Engineering
Fundamentals

Engineering Fundamentals

EXAMINATION REVIEW

Donald G. Newnan, Ph.D., P.E.
San Jose State University

Bruce E. Larock, Ph.D., P.E.
University of California, Davis

A WILEY–INTERSCIENCE PUBLICATION, SECOND EDITION

JOHN WILEY & SONS, New York · Chichester · Brisbane · Toronto

Copyright © 1978 by John Wiley & Sons, Inc.

All rights reserved. Published simultaneously in Canada.

Reproduction or translation of any part of this work
beyond that permitted by Sections 107 or 108 of the
1976 United States Copyright Act without the permission
of the copyright owner is unlawful. Requests for
permission or further information should be addressed to
the Permissions Department, John Wiley & Sons, Inc.

Library of Congress Cataloging in Publication Data

Newnan, Donald G
  Engineering fundamentals.

  "A Wiley–Interscience publication."
  Includes index.
  1. Engineering. 2. Engineering—Problems,
exercises, etc. I. Larock, Bruce E., 1940—
joint author. II. Title.

TA153.N47 1978 620'.0076 77–12592
ISBN 0-471-01900-3

Printed in the United States of America

10 9 8 7 6 5 4 3

Acknowledgments

Many individuals and groups have aided us in one way or another during the preparation of the two editions of this book. We were materially assisted by James H. Sams, then executive secretary of the National Council of Engineering Examiners, and by a number of the State Boards of Registration for Engineers. In preparing the second edition we were assisted by Ralph W. Shoemaker and by C. Dean Newnan. Joanne Weigt ably typed a portion of the manuscript. Finally, our Wiley editor, Beatrice Shube, has been a constructive part of this project from the beginning. The final responsibility for the correctness of the text resides with the authors, however.

DONALD G. NEWNAN
BRUCE E. LAROCK

San Jose, California
Davis, California

Contents

1 Introductory Comments

This book provides the engineer with a systematic review of the fundamental principles of engineering science. An orderly program of continuing education and professional registration are desirable and, in many cases, necessary elements in the professional development of the engineer. Ready access to a compact collection of basic engineering principles, to a large number of problems illustrating the use of these principles, and to a complete solution to every one of these many problems should materially assist the engineer in the review process.

Registration is uniformly recognized in the United States as a landmark along any engineer's path of professional development. In each of the 50 states, and the five other jurisdictions, laws regulate the practice of engineering and the right of an individual to designate himself as an engineer. These laws were enacted to protect the public from people who might call themselves engineers but who in reality are not qualified to perform competent engineering work. Thus registration as a professional engineer (PE) is a desirable and often mandatory goal.

There are normally four steps in becoming a registered professional engineer. Graduation from an engineering school, or practical engineering experience, is the first step. The individual then passes an eight-hour examination on the fundamentals of engineeering. The next step is the professional practice of engineering under responsible supervision. The minimum time varies from two to three years, depending on the state or jurisdiction. Finally, it is necessary to pass another eight-hour examination on the professional practice of a specific branch of engineering.

The fundamentals of engineering examination is prepared by the National Council of Engineering Examiners (NCEE) and is known by various names.

In many states it is called the EIT or Engineer-In-Training examination. Elsewhere it may be called the Engineering Fundamentals exam or the Intern Engineer exam. Engineers in 50 of the 55 states and other jurisdictions take the *same* nationally prepared examination at the same time. Engineers from all branches of engineering (CE, ME, EE, and so on) take the same engineering fundamentals examination. The completed examinations are sent to NCEE in South Carolina for scoring.

The engineering fundamentals examination is a test of basic engineering concepts. The eight-hour examination is split into two four-hour sessions. The first session presently consists of 150 multiple-choice questions with five choices. The topics and the approximate number of questions are as follows:

	No. of questions
Mathematics	22
Nucleonics and wave phenomena	5
Chemistry	14
Statics	14
Dynamics	14
Mechanics of materials	14
Fluid mechanics	14
Thermodynamics	18
Electrical theory	16
Materials science	9
Economic analysis	10
	150

A perfect score of 50 points is given for 100 correct answers.

The second four-hour session presently consists of three problem sets in each of the following subjects: statics, dynamics, mechanics of materials, fluid mechanics, thermodynamics, electrical theory, and economic analysis. Each of the problem sets consists of introductory information, followed by 10 multiple-choice problems that relate to the introductory material. The applicant must select and solve five of the 21 problem sets, making a total of 50 multiple-choice problems. The problems must be selected from at least four different subject areas, with the further restriction that no more than one problem set may be chosen from economic analysis.

This book is devoted exclusively to an orderly review of the content of the engineering fundamentals examination. (Other books have been written by one of the authors to cover the professional practice examinations.) Each chapter begins with a short review of the fundamental principles that are conceptually important and practically useful in the field. The goal is to present basic ideas compactly and directly; lengthy, detailed derivations are avoided. The result is a selective overview of the discipline rather than an encyclopedic coverage. The remainder of each chapter consists of multiple-choice problems. An important feature of the book is the presentation of a complete solution for each problem. The International System of Units (SI units) are used in some of the problems. (Conversion factors and other information on SI units are provided in Appendix A.) The goal has been to provide an orderly review of engineering fundamentals and typical examination problems for study and solution.

2 Mathematics

According to some, mathematics is *the* most fundamental branch of all science. Indeed, the goal of much scientific and technical work is to express in precise mathematical terms the behavior of our universe and its smaller component parts. In varying degrees mathematics is used in all the disciplines that together make up engineering fundamentals. For this reason it is highly desirable to have a working knowledge of some of the basic relations of algebra, trigonometry, geometry, and calculus. We present here a *brief* review of fundamental principles; a thorough review, including proofs, is outside the scope of this volume.

ALGEBRA

The basic rules of algebra apply equally well to real and complex numbers, that is, numbers expressible in the form $a_1 + ia_2$, where a_1 and a_2 are real numbers, zero or nonzero, and $i^2 = -1$. The basic rules, given in additive and multiplicative form, are three:

Commutative: $\qquad a + b = b + a \qquad\qquad ab = ba$

Distributive: $\qquad a(b + c) = ab + ac$

Associative: $\qquad a + (b + c) = (a + b) + c \qquad a(bc) = (ab)c$

The laws of exponents and logarithms are intimately related. For positive numbers a and b and any positive or negative exponents x and y, the rules for exponents are as follows:

$$b^{-x} = \frac{1}{b^x} \qquad b^x b^y = b^{x+y}$$

$$(ab)^x = a^x b^x \qquad b^{xy} = (b^x)^y$$

If $b^y = x$ for positive b and x, then $y = \log_b x$ is the definition of the logarithm of x to the base b. The logarithm is therefore a kind of exponent. The most commonly used base numbers are $b = 10$ for common logarithms and $b = e = 2.718 \cdots$ for natural logarithms. (When $b = 10$ it is often not written down; when $b = e$ often $\log_e = \ln$ is written.) Regardless of the value of b these laws hold for logarithms:

$$b^{\log_b x} = x \qquad \log_b b^x = x$$

$$\log_b (xy^n) = \log_b x + n \log_b y \qquad \text{for any value of } n$$

To change, for example, the base of a logarithm from any base b to the base e,

$$\log_b x = \frac{\log_e x}{\log_e b} = \frac{\ln x}{\ln b} = \log_b e \times \ln x$$

since $(\log_b e)(\log_e b) = \log_e (b^{\log_b e}) = \log_e e = 1$.

An entire branch of mathematics, linear algebra, has grown out of an interest in solving sets of linear, simultaneous algebraic equations. The field is a generalization of solving the equation $ax = b$, which is linear in the one unknown x and has the obvious solution $x = b/a$. For two simultaneous equations in the unknowns x and y, the equations are often solved by eliminating y and solving for x:

$$a_{11}x + a_{12}y = b_1$$

$$a_{21}x + a_{22}y = b_2$$

From the first equation $y = (1/a_{12})(b_1 - a_{11}x)$. Insertion of this expression for y into the second equation yields an equation of the form of the single linear equation, which is easily solved.

The foregoing problem and also larger sets of simultaneous linear equations can be solved by using determinants. The determinant of the coefficients in the problem is

$$D = \begin{vmatrix} a_{11} & a_{12} \\ a_{21} & a_{22} \end{vmatrix} = a_{11}a_{22} - a_{12}a_{21}$$

If D is nonzero, then Cramer's rule gives the solution for x and y as

$$x = \frac{D_1}{D} \qquad y = \frac{D_2}{D}$$

D_1 is formed from D by replacing a_{11} and a_{21} by b_1 and b_2, respectively. To find D_2, replace the second column of a's by the b's. The same procedure is followed for three or more unknown variables. A 3×3 determinant can be reduced to a 2×2 determinant by expanding it in terms of minors along one column or row:

$$\begin{vmatrix} a_{11} & a_{12} & a_{13} \\ a_{21} & a_{22} & a_{23} \\ a_{31} & a_{32} & a_{33} \end{vmatrix} = a_{11} \begin{vmatrix} a_{22} & a_{23} \\ a_{32} & a_{33} \end{vmatrix} - a_{12} \begin{vmatrix} a_{21} & a_{23} \\ a_{31} & a_{33} \end{vmatrix} + a_{13} \begin{vmatrix} a_{21} & a_{22} \\ a_{31} & a_{32} \end{vmatrix}$$

Systems of three linear equations in three unknowns can still be solved by successive elimination, but the use of Cramer's rule and determinants is often more efficient. For four or more unknowns the required bookkeeping becomes formidable by either method, but it is then preferable to use Cramer's rule because it is more systematic.

Quadratic equations are always solvable by algebra. If $ax^2 + bx + c = 0$, then the two solutions are

$$x = \frac{1}{2a}[-b \pm (b^2 - 4ac)^{1/2}]$$

If $b^2 < 4ac$, the roots of the equation are complex numbers. Formulas also exist that give the solutions to third- and fourth-order equations, but it is usually easier to try solving the equation by (a) attempting to factor the equation algebraically (often not successful), (b) graphing the equation and noting the points of intersection with the x-axis, or (c) substituting numerically by trial and error.

Another useful formula in algebra is the binomial theorem, which is a special form of the Taylor's series of calculus:

$$(a + b)^n = a^n + \frac{n}{1!}a^{n-1}b + \frac{n(n-1)}{2!}a^{n-2}b^2 + \cdots$$

$$+ \frac{n(n-1) \cdots (n-r+1)}{r!}a^{n-r}b^r + \cdots + b^n$$

For a positive integer n this expansion has $(n + 1)$ terms. In the formula the convenient "factorial" notation $n! = n(n-1)(n-2) \cdots (3)(2)(1)$ has been used.

TRIGONOMETRY

Trigonometry deals with the relations between the angles and the sides of triangles. The periodic functions defined by these relations, however, have vastly wider applications. In using these functions we often deal with angle measurement, of which there are two kinds. One system divides one revolution into 360° (degrees). Each degree is further divisible into 60′ (minutes), and each minute into 60″ (seconds), although often a fraction of a degree is written as a decimal (e.g., $30' = 0.5°$). The unit of measurement in the second system is the radian (rad); 2π rad equals one revolution, or $180° = \pi$ rad $= 3.14159 \cdots$ rad.

The two most basic trigonometric functions are the sine and cosine (Fig. 2-1), which are defined as follows:

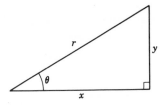

Figure 2-1

$$\sin \theta = \frac{y}{r} \qquad \cos \theta = \frac{x}{r}$$

Here x and y may assume any value, but r is always positive. The other four basic functions are

$$\tan \theta = \frac{\sin \theta}{\cos \theta} = \frac{y}{x} \qquad \cot \theta = \frac{1}{\tan \theta} = \frac{x}{y}$$

$$\sec \theta = \frac{1}{\cos \theta} = \frac{r}{x} \qquad \csc \theta = \frac{1}{\sin \theta} = \frac{r}{y}$$

The sine and cosine are odd and even periodic functions, respectively, with periods of 2π (Fig. 2-2). By learning the variations of these two functions, one can easily deduce the variation of the other functions.

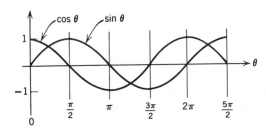

Figure 2-2

Much can be done in trigonometry by remembering a few fundamental identities. Among them are these:

$$\sin^2 \theta + \cos^2 \theta = 1 \qquad 1 + \tan^2 \theta = \sec^2 \theta \qquad 1 + \cot^2 \theta = \csc^2 \theta$$

$$\sin (\theta \pm \phi) = \sin \theta \cos \phi \pm \cos \theta \sin \phi$$

$$\cos (\theta \pm \phi) = \cos \theta \cos \phi \mp \sin \theta \sin \phi$$

From these last two identities the double-angle (sin 2θ, cos 2θ) formulas can be derived by letting $\theta = \phi$. The half-angle (sin $\theta/2$, cos $\theta/2$) formulas can also be derived by replacing θ and ϕ by $\theta/2$ and rearranging the resulting expressions.

In solving for the unknown parts of a plane triangle (Fig. 2-3), two or three basic formulas are often useful. These are

Sum of angles: $\qquad \alpha + \beta + \gamma = 180°$

Law of sines: $\qquad \dfrac{a}{\sin \alpha} = \dfrac{b}{\sin \beta} = \dfrac{c}{\sin \gamma}$

Law of cosines: $\qquad a^2 = b^2 + c^2 - 2bc \cos \alpha$

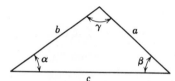

Figure 2-3

Note that if $\alpha = 90°$ the triangle is a right triangle, and the law of cosines then becomes a statement of the Pythagorean formula.

GEOMETRY

Here we group together some elements of elementary plane geometry, which describes some spatial properties of objects of various shapes, and analytic geometry, which employs algebraic notation in its more detailed description of some of these same objects.

The triangle and rectangle are basic geometric figures; they may also be considered as special cases of the trapezoid. From Fig. 2-4 we can consider

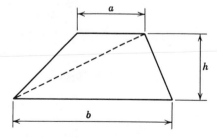

Figure 2-4

the trapezoid as the sum of two triangles. The area A of each shape is as follows:

$$\text{Trapezoid:} \quad A = \frac{h}{2}(a+b)$$

$$\text{Rectangle:} \quad A = hb \quad (a = b)$$

$$\text{Triangle:} \quad A = \frac{hb}{2} \quad (a = 0)$$

Another important shape is the regular polygon having n sides. The central angle subtended by one side is the vertex angle; its value is $2\pi/n$. The included angle between two successive sides of the polygon is $(n-2)\pi/n$.

The most important nonpolygonal geometric shape is the circle. For a circle of radius r and diameter $d = 2r$, the circumference is $c = \pi d$ and the enclosed area $A = \pi r^2$. In three dimensions its counterpart is the sphere which has a surface area $S = 4\pi r^2$ and an enclosed volume $V = \frac{4}{3}\pi r^3$.

Plane analytic geometry describes algebraically the properties of one- and two-dimensional geometric forms in an (x, y) plane.

The general equation of a straight line is $Ax + By + C = 0$. This equation is often more usefully written in one of the three following forms:

$$\text{Point-slope:} \quad y - y_1 = m(x - x_1)$$

$$\text{Slope-intercept:} \quad y = mx + b$$

$$\text{Two-intercept:} \quad \frac{x}{a} + \frac{y}{b} = 1$$

For a straight line passing through the points $P_1 = (x_1, y_1)$ and $P_2 = (x_2, y_2)$, the slope $m = (y_2 - y_1)/(x_2 - x_1)$; the intercepts a and b are the coordinate values occurring where the line intersects the x- and y-axes, respectively. The distance between points P_1 and P_2 is $D = [(x_2 - x_1)^2 + (y_2 - y_1)^2]^{1/2}$. Also, parallel lines have equal slopes, whereas perpendicular lines have negative reciprocal slopes.

Included in the general equation of second degree $Ax^2 + Bxy + Cy^2 + Dx + Ey + F = 0$ are a set of geometric shapes called the conic sections. The different conic sections can be recognized by investigating $B^2 - 4AC$:

If $B^2 - 4AC > 0$, the section is a hyperbola

If $B^2 - 4AC = 0$, the section is a parabola

If $B^2 - 4AC < 0$, the section is an ellipse

In the last case if $A = C$ and they are not zero, the section is a circle. If A, B, and C are all zero, the straight line again results.

The basic equation for the hyperbola can be found from the general second-degree equation. For a hyperbola centered at the coordinate origin with limbs opening left and right (Fig. 2-5a), the equation is

$$\frac{x^2}{a^2} - \frac{y^2}{b^2} = 1$$

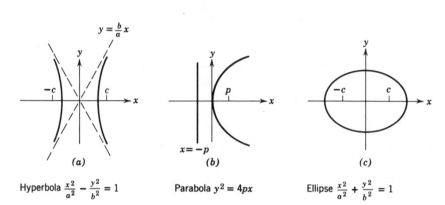

(a)

Hyperbola $\frac{x^2}{a^2} - \frac{y^2}{b^2} = 1$

(b)

Parabola $y^2 = 4px$

(c)

Ellipse $\frac{x^2}{a^2} + \frac{y^2}{b^2} = 1$

Figure 2-5

The difference of distances from the two focuses ($\pm c$, 0) to a point on the hyperbola is always constant, where $c^2 = a^2 + b^2$. The limbs are asymptotic to the straight lines $y = \pm(b/a)x$. For a hyperbola centered at the point (h, k), replace x by $(x - h)$ and y by $(y - k)$. (This procedure for shifting the location of figures applies generally for all conic sections.)

The parabola, which geometrically is the locus of points equidistant from a point and a line (Fig. 2-5b), may be written in type form as

$$y^2 = 4px$$

when the vertex of the parabola is at the coordinate origin and the parabola opens to the right. Here the parabola is equidistant from the focus point $(p, 0)$ and the directrix line $x = -p$. Change the sign of p to obtain a parabola opening to the right; interchange the roles of y and x to obtain parabolas opening upward or downward.

The type equation for an ellipse, centered on the coordinate origin (Fig. 2-5c), is

$$\frac{x^2}{a^2} + \frac{y^2}{b^2} = 1$$

The semimajor and semiminor axes are a and b. The focuses of the ellipse are at $(\pm c, 0)$, where $c^2 = a^2 - b^2$. Any point on the ellipse is such that the sum of the distances to that point from the two focuses is a constant. If $a = b = r$, the ellipse then becomes a circle of radius r.

Sometimes it is more convenient to use the polar (r, θ), cylindrical (r, θ, z), or spherical (ρ, θ, ϕ) coordinate system in place of the two- or three-dimensional Cartesian (x, y, z) coordinate system. By reference to Fig. 2-6,

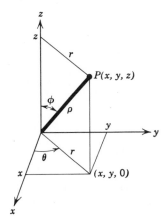

Figure 2-6

these coordinate systems can be related to one another. The relations between the polar, cylindrical, and Cartesian coordinate systems are

$$x = r \cos \theta \qquad y = r \sin \theta \qquad z = z$$

Since we also have

$$z = \rho \cos \phi \qquad r = \rho \sin \phi$$

the relations between the spherical and Cartesian coordinate systems are

$$x = \rho \sin \phi \cos \theta \qquad y = \rho \sin \phi \sin \theta \qquad z = \rho \cos \phi$$

CALCULUS

At a point on the curve $y = f(x)$ the slope of the curve is the ratio of the change in $f(x)$ to the change in x when the change in x approaches zero in the limit; mathematically,

$$\text{Slope} = \frac{dy}{dx} = \frac{df(x)}{dx} = f'(x)$$

This is also called the rate of change of y with respect to x or the first derivative of $f(x)$. Second and higher derivatives are in turn defined as the rate of change of the next lower ordered derivative.

Derivatives of basic functions of x are given now. Let f and g be functions of x; c, m, and n are constants.

$$\frac{d}{dx}(c) = 0$$

$$\frac{d}{dx}(cx^n) = cnx^{n-1} \qquad \text{for} \quad n \neq 0$$

$$\frac{d}{dx}(f \pm g) = \frac{df}{dx} \pm \frac{dg}{dx}$$

$$\frac{d}{dx}(f^m) = mf^{m-1}\frac{df}{dx}$$

$$\frac{d}{dx}(f^m g^n) = f^m \frac{d}{dx}(g^n) + g^n \frac{d}{dx}(f^m)$$

Here m and n assume any value. If $m = 1$, $n = 1$ this rule treats the simple product of two functions; if $m = 1$, $n = -1$ the rule governs the differentiation of the quotient of two functions. If $x = x(t)$,

$$\frac{df}{dt} = \frac{df}{dx}\frac{dx}{dt}$$

This is the chain rule of differentiation.

$$\frac{d}{dx}(\sin f) = \cos f \frac{df}{dx}$$

$$\frac{d}{dx}(\cos f) = -\sin f \frac{df}{dx}$$

At a maximum or a minimum of the function $f(x)$, the rate of change of f is zero; that is, $f'(x) = 0$. At that point f is a maximum if $f'' < 0$; it is a minimum if $f'' > 0$. Often, however, other physical considerations indicate whether the function has a maximum or minimum, and the second derivative test is not actually needed. If $f'' = 0$, the point is usually (but not always) a point of inflection, a point where the curvature of the function changes from concave upward to concave downward or vice versa.

If the function $f(x)$ approaches the value c as x approaches the value x_0, then we say c is the limiting value of $f(x)$ at the point x_0 and express this mathematically as

$$\lim_{x \to x_0} f(x) = c$$

The algebra of limits is no different from ordinary algebra:

$$\lim (f+g) = \lim f + \lim g$$

$$\lim (fg) = (\lim f)(\lim g)$$

$$\lim \frac{f}{g} = \frac{\lim f}{\lim g} \qquad \text{if} \quad \lim g \neq 0$$

In the use of this last equation, however, the indeterminate forms $0/0$ or ∞/∞ may be encountered. In this case L'Hospital's rule is useful: Let f and g be functions having continuous derivatives with $g'(x_0) \neq 0$. If on approaching the limit point $x = x_0$ we have

$$\lim_{x \to x_0} f(x) = \lim_{x \to x_0} g(x) = 0$$

or

$$\lim_{x \to x_0} f(x) = \lim_{x \to x_0} g(x) = \pm\infty$$

then

$$\lim_{x \to x_0} \left[\frac{f(x)}{g(x)} \right] = \lim_{x \to x_0} \left[\frac{f'(x)}{g'(x)} \right]$$

An example is

$$\lim_{x \to 0} \frac{\sin x}{x} = \lim_{x \to 0} \frac{\cos x}{1} = 1$$

Continuous functions may be expressed as power series expansions around a point $x = c$ by use of the Taylor's series

$$f(x) = f(c) + f'(c)\frac{(x-c)}{1!} + f''(c)\frac{(x-c)^2}{2!} + \cdots + f^{(n)}(c)\frac{(x-c)^n}{n!} + \cdots$$

This series is particularly useful when a polynomial representation for a function is desired for x near c. The series can often then be truncated after only a few terms with little loss in accuracy.

Integration is the inverse of the process of differentiation. It may also be defined as the limit of a sequence; by this process the integral may be used to evaluate the exact area under a curve (Fig. 2-7). The area under this curve between a and b is $A = \int_a^b f(x)\,dx$.

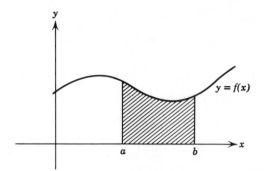

Figure 2-7

Some basic integration formulas follow, using the same notation as was used for derivatives.

$$\int \frac{df(x)}{dx}\,dx = f(x) + C \qquad (C = \text{constant of integration})$$

$$\int 0 \times dx = C$$

$$\int cf(x)\,dx = c\int f(x)\,dx$$

$$\int (f \pm g)\,dx = \int f\,dx \pm \int g\,dx$$

$$\int x^n\,dx = \frac{x^{n+1}}{n+1} + C \qquad (n \neq -1)$$

$$\int \frac{dx}{x} = \ln x + C$$

$$\int e^x\,dx = e^x + C$$

$$\int \sin x\,dx = -\cos x + C$$

$$\int \cos x\,dx = \sin x + C$$

$$\int f\,dg = fg - \int g\,df$$

This last formula, involving the two functions $f(x)$ and $g(x)$, is called integration by parts and is a powerful tool of integration. Evaluating any integral between definite limits will eliminate the constant of integration. We should also note that, for $a \leq b \leq c$,

$$\int_a^c f(x)\, dx = \int_a^b f(x)\, dx + \int_b^c f(x)\, dx$$

and

$$\int_a^b f(x)\, dx = -\int_b^a f(x)\, dx$$

DIFFERENTIAL EQUATIONS

Only the barest introduction to ordinary differential equations can be given here. A few problems in this chapter give examples of solving differential equations, but for a comprehensive review of the subject the reader should consult an appropriate mathematics text.

First-order ordinary differential equations are separable if they can be put in the form

$$M(x)\, dx + N(y)\, dy = 0$$

The equation is solved by direct integration; as in all differential equations a boundary condition must be specified before the constant of integration can be evaluated. If $M = M(x, y)$ and $N = N(x, y)$, a solution $F(x, y)$ can be found if the differential equation is exact, that is, if $\partial M/\partial y = \partial N/\partial x$. Then the solution $F(x, y)$ must satisfy the requirements $M = \partial F/\partial x$, $N = \partial F/\partial y$.

Many additional terms and types of basic ordinary differential equations are important to engineering but cannot be reviewed here. Just one representative important differential equation will be mentioned, however, to illustrate some concepts. The equation is

$$\frac{d^2 y}{dx^2} + p^2 y = f(x) \qquad (p = \text{constant})$$

plus two boundary conditions since this equation involves a second derivative. The equation is linear (y and/or derivatives of y are not multiplied together to form quadratic or higher order terms) and nonhomogeneous (the right term does not involve y; if it were not present, the equation would be homogeneous). The general solution is the sum of a complementary solution [solution of the equation with $f(x) = 0$] plus one particular solution [solution with $f(x)$ present]. If, for example, $f(x) = x$, a particular solution is $y = x/p^2$.

Since $p^2 > 0$, the general complementary solution is $y = A \sin px + B \cos px$. If the plus sign in the equation were a minus sign, the solution would involve terms of the form $e^{\pm px}$.

STATISTICS AND PROBABILITY

Statistics is used as an aid in drawing conclusions from masses of data. The computation of certain properties of the data is useful in answering the questions "How big is it?" and "How much variation in size is there?"

The arithmetic mean, the median, and the mode are valuable tools in ascertaining the answer to the first question. The arithmetic mean $\bar{x}$ is the average value of N individual values x_i:

$$\bar{x} = \frac{1}{N} \sum_{i=1}^{N} x_i$$

The median value is the middle value when all data are arranged in order by magnitude; half the values are larger than the median value, half are smaller. The mode, on the other hand, is the value that occurs most frequently. The standard deviation σ is a statistic that helps to answer the second question; it gives the rms deviation from the mean value $\bar{x}$:

$$\sigma = \left[\frac{1}{N-1} \sum_{i=1}^{N} (x_i - \bar{x})^2 \right]^{1/2}$$

This expression gives a conservative and unbiased value for σ, but it is not unusual to find the divisor $N - 1$ replaced by N. The square of the standard deviation σ^2 is called the *variance*.

Much of basic probability theory deals with mutually exclusive events or independent events. If only one of a set of possible events can occur, the events are mutually exclusive. Events are independent if the occurrence or nonoccurrence of one event does not affect the probability of occurrence of the other events. In this connection it is sometimes necessary to compute the number of permutations or set arrangements of n things taken r at a time, which is $n!/(n-r)!$, and the number of combinations (no set arrangement) of n things taken r at a time, which can be expressed in the equivalent forms

$$\binom{n}{r} = \frac{n!}{r!(n-r)!}.$$

The probability of success of an event plus the probability of failure of an event is obviously unity, a fact of great simplicity and use. The probability of success itself is the ratio of the number of ways of achieving success to the total number of possible events (successes and failures). For independent events the following two rules are also useful:

1. The probability of A *or* B occurring equals the *sum* of the probability of occurrence of A and the probability of occurrence of B.

2. The probability of A *and* B occurring equals the *product* of the two individual probabilities.

PROBLEM 2-1

The number below that has four significant figures is

   (A) 1414.0
   (B) 1.4140
   (C) 0.141
   (D) 0.01414
   (E) 0.0014

Solution. A significant figure is any of the digits 1 through 9 as well as 0 except when 0 is used to fix the decimal point or to fill the places of unknown or discarded digits.

Part	No. of Significant Figures
A	5
B	5
C	3
D	4
E	2

Answer is (D)

PROBLEM 2-2

The fourth term in the expansion of $(1 - 2x^{-1/2})^{-2}$ is

   (A) $-6x^{-1}$
   (B) $12x^{-1}$
   (C) $32x^{-3/2}$
   (D) $64x^{-3/2}$
   (E) $80x^{-2}$

Solution.   Using the binomial expansion

$$(1-A)^{-n} = 1 + nA + \frac{n(n+1)}{2!}A^2 + \frac{n(n+1)(n+2)}{3!}A^3$$

$$+ \frac{n(n+1)(n+2)(n+3)}{4!}A^4 + \cdots$$

in which we have $A = 2x^{-1/2}$ and $n = 2$ here, we obtain

$$(1-2x^{-1/2})^{-2} = 1 + 2(2x^{-1/2}) + \frac{(2)(3)}{2}(2x^{-1/2})^2$$

$$+ \frac{(2)(3)(4)}{(3)(2)}(2x^{-1/2})^3 + \frac{(2)(3)(4)(5)}{(4)(3)(2)}(2x^{-1/2})^4 + \cdots$$

$$= 1 + 4x^{-1/2} + 12x^{-1} + 32x^{-3/2} + 80x^{-2} + \cdots$$

Answer is (C)

PROBLEM 2-3

If $x = +6$ and -4, the equation satisfying both these values would be

(A) $2x^2 + 3x - 24 = 0$
(B) $x^2 + 10x - 24 = 0$
(C) $x^2 - 2x - 24 = 0$
(D) $x^2 - 4x - 32 = 0$
(E) $x^2 - 3x + 18 = 0$

Solution.   If a quadratic equation has solutions of $+6$, -4, it must have the factored form

$$(x-6)(x+4) = 0$$

Expanding this, we obtain

$$x^2 - 6x + 4x - 24 = 0$$

$$x^2 - 2x - 24 = 0$$

Answer is (C)

PROBLEM 2-4

If $i = \sqrt{-1}$, the quantity i^{27} is equal to

(A) 0
(B) i
(C) $-i$
(D) 1
(E) -1

Solution. $(i)^n$ is a periodic function; that is,

$$i^1 = i \qquad i^2 = -1 \qquad i^3 = -i \qquad i^4 = +1 \qquad i^5 = i \qquad \cdots$$

More generally,

$$i^{4n+1} = i \qquad i^{4n+2} = -1 \qquad i^{4n+3} = -i \qquad i^{4n+4} = +1$$

for any integer n.
  Here $27 = 4n + 3$ and $i^{27} = -i$.

Answer is (C)

PROBLEM 2-5

If $x = \log_a N$, then

(A) $x = N^a$
(B) $a = x^N$
(C) $x = a^N$
(D) $N = a^x$
(E) $N = x^a$

Solution. If x is the logarithm of N to the base a ($x = \log_a N$), then by definition $N = a^x$.

Answer is (D)

PROBLEM 2-6

The $\log_{10}$ of 2 is 0.30103. The log of $\frac{1}{2}$ is

    (A) $9.30103 - 10$
    (B) $0.30103 \div 2$
    (C) $1 - 0.30103$
    (D) $9.69897 - 10$
    (E) $1 \div 0.30103$

Solution.

$$\log \tfrac{1}{2} = \log 1 - \log 2 = 0.00000 - 0.30103$$

which may be written as

$$= 10. - 0.30103 - 10$$
$$= 9.69897 - 10$$

Answer is (D)

PROBLEM 2-7

If $\log_a 10 = 0.250$, $\log_{10} a$ equals

    (A) 4
    (B) 0.50
    (C) 2
    (D) 0.25
    (E) 1000

Solution. $\text{Log}_a 10 = 0.250$ can be written as $10 = a^{0.250}$. Taking $\log_{10}$,

$$\log_{10} 10 = \log_{10} a^{0.250}$$
$$1 = 0.250 \log_{10} a$$

Since $1 = 0.250 \log_{10} a$,

$$\log_{10} a = \frac{1}{0.250} = 4$$

Answer is (A)

PROBLEM 2-8

If $x = \frac{1}{2}\ln\dfrac{1+u}{1-u}$ (ln = natural logarithm), then

(A) $u = e^x$
(B) $u = \tanh x$
(C) $u = (e^x - 1)/(e^x + 1)$
(D) $u = (2x - 1)/(2x + 1)$
(E) $u = \ln(\sin x)$

Solution.

$$2x = \ln\frac{1+u}{1-u}$$

$$e^{2x} = \exp\left(\ln\frac{1+u}{1-u}\right) = \frac{1+u}{1-u}$$

$$(1-u)e^{2x} = 1+u$$

$$u(e^{2x} + 1) = e^{2x} - 1$$

$$u = \frac{e^{2x}-1}{e^{2x}+1} = \frac{e^x - e^{-x}}{e^x + e^{-x}} = \tanh x$$

Answer is (B)

PROBLEM 2-9

In the equation

$$\log_{10}(X-1) + \log_{10} X = 1$$

the value of X is most nearly

(A) −2.7
(B) 1.6
(C) 3.7
(D) −0.6
(E) 5.5

Solution. Since the sum of logarithms of numbers is their product and $\log 10 = 1$, then

$$(X-1)(X) = 10 \quad \text{or} \quad X^2 - X = 10$$

$$X^2 - X - 10 = 0$$

$$X = \frac{1 \pm \sqrt{1+40}}{2} = \frac{1 \pm 6.4}{2} = 3.7$$

Note that there is only one real root. The other answer from the quadratic equation (-2.7) is not real, for the log of a negative number has no meaning.

Answer is (C)

PROBLEM 2-10

Each interior angle of a regular polygon with eight sides is nearest to

(A) 100°
(B) 80°
(C) 150°
(D) 125°
(E) 135°

Solution. The sum of the interior angles of a polygon is equal to $(n-2) \times 180°$. In a regular polygon all sides are equal, hence all angles are equal.

$$\frac{(n-2) \times 180°}{8} = \frac{6}{8}(180) = 135°$$

Answer is (E)

PROBLEM 2-11

If the sine of angle α is given as K, the tangent of angle α is equal to

(A) $1 - K$
(B) $1/K$
(C) $\sqrt{1-K^2}$
(D) $\dfrac{1}{\sqrt{1-K^2}}$
(E) $\dfrac{K}{\sqrt{1-K^2}}$

Solution.

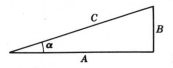

Figure 2-8

Sin $\alpha = B/C = K$. If we let $C = 1$, then $B = K$. Since $A^2 + B^2 = C^2$, $A^2 + K^2 = 1^2$ or $A = \sqrt{1 - K^2}$. Then the triangle appears as:

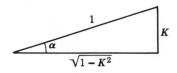

Figure 2-9

$$\tan \alpha = \frac{K}{\sqrt{1 - K^2}}$$

Answer is (E)

PROBLEM 2-12

Which of the following is incorrect?

(A) $\cos^2 A = \tan A \cot A$
(B) $\sin A = \cos A \tan A$
(C) $\cos 2A = \cos^2 A - \sin^2 A$
(D) $2 \sin^2 A = 1 - \cos 2A$
(E) $\sin (A + B) = \sin A \cos B + \cos A \sin B$

Solution. An identity is an equality that is valid for all values of the variable(s) in the equation. Considering the first expression,

$$\tan A \cot A = 1 \quad \text{or} \quad \cos^2 A = 1$$

which is not true for all values of A.
 The four other expressions are correct.

Answer is (A)

PROBLEM 2-13

As shown, a circle can be circumscribed around a triangle *ABC. AC* is a diameter of the circle.

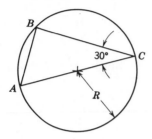

Figure 2-10

In terms of the radius R, the area of the triangle is most nearly

(A) $0.750R^2$
(B) $0.785R^2$
(C) $0.866R^2$
(D) $1.000R^2$
(E) $1.414R^2$

Solution. A geometric theorem states that a triangle inscribed in a semicircle is a right triangle, so angle $ABC = 90°$.

$\mathrm{Sin}\ 30° = \dfrac{AB}{2R} = \dfrac{1}{2}$. Therefore $AB = R$. $\mathrm{Cos}\ 30° = \dfrac{BC}{2R} = \dfrac{\sqrt{3}}{2}$. Therefore $BC = \sqrt{3}R$.

$$\mathrm{Area} = \tfrac{1}{2}(AB)(BC) = \tfrac{1}{2}R\sqrt{3}R$$
$$= 0.866R^2$$

Answer is (C)

PROBLEM 2-14

Given the following relations:

$$r \cos \phi \cos \theta = 4 \qquad\qquad (1)$$
$$r \cos \phi \sin \theta = 3 \qquad\qquad (2)$$
$$r \sin \phi \quad\ \ = 5 \qquad\qquad (3)$$

The value of r is most nearly

(A) 6.00
(B) 7.07
(C) 7.50
(D) 8.66
(E) 10.00

Solution. Divide Eq. (2) by Eq. (1):

$$\frac{r \cos \phi \sin \theta}{r \cos \phi \cos \theta} = \frac{3}{4} = \tan \theta$$

Therefore

$$\theta = \tan^{-1}\left(\tfrac{3}{4}\right) = 36.9°$$

Divide Eq. (3) by Eq. (1):

$$\frac{r \sin \phi}{r \cos \phi \cos \theta} = \frac{5}{4} \qquad \tan \phi \frac{1}{\cos \theta} = \frac{5}{4}$$

but $(1/\cos \theta) = (1/\cos 36.9°) = (1/0.8) = 1.25$, $1.25 \tan \phi = \tfrac{5}{4}$, $\tan \phi = 1$, $\phi = 45°$, $r \sin \phi = 5$, and $r \sin 45° = 5$. Therefore

$$r = \frac{5}{0.707} = 7.07$$

This result is easily checked by squaring and adding the three original equations and recalling that $\cos^2 \theta + \sin^2 \theta = 1$ for both θ and ϕ; $r^2[\cos^2 \phi(\cos^2 \theta + \sin^2 \theta) + \sin^2 \phi] = r^2 = 50$.

Answer is (B)

PROBLEM 2-15

It is 3.8 km from point A to the north end of the lake and 5.3 km from A to the south end of the lake. The lake subtends an angle of 110° at A. The length of the lake from north to south is nearest to

(A) 5.4 km
(B) 6.5 km
(C) 7.5 km
(D) 8.1 km
(E) 9.1 km

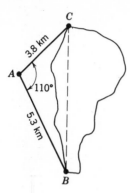

Figure 2-11

Solution.

The law of cosines applies directly:

$$a^2 = b^2 + c^2 - 2bc \cos \alpha$$
$$a^2 = (3.8)^2 + (5.3)^2 - 2(3.8)(5.3) \cos 110°$$

Since $\cos 110° = -0.342$,

$$a^2 = 14.4 + 28.1 - (-13.8) = 56.3$$
$$a = 7.50 \text{ km}$$

Answer is (C)

PROBLEM 2-16

Two points lie on a horizontal line directly south of a tower 100 ft high. The angles of depression to the points are 28°10′ and 42°50′. The distance between the points is closest to

(A) 39.7 ft
(B) 64.8 ft
(C) 70.4 ft
(D) 78.9 ft
(E) 104.0 ft

Solution. In this problem it is necessary to know that zero factorial is equal to one.

$$\frac{7! \times 6!}{8! \times 0!} = \frac{\cancel{7!} \times 6 \times 5 \times \cancel{4} \times 3 \times \cancel{2} \times 1}{\cancel{8} \times \cancel{7!} \times 1} = 90$$

Answer is (E)

PROBLEM 2-19

When the determinant D is

$$D = \begin{vmatrix} 1 & 1 & 1 \\ 2 & -1 & 1 \\ 1 & 2 & -1 \end{vmatrix}$$

its value is

- (A) −5
- (B) −3
- (C) +1
- (D) +3
- (E) +7

Solution. Expanding by minors along the top row,

$$D = \begin{vmatrix} 1 & 1 & 1 \\ 2 & -1 & 1 \\ 1 & 2 & -1 \end{vmatrix} = 1 \begin{vmatrix} -1 & 1 \\ 2 & -1 \end{vmatrix} - 1 \begin{vmatrix} 2 & 1 \\ 1 & -1 \end{vmatrix} + 1 \begin{vmatrix} 2 & -1 \\ 1 & 2 \end{vmatrix}$$

$$= 1(1-2) - 1(-2-1) + 1(4+1)$$
$$= 1(-1) - 1(-3) + 1(+5)$$
$$= -1 + 3 + 5$$
$$= +7$$

Answer is (E)

PROBLEM 2-20

We are given the following three equations:

$$\frac{L}{2}+\frac{m}{3}+\frac{n}{4}=62$$

$$\frac{L}{4}+\frac{m}{5}+\frac{n}{6}=38$$

$$\frac{L}{3}+\frac{m}{4}+\frac{n}{5}=47$$

The value of L is

(A) 12
(B) 24
(C) 60
(D) 72
(E) 120

Solution.   The problem could also be solved by successive elimination, but we choose to use Cramer's rule and determinants here:

$$D=\begin{vmatrix} \frac{1}{2} & \frac{1}{3} & \frac{1}{4} \\ \frac{1}{4} & \frac{1}{5} & \frac{1}{6} \\ \frac{1}{3} & \frac{1}{4} & \frac{1}{5} \end{vmatrix}=\frac{1}{2}\begin{vmatrix} \frac{1}{5} & \frac{1}{6} \\ \frac{1}{4} & \frac{1}{5} \end{vmatrix}-\frac{1}{3}\begin{vmatrix} \frac{1}{4} & \frac{1}{6} \\ \frac{1}{3} & \frac{1}{5} \end{vmatrix}+\frac{1}{4}\begin{vmatrix} \frac{1}{4} & \frac{1}{5} \\ \frac{1}{3} & \frac{1}{4} \end{vmatrix}$$

$$D = -\frac{1}{1200}+\frac{1}{540}-\frac{1}{960}=-\frac{1}{43{,}200} \qquad \frac{1}{D}=-43{,}200\neq0$$

$$D_L=\begin{vmatrix} 62 & \frac{1}{3} & \frac{1}{4} \\ 38 & \frac{1}{5} & \frac{1}{6} \\ 47 & \frac{1}{4} & \frac{1}{5} \end{vmatrix}=-\frac{62}{600}-\frac{38}{240}+\frac{47}{180}=-\frac{1}{1800}$$

$$L=\frac{D_L}{D}=\frac{43{,}200}{1800}=24$$

Answer is (B)

Additional calculations would show that $m = 60$ and $n = 120$.

PROBLEM 2-21

We are given this set of three simultaneous equations:

$$5X + 2Y + 4Z = \quad 4 \tag{1}$$
$$3X - \ Y + 2Z = -11 \tag{2}$$
$$7X - 3Y - 3Z = \quad 8 \tag{3}$$

The correct value of Y is nearest to

(A) −5
(B) −1
(C) 2
(D) 7
(E) 10

Solution. Although determinants could be used here, we use the successive elimination approach.

$$
\begin{array}{rl}
(1) & 5X + \ 2Y + 4Z = \quad 4 \\
-2 \times (2) & \underline{-6X + \ 2Y - 4Z = \quad 22} \\
(4) & -X + \ 4Y \qquad\quad = \quad 26 \\[6pt]
3 \times (2) & 9X - \ 3Y + 6Z = -33 \\
2 \times (3) & \underline{14X - \ 6Y - 6Z = \quad 16} \\
(5) & 23X - \ 9Y \qquad\quad = -17 \\[6pt]
23 \times (4) & -23X + 92Y \qquad = 598 \\
(5) & \underline{23X - \ 9Y \qquad = -17} \\
& 83Y \qquad\quad = 581 \qquad Y = 7
\end{array}
$$

Answer is (D)

Additional calculations show that $X = 2$ and $Z = -5$.

PROBLEM 2-22

A certain job can be performed by group X in 100 hr. Group Y can perform the same job in 25 hr, and group Z requires 20 hr. If the three groups, X, Y, and Z, work together, the number of hours required to complete the job is nearest

(A) 8
(B) 10
(C) 12
(D) 14
(E) 16

Solution. Let N = number of hours to complete the job. The hourly progress is:

$$\text{Group } X = \frac{N}{100} \qquad \text{Group } Y = \frac{N}{25} \qquad \text{Group } Z = \frac{N}{20}$$

The combined effort is:

$$\frac{N}{100} + \frac{N}{25} + \frac{N}{20} = 1 \qquad (0.01 + 0.04 + 0.05)N = 1$$

$$N = \frac{1}{0.10} = 10 \text{ hr}$$

Answer is (B)

PROBLEM 2-23

The surface area of a tetrahedron is described by

(A) 4 equilateral triangles
(B) 6 squares
(C) 12 pentagons
(D) 3 trapeziums
(E) 8 pentagons

Solution. A tetrahedron (triangular pyramid) is bounded by four equilateral triangles.

Answer is (A)

PROBLEM 2-24

The equation of a straight line that has a slope of $+2$ and passes through a point with x and y coordinates of 4 and 5, respectively, is

(A) $x+2y=14$
(B) $xy=20$
(C) $2x+y=13$
(D) $4y=5x$
(E) $y=2x-3$

Solution. The point-slope equation for a straight line is $y-y_1=m(x-x_1)$, where m is the slope and x_1 and y_1 are the coordinates of a point on the line.

$$y-5=2(x-4) \qquad y-5=2x-8 \qquad y=2x-3$$

Answer is (E)

PROBLEM 2-25

The equation (in rectangular coordinates) of the plane passing through the three points $(1, 3, 5)$, $(2, 4, 4)$, and $(3, 4, 2)$ is

(A) $xyz=-2$
(B) $y-z=2(x+2)$
(C) $2x-y+z-4=0$
(D) $x-2=yz$
(E) $x+2(y+z)=17$

Solution. The general equation of a plane may be written

$$Ax+By+Cz+D=0$$

Substituting the three points:

$$A+3B+5C+D=0 \tag{1}$$
$$2A+4B+4C+D=0 \tag{2}$$
$$3A+4B+2C+D=0 \tag{3}$$

Solving for A, B, and D in terms of C, we have:

Equation $(3)-(2)$,

$$A-2C=0 \qquad A=2C \tag{4}$$

Equation $(3)-(1)$,

$$2A+B-3C=0 \tag{5}$$

Substituting (4) into (5),

$$2(2C)+B-3C=0 \qquad B=-C \tag{6}$$

Substituting (4) and (6) back into (1) gives

$$2C+3(-C)+5C+D=0 \qquad D=-4C$$

From the general equation

$$2Cx-Cy+Cz-4C=0$$
$$C(2x-y+z-4)=0$$

The equation of the plane is $2x-y+z-4=0$.

Answer is (C)

PROBLEM 2-26

The curve represented by the equation $\dfrac{x^2}{a^2}-\dfrac{y^2}{b^2}=1$ is a

    (A) straight line
    (B) circle
    (C) ellipse
    (D) parabola
    (E) hyperbola

Solution. The simple equations for the various curves are

straight line $y=mx+b$

circle $x^2+y^2=a^2$

ellipse $\dfrac{x^2}{a^2}+\dfrac{y^2}{b^2}=1$

parabola $y^2=ax$

$$\text{hyperbola} \qquad \frac{x^2}{a^2} - \frac{y^2}{b^2} = 1$$

Answer is (E)

PROBLEM 2-27

The equation of the largest circle that is tangent to both coordinate axes and has its center on the line $2X + Y - 6 = 0$ is

(A) $X^2 + Y^2 = 36$
(B) $(X-2)^2 + (Y-2)^2 = 4$
(C) $(X+6)^2 + (Y-6)^2 = 36$
(D) $(X-6)^2 + (Y+6)^2 = 36$
(E) $X^2 + Y^2 = 4(X + Y - 1)$

Solution.   The center of the circle must be at the intersection of two straight lines: $2X + Y - 6 = 0$, and $X - Y = 0$ or $X + Y = 0$. Solving the two cases we get:

$$\begin{array}{r} 2X + Y - 6 = 0 \\ \underline{X - Y \quad = 0} \\ 3X \quad\quad -6 = 0 \\ X = 2 \\ Y = 2 \end{array}$$

The circle has its center at (2, 2) with radius $= 2$.

$$\begin{array}{r} 2X + Y - 6 = 0 \\ \underline{-X - Y \quad = 0} \\ X \quad\quad -6 = 0 \\ X = 6 \\ Y = -6 \end{array}$$

The circle has its center at (6, −6) with radius $= 6$.
   The second circle is the correct one, since the largest circle is desired. The general equation for a circle is $(X-a)^2 + (Y-b)^2 = R^2$. Substituting, we get $(X-6)^2 + (Y+6)^2 = 6^2$.

Answer is (D)

PROBLEM 2-28

The cable of a suspension bridge hangs in the shape of an arc of a parabola *AB*. The supporting towers are 70 ft high and 200 ft apart and the lowest point on the cable is 20 ft above the roadway. The length of the supporting rod *L* 50 ft from the middle of the bridge is nearest

(A) 30.0 ft
(B) 32.5 ft
(C) 36.7 ft
(D) 38.2 ft
(E) 45.0 ft

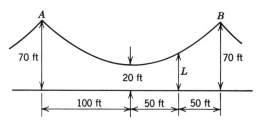

Figure 2-14

Solution.

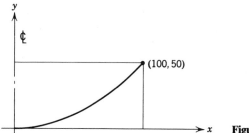

When $x = \pm 100$, $y = 70 - 20 = 50$

$$y = kx^2$$

$$50 = k(100)^2$$

Therefore $k = 0.005$ and $y = 0.005x^2$.

When $x = 50$ ft, $y = 0.005(50)^2 = 0.005(2500) = 12.5$ ft. The length of the supporting rod is $L = 12.5 + 20 = 32.5$ ft.

Answer is (B)

PROBLEM 2-29

In the equation $x = \dfrac{t^2 + t}{2t^2 + 1}$, the limit of x as t approaches infinity is

- (A) ∞
- (B) 2
- (C) 1
- (D) $\frac{1}{2}$
- (E) 0

Solution.

$$\lim_{t \to \infty} \frac{t^2 + t}{2t^2 + 1} = ?$$

The usual rule is to divide both the numerator and denominator by the highest power of the variable occurring in either. In this case, divide by t^2. This step is valid since $t \neq 0$.

$$\lim_{t \to \infty} \frac{t^2 + t}{2t^2 + 1} = \lim_{t \to \infty} \frac{1 + (1/t)}{2 + (1/t^2)} = \frac{1}{2}$$

The limit of each term in the numerator and denominator containing t is zero. $(1/\infty = 0.)$

Answer is (D)

PROBLEM 2-30

In the equation $y = \dfrac{-x^3 + 3x + 2}{x^2 + 2x + 1}$, the limit of y as x approaches a value of -1 is

- (A) 0
- (B) 1
- (C) 2
- (D) 3
- (E) ∞

Solution.

$$\lim_{x\to-1}\frac{-x^3+3x+2}{x^2+2x+1}=\frac{-(-1)^3+3(-1)+2}{(-1)^2+2(-1)+1}=\frac{+1-3+2}{+1-2+1}=\frac{0}{0}$$

which is indeterminate.

Not obtaining a solution above, we try factoring the numerator and denominator:

$$\lim_{x\to-1}\frac{-x^3+3x+2}{x^2+2x+1}=\lim_{x\to-1}\frac{(x+1)(x+1)(-x+2)}{(x+1)(x+1)}$$

$$=\lim_{x\to-1}(-x+2)=+3$$

We find that y is not continuous at $x=-1$ but does approach a limit of $+3$ as x approaches -1.

Answer is (D)

PROBLEM 2-31

In the equation of $y=\dfrac{\ln(1-z)}{z}$, the limit of y as z approaches a value of zero is

(A) ∞

(B) 3

(C) 1

(D) 0

(E) -1

Solution.

$$\lim_{z\to0}y=\frac{\ln(1-z)}{z}=\frac{\ln(1-0)}{0}=\frac{0}{0}$$

The form is indeterminate. Applying L'Hospital's rule,

$$\lim_{z\to0}\frac{\ln(1-z)}{z}=\lim_{z\to0}\frac{\dfrac{d}{dz}[\ln(1-z)]}{\dfrac{d}{dz}(z)}$$

$$=\lim_{z\to0}\frac{-1}{1-z}=-1$$

Answer is (E)

PROBLEM 2-32

The slope of the curve $y = x^3 - 4x$ as it passes through the origin ($x = 0$; $y = 0$) is equal to

    (A) +4
    (B) +2
    (C) 0
    (D) −2
    (E) −4

Solution.

$$\frac{dy}{dx}\bigg|_{x=0} = [3x^2 - 4]_{x=0} = -4$$

Therefore the slope of the curve at $x = 0$ is −4.

Answer is (E)

PROBLEM 2-33

For the position-time function $x = 3t^2 + 2t$, the velocity in the x direction at $t = 1$ is

    (A) 9
    (B) 8
    (C) 7
    (D) 6
    (E) 5

Solution.

$$\text{Velocity} = \frac{dx}{dt} = [6t + 2]_{t=1} = 6 + 2 = 8$$

Answer is (B)

PROBLEM 2-34

The stiffness of a rectangular timber is proportional to the width and the cube of the depth. The width of the stiffest beam that can be made of a circular log whose diameter is 20 in. is closest to

(A) 10 in.
(B) 12 in.
(C) 14 in.
(D) 16 in.
(E) 17 in.

Solution.

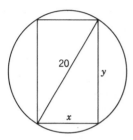

Figure 2-16

Here x = beam width, y = beam depth, and log diameter = 20 in.

To maximize stiffness, the function xy^3 must be maximized. From the figure, $y = (20^2 - x^2)^{1/2}$. Writing xy^3 as a function of one variable,

$$xy^3 = x(20^2 - x^2)^{3/2}$$
$$f(x) = x(20^2 - x^2)^{3/2} \qquad \text{for } 0 \le x \le 20$$

Since $f(0) = f(20) = 0$ and $f(x) > 0$ for intermediate values of x, it is clear that $f(x)$ must attain a maximum value. To find the maximum, we set $f'(x) = 0$ and solve for x. [Alternatively, we could check that $f(x)$ is indeed maximized by verifying that $f''(x) < 0$ at that point.] Using the chain rule for differentiation of a product,

$$f'(x) = x\tfrac{3}{2}(20^2 - x^2)^{1/2} \times (-2x) + (20^2 - x^2)^{3/2} \times (1)$$
$$= -3x^2(20^2 - x^2)^{1/2} + (20^2 - x^2)^{3/2} = 0$$

Dividing by $(20^2 - x^2)^{1/2}$, we get $-3x^2 + 20^2 - x^2 = 0$, $\quad -4x^2 + 20^2 = 0$, and $x^2 = 400/4 = 100$. Then

$$x = 10 \text{ in.}$$

Answer is (A)

PROBLEM 2-35

Circular cylindrical cans of volume V_0 are to be manufactured with both ends closed. The ratio between the diameter and height that will require the minimum amount of metal to make each can is nearest to

(A) $d/h = 0.6$
(B) $d/h = 0.8$
(C) $d/h = 1.0$
(D) $d/h = 1.2$
(E) $d/h = 1.4$

Solution.

$$\text{Total surface area} = 2 \text{ end areas} + \text{side surface area}$$

$$= 2\left(\frac{\pi}{4}d^2\right) + \pi\,dh$$

The surface area is to be a minimum. To find extreme values we must find the first derivative of the function. But the function contains two variables (d and h), so one must be defined in terms of the other to eliminate one variable.

$$V_0 = \frac{\pi}{4}d^2 h$$

Therefore

$$h = \frac{4V_0}{\pi d^2}$$

Hence

$$\text{Surface area} = 2\left(\frac{\pi}{4}d^2\right) + \pi d\left(\frac{4V_0}{\pi d^2}\right) = \frac{\pi}{2}d^2 + \frac{4V_0}{d}$$

Taking the first derivative and equating it to zero,

$$f'(d) = \frac{\pi}{2}(2d) + 4V_0\left(\frac{-1}{d^2}\right) = 0$$

$$\pi d = \frac{4V_0}{d^2} \quad \text{and} \quad d^3 = \frac{4V_0}{\pi}$$

[Check: $f''(d) = \pi + 8V_0/d^3 > 0$ and therefore is a minimum.]

Since $V_0 = (\pi/4) d^2 h$, we also have $d^2 h = 4 V_0/\pi$. Therefore

$$d^2 h = d^3 \quad \text{or} \quad h = d$$

$$\text{Ratio } d/h = 1$$

Answer is (C)

PROBLEM 2-36

The area under the curve $y = x^2$ between the values $x = +1$ ft and $x = +7$ ft is nearest to

(A) 96 ft^2
(B) 114 ft^2
(C) 147 ft^2
(D) 171 ft^2
(E) 342 ft^2

Solution.

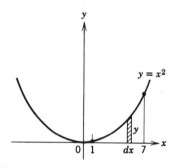

Figure 2-17

$$dA = y \, dx$$

$$A = \int dA = \int y \, dx$$

Substituting for y in terms of x,

$$A = \int_1^7 x^2 \, dx = \frac{x^3}{3} \Big|_1^7$$

$$= \frac{343}{3} - \frac{1}{3} = \frac{342}{3}$$

$$= 114 \text{ ft}^2$$

Answer is (B)

PROBLEM 2-37

The area formed by the boundaries $y = 1$, $x = 1$, and $y = e^{-x}$ is closest to

(A) 0.50
(B) 0.46
(C) 0.42
(D) 0.38
(E) 0.34

Solution.

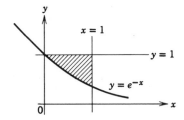

Figure 2-18

$$\text{Area} = 1^2 - \int_0^1 e^{-x}\, dx = 1 - [-e^{-x}]\Big|_0^1 = 1 - [-e^{-1} - (-e^{-0})]$$

$$= 1 + e^{-1} - 1 = e^{-1}$$

$$= \frac{1}{2.718} = 0.368$$

Answer is (D)

PROBLEM 2-38

The rate of decay of radioactive elements is usually assumed to be proportional to the number of atoms that have not decayed, where λ is the proportionality constant. If at time $t = 0$ there are X_0 atoms of a given element, the expression for the number of atoms, X, that have not decayed (as a function of time t, λ, and X_0) is

(A) $X_0(1 - \lambda t)$
(B) $X_0 e^{-\lambda t}$
(C) $X_0(1 - e^{-\lambda t})$
(D) $X_0/(1 + \lambda t)$
(E) $X_0(1 - \lambda t^{1/2})$

Solution. X = number of atoms that have not decayed. The rate of decay of X is proportional to X, or $dX/dt = -\lambda X$. Rearranging this equation,

$$\frac{dX}{X} = -\lambda \, dt$$

Integrating,

$$\ln X = -\lambda t + A$$
$$X = e^{-\lambda t} e^A$$

When $t = 0$, $X = X_0 = e^A$. Therefore $X = X_0 e^{-\lambda t}$.

Answer is (B)

PROBLEM 2-39

A 1000-ft^3 storage tank is filled with natural gas at 80°F and 1 atm pressure. The tank is flushed out with nitrogen gas at 80°F and 1 atm pressure, at a constant rate of 300 cfm. The flushing process is carried out at constant temperature and pressure, under conditions of perfect mixing in the tank at all times. The time required to reach a gas composition of 95 vol. % nitrogen in the tank is nearest to

    (A) 3 min
    (B) 5 min
    (C) 7 min
    (D) 10 min
    (E) 15 min

Solution.   Let

    g = quantity of pure natural gas in the tank at any time
    x = quantity of nitrogen added to the tank = quantity of mixture removed
       from the tank
    volume of tank = 1000 ft^3
    5% of volume of tank = $0.05 \times 1000 = 50$ ft^3

Suppose a volume Δx of the mixture is removed from the tank. The amount of natural gas thus removed will be $(g/1000)\,\Delta x$. Hence the change in the amount of natural gas in the tank is given by $\Delta g = -(g/1000)\,\Delta x$. Then the ratio of the quantity of natural gas removed to the volume of nitrogen added is

$$\frac{\Delta g}{\Delta x} = -\frac{g}{1000}$$

When $\Delta x \rightarrow 0$ we obtain the instantaneous rate of change of g with respect to x:

$$\frac{dg}{dx} = \frac{-g}{1000}$$

Separating the variables of the differential equation,

$$\frac{dg}{g} + \frac{dx}{1000} = 0$$

Integrating,

$$\int \frac{dg}{g} + \int \frac{dx}{1000} = C \qquad \ln g + \frac{x}{1000} = C$$

When $g = 1000$, $x = 0$. Therefore $C = \ln 1000$.

$$\frac{x}{1000} = \ln 1000 - \ln g = \ln \frac{1000}{g}$$

We want to find x when g is 5% by volume or 50 ft^3. Hence

$$\frac{x}{1000} = \ln \frac{1000}{50} = \ln 20 = 3.0$$

$$x = 3000 \text{ ft}^3 \text{ nitrogen}$$

Since nitrogen flows in at 300 cfm, the time required to reach 5% by volume natural gas = 3000/300 = 10 min.

Answer is (D)

PROBLEM 2-40

A factory has measured the diameter of 100 random samples of its product. The results, arranged in ascending order, were:

(45 results between 0.859 and 0.900, inclusive) · · ·
0.901   0.902   0.902   0.902   0.903   0.903
0.904   0.904   0.904   0.904 · · ·
(45 more different results from 0.905 to 0.958, inclusive)

The sum of all 100 observations is 91.170. No observed value among those not numerically shown above occurred more than twice. The smallest value observed was 0.859; the largest, 0.958. From the information given, the median for the data is

(A) 0.901

(B) 0.902

(C) 0.903

(D) 0.904

(E) 0.912

Solution. Mean: The arithmetic mean is what is commonly called the *average*, that is, the sum of the values divided by the number of values.

$$\bar{X} = \frac{1}{N} \sum_{i=1}^{N} X_i = \frac{91.170}{100} = 0.9117 = 0.912$$

Median: The median of a set of data is the middle value in order of size if N is odd or the value midway between the two middle items if N is even. Here the median is halfway between 0.903 and 0.903, or 0.903.

Mode: This is the most frequent value. In this case it is 0.904, which occurred four times.

Answer is (C)

PROBLEM 2-41

The geometric mean of the numbers 4 and 49 is

(A) 14.0

(B) 22.0

(C) 26.5

(D) 33.0

(E) 49.3

Solution. The geometric mean (G) may be defined as the Nth root of the product of N values:

$$G = \sqrt[N]{x_1 x_2 \cdots x_n}$$

In this case

$$G = \sqrt[2]{(4)(49)} = \sqrt{4} \times \sqrt{49} = 2(7) = 14$$

Answer is (A)

PROBLEM 2-42

The probability of obtaining at least one 6 in three throws of a die is nearest to

(A) 0.17
(B) 0.42
(C) 0.50
(D) 0.58
(E) 0.83

Solution. Let P_1 be the probability of getting one or more 6's and P_2 be the probability of getting no 6's. For any situation the sum of probabilities for all possibilities is 1:

$$P_1 + P_2 = 1$$

The probability of *not* getting a 6 on any single roll is $\frac{5}{6}$; thus the probability that a 6 will not turn up in three rolls is

$$P_2 = \left(\tfrac{5}{6}\right)\left(\tfrac{5}{6}\right)\left(\tfrac{5}{6}\right) = \tfrac{125}{216} = 0.58$$
$$P_1 = 1 - P_2 = 1 - 0.58 = 0.42$$

Therefore the probability of obtaining at least one 6 in three throws of a die is 0.42.

Answer is (B)

PROBLEM 2-43

An elementary game is played by rolling a die and drawing a ball from a bag containing three white and seven black balls. The player wins whenever he rolls a number less than 4 and draws a black ball. What is the probability of winning in the first attempt?

(A) 7/20
(B) 12/10
(C) 1/2
(D) 7/10
(E) 13/20

Solution. The probability of rolling a number less than 4 with a die $=\frac{3}{6}=\frac{1}{2}$. The probability of drawing a black ball $= 7/(7+3)=7/10$. The probability of a series of independent events equals the product of probabilities of the individual events:

$$\tfrac{1}{2}\times\tfrac{7}{10}=\tfrac{7}{20}$$

Answer is (A)

PROBLEM 2-44

The table gives the values of x and the frequency f with which they occur.

x	f
2	4
4	6
7	6
12	4

(1) The arithmetic mean $\bar{x}$ is nearest to

(A) 6.55
(B) 6.40
(C) 6.25
(D) 6.10
(E) 5.95

Solution. When individual values recur, it is convenient to compute the mean as the sum of each distinct value times its frequency of occurrence f, divided by the total number of values.

x	f	fx
2	4	8
4	6	24
7	6	42
12	4	48
	$N=20$	$122=\Sigma\,(fx)$

$$\bar{x}=\frac{\Sigma\,(fx)}{N}$$

$$\bar{x}=\tfrac{122}{20}=6.1$$

This procedure is equivalent to summing 20 individual values of x and dividing by 20.

$$\text{Answer is (D)}$$

(2) The standard deviation σ is nearest to

(A) 4.7
(B) 4.4
(C) 4.1
(D) 3.8
(E) 3.5

Solution. The standard deviation σ is a measure of the dispersion or scatter of a set of values. It is sometimes called the rms deviation, as this describes its method of calculation. First, square the deviations of individual values from the arithmetic mean. Then take the mean of these squares and extract the square root.

$$\sigma = \left[\frac{\Sigma f(x-\bar{x})^2}{N-1}\right]^{1/2}$$

x	f	$x-\bar{x}$	$(x-\bar{x})^2$	$f(x-\bar{x})^2$
2	4	−4.1	16.81	67.24
4	6	−2.1	4.41	26.46
7	6	0.9	0.81	4.86
12	4	5.9	34.81	139.24
	$N=20$			$\Sigma=237.80$

$$\sigma = \left[\frac{\Sigma f(x-\bar{x})^2}{N-1}\right]^{1/2} = \left(\frac{237.80}{19}\right)^{1/2} = (12.52)^{1/2} = 3.54$$

$$\text{Answer is (E)}$$

(If N rather than $N-1$ were used in the denominator of σ, the result would be $\sigma=3.45$.)

PROBLEM 2-45

Assume a group of nine people consists of four men and five women. The probability that a committee of three, selected at random, would consist of two men and one woman is nearest to

(A) 0.30
(B) 0.35
(C) 0.40
(D) 0.45
(E) 0.50

Solution.   Here we use the notation

$$\binom{n}{r} = \frac{n!}{r!(n-r)!}$$

which gives the total number of ways r objects can be chosen from n objects.
      There are nine people (four men and five women). The total number of committees of three people would be equal to the total number of committees consisting of

0 men	3 women	or	$\binom{4}{0} \times \binom{5}{3} = 1 \times \dfrac{5!}{3!\,2!} = 10$
1 man	2 women		$\binom{4}{1} \times \binom{5}{2} = \dfrac{4!}{3!} \times \dfrac{5!}{2!\,3!} = 40$
2 men	1 woman		$\binom{4}{2} \times \binom{5}{1} = \dfrac{4!}{2!\,2!} \times \dfrac{5!}{4!} = 30$
3 men	0 women		$\binom{4}{3} \times \binom{5}{0} = \dfrac{4!}{3!} \times 1 \quad = \underline{4}$

$$\text{Total } 84$$

The number of committees consisting of two men and one woman is 30 out of 84 possible combinations. Hence the probability of this happening is

$$P = \tfrac{30}{84} = 0.357$$

Answer is (B)

PROBLEM 2-46

The sum of all whole numbers from 1 to 100 inclusive is nearest to

(A) 6500
(B) 6000
(C) 5500
(D) 5050
(E) 5005

Solution. We want to determine the sum S of an arithmetic progression. Let a be the first term, d the difference between successive terms, l the last term, and n the number of terms. Then

$$S = a + (a+d) + (a+2d) + \cdots + [a+(n-1)d]$$

or, in reverse order,

$$S = [a+(n-1)d] + [a+(n-2)d] + \cdots + a$$

Adding these two equations,

$$2S = n[2a + (n-1)d]$$

or, since $[a+(n-1)d] = l$,

$$S = \frac{n}{2}(a+l)$$

In this problem $a = 1$, $n = l = 100$, and $S = \frac{100}{2}(101)$.

$$S = 5050$$

Answer is (D)

PROBLEM 2-47

Given a universe $= (1, 2, 3, 4, 5, 6, 7)$, set $M = (1, 3, 6)$, and set $N = (1, 2, 6, 7)$, the set $M \cap N$ is

(A) (2, 4, 5, 7)
(B) (3, 4, 5)
(C) (1, 6)
(D) (3)
(E) (1, 2, 4, 5, 6, 7)

Solution. $\bar{M}$ represents non-*M*, or all elements of the universe that are not in set *M*.

(A) $\bar{M} = (2, 4, 5, 7)$

$\bar{N}$ represents non-*N*.

(B) $\bar{N} = (3, 4, 5)$

$M \cap N$ denotes the intersection of sets *M* and *N*, which includes all elements in *both* sets *M* and *N*.

(C) $M \cap N = (1, 6)$

$M \cap \bar{N}$ is the intersection of sets *M* and $\bar{N}$.

(D) $M \cap \bar{N} = (3)$

$\bar{M} \cup N$ represents the union of sets $\bar{M}$ and *N*; it includes all elements that are members of *either $\bar{M}$ or N*.

(E) $\bar{M} \cup N = (1, 2, 4, 5, 6, 7)$

Answer is (C)

3 Statics

Statics is the subdivision of mechanics that considers the equilibrium of stationary or uniformly translating particles or rigid bodies. By this definition a unifying characteristic of *all* statics problems is the absence of any acceleration.

A variety of forces that may act on a body are considered when statics is studied. The forces may either act at specific points on the body or be distributed over a region in space; the weight of a body or a distributed pressure are examples of the latter. In connection with distributed forces it is common to discuss the determination of centroids, centers of gravity, and moments of inertia. This is done here. We defer the subject of shear and bending moment diagrams, which are related to the internal static equilibrium of a body, to Chapter 5. Fluid statics problems are covered in Chapter 6.

EQUILIBRIUM

A body is in a state of static equilibrium when no net or unbalanced resultant force $\mathbf{R}$ acts on the body and, in addition, the forces on the body create no net tendency toward rotation or couple $\mathbf{M}$. These requirements for equilibrium, restated, are

$$\mathbf{R} = 0 \qquad \mathbf{M} = 0 \tag{3-1}$$

The sense of the vector that represents a moment is determined by use of the right-hand rule.

For the Cartesian (x, y, z) coordinate system shown in Fig. 3-1, Eq. (3-1) become

$$|\mathbf{R}| = [(\textstyle\sum F_x)^2 + (\sum F_y)^2 + (\sum F_z)^2]^{1/2} = 0$$

$$|\mathbf{M}| = [(\textstyle\sum M_x)^2 + (\sum M_y)^2 + (\sum M_z)^2]^{1/2} = 0 \tag{3-2}$$

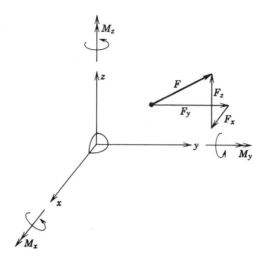

Figure 3-1 Cartesian coordinate system

which is equivalent to the requirements

$$\sum F_x = 0 \qquad \sum M_x = 0$$
$$\sum F_y = 0 \qquad \sum M_y = 0 \qquad (3\text{-}3)$$
$$\sum F_z = 0 \qquad \sum M_z = 0$$

For a three-dimensional statically determinate problem we thus have six independent equations for equilibrium. For the corresponding two-dimensional situation, the three equations are

$$\sum F_x = 0 \qquad \sum F_y = 0 \qquad \sum M_z = 0 \qquad (3\text{-}4)$$

Commonly encountered examples of bodies in static equilibrium are the two- and three-force bodies.

The two-force body is one that is acted on by concentrated forces that are applied at only two points on the body. Equations (3-4) can be used to show that these two forces have the same line of action, the same magnitude, and act in opposite directions. This information is particularly useful in the solution of statically determinate truss problems.

Concentrated forces acting at three points on a body cause the body to be called a three-force body. For equilibrium these forces must either (*a*) have lines of action that all pass through the same point so that no couple acts on the body or (*b*) act along parallel lines of action.

Example 1

A 4-ft × 10-ft block that weighs 1600 lb is held in a horizontal position by a force P and hinge A, as shown in Fig. 3-2a. Determine the force P and the resultant hinge reaction R_A.

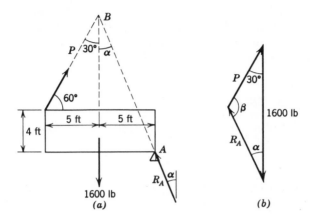

Figure 3-2

Solution. The block, which weighs 1600 lb, is held in equilibrium by forces P and R_A, thus forming a three-force body. The point common to all three lines of action is point B, which is directly above the location of the equivalent concentrated 1600-lb weight. A force triangle, shown in Fig. 3-2b, can easily be drawn.

From Fig. 3-2a we have

$$\tan \alpha = \frac{5}{4 + 5 \tan 60°} = \frac{5}{4 + 5\sqrt{3}} = 0.395$$

$$\alpha = 21°33'$$

Hence

$$\beta = 180° - 30° - \alpha = 128°27'$$

Using the law of sines,

$$\frac{P}{\sin \alpha} = \frac{R_A}{\sin 30°} = \frac{1600}{\sin \beta} = 2046$$

$$P = 2046 \sin 21°33' = 752 \text{ lb}$$

$$R_A = 2046 \sin 30° = 1023 \text{ lb} \qquad \text{acting at an angle } \alpha$$

FREE-BODY DIAGRAM

In Fig. 3-2*a* we have drawn a free-body diagram of the block. Relevant dimensions and angles are also shown. It is helpful to prepare a free-body diagram during the course of solving most problems in mechanics. The diagram should clearly show the essential elements of the problem and no other information. Sometimes, in the interest of clarity, dimensions are also excluded from the free-body diagram when they would tend to clutter the drawing. The use of a free-body diagram in finding the reactions for simple bodies is further shown in the solved problems.

FRAMES AND TRUSSES

Certain useful procedures have evolved from the analysis of the forces in frames and trusses. A truss is an assemblage of pieces that may be accurately represented as two-force members; a frame is composed of members that each are usually acted on by applied loads at more than two points. The analysis of a statically determinate frame is usually accomplished by drawing a free-body diagram of each member and then writing the appropriate equilibrium equations for each diagram. These equations are then solved for the unknown forces and reactions. The analysis of trusses, however, usually proceeds by some combination of the methods known as the method of joints or the method of sections.

Example 2

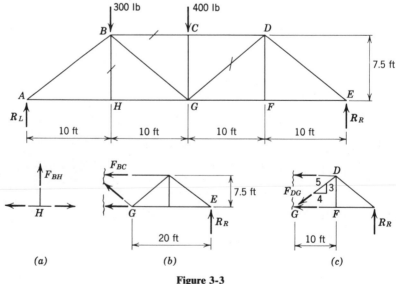

Figure 3-3

Determine the force in members *BH*, *BC*, and *GD* of the truss shown in Fig. 3-3. Note that the truss is composed of triangles 7.5 ft : 10.0 ft : 12.5 ft, so that they are 3 : 4 : 5 right triangles.

Solution.   First we solve for the reactions R_L and R_R:

$$\sum M_E = 0 \qquad 40R_L = 30(300) + 20(400)$$

$$R_L = 425 \text{ lb}$$

$$\sum F_v = 0 \qquad R_R = 300 + 400 - 425 = 275 \text{ lb}$$

The method of joints considers the equilibrium of each pinned connection between members. Because of the pin, no moment can be transmitted through the joint. For a two-dimensional truss, joint equilibrium requires that $\sum F_h = 0$ and $\sum F_v = 0$ be satisfied for the forces acting at each joint. We have an especially simple case in this problem, as shown in Fig. 3-3a. All bar forces are shown to be in tension. Summing forces vertically,

$$\sum F_v = 0 = F_{BH}$$

The method of joints is most efficiently used (*a*) when the forces in all members of a truss are desired or (*b*) when special cases such as the example arise.

The method of sections is normally more efficient than the method of joints when only a few selected bar forces must be found. If the truss as a whole is in equilibrium, then any segment or section of the structure must also be in equilibrium. This principle is put to use by selecting an "appropriate" section of a structure and applying the principles of statics. In most cases the appropriate section is one that (*a*) severs the member of interest and (*b*) results in a free body acted on by only three unknown forces. Numerous exceptions to the second requirement exist, however.

Selecting a section as shown in Fig. 3-3b, we write

$$\sum M_G = 0 \qquad 7.5F_{BC} = -20R_R$$

$$F_{BC} = \frac{-20(275)}{7.5} = 733 \text{ lb compression}$$

Referring to Fig. 3-3c,

$$\sum F_v = 0 \qquad \tfrac{3}{5}F_{DG} = R_R$$

$$F_{DG} = \tfrac{5}{3}(275) = 458 \text{ lb tension}$$

FRICTION

In many situations forces due to friction cause a body to remain in static equilibrium. In these problems it is important to recall that frictional forces always act to oppose any actual or impending motion. For most cases of dry friction the frictional force F is simply proportional to the normal force N, or $F = \mu N$ where μ is the coefficient of either static or kinetic friction. (A few more specialized friction problems, involving rolling friction or belt friction, are presented only in the statics problem section.)

Example 3

A 500-lb block rests on a 30° plane (Fig. 3-4). If the coefficient of static friction is 0.30 and the coefficient of kinetic friction is 0.20, what is the value of P needed to

(*a*) prevent the block from sliding down the plane?
(*b*) start the block moving up the plane?
(*c*) keep the block moving up the plane?

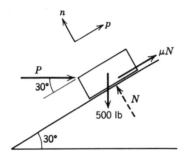

Figure 3-4

Solution.   Normal to the plane, $\sum F_n = 0$

$$N - P \sin 30° - 500 \cos 30° = 0$$

$$N = \frac{P}{2} + 433$$

(*a*) Parallel to the plane, $\sum F_p = 0$

$$\mu N + P \cos 30° - 500 \sin 30° = 0$$

$$0.3\left(\frac{P}{2} + 433\right) + 0.866P - 250 = 0$$

$$P = 118.2 \text{ lb}$$

(*b*) Here we have impending motion *up* the plane so the direction of μN in Fig. 3-4 must be directed down the plane to oppose the motion. This can be accomplished mathematically by replacing μN by $-\mu N$ in the earlier equation. Thus

$$-0.3\left(\frac{P}{2}+433\right)+0.866P-250=0$$

$$P = 531\,\text{lb}$$

(c) The problem is unchanged from (b) except that we now consider the kinetic ($\mu = 0.2$) rather than the stationary ($\mu = 0.3$) case.

$$-0.2\left(\frac{P}{2}+433\right)+0.866P-250=0$$

$$P = 439\,\text{lb}$$

CENTROID, CENTER OF GRAVITY, MOMENT OF INERTIA

The center of gravity (also called center of mass) and mass moment of inertia are physical properties of a body. If the density is uniform throughout the body, these properties exactly coincide with the associated, but purely geometric, properties of a body which are called the centroid and moment of inertia. The equations for the center of gravity of a body are

$$\bar{X}=\frac{1}{M}\int_{\text{V}} \rho x\, d\text{V} \qquad \bar{Y}=\frac{1}{M}\int_{\text{V}} \rho y\, d\text{V} \qquad \bar{Z}=\frac{1}{M}\int_{\text{V}} \rho z\, d\text{V} \quad (3\text{-}5)$$

where $M = \int_{\text{V}} \rho\, d\text{V}$ is the mass of the body composed of volume elements $d\text{V}$ that have density ρ. These equations apply equally well for rods, plane areas, and volumes of any shape; one need only take care to select an appropriate volume element. The integrals may be replaced by finite sums for single or composite bodies whenever the component volume and the centroid or center of gravity of each individual element are already known. Note that the density ρ will cancel in Eq. (3-5) when it is constant throughout a body, and we then have the equations for the centroid of the body. These principles are illustrated in Example 4. Appendix B gives the centroidal coordinates for some common geometric shapes.

Example 4

Locate the X, Y, Z coordinates of the centroid for the slender rod $ABCD$ of constant density shown in Fig. 3-5. The semicircular portion ABC has a radius of π m and lies in the Y-Z plane. The straight portion is 10 meters long and lies in the X-Y plane. The point D is located at $X = 6$, $Y = 8$.

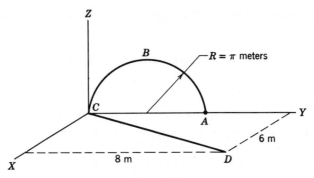

Figure 3-5

Solution. Length $L = \pi R + \sqrt{6^2 + 8^2} = 9.87 + 10 = 19.87\,\text{m}$

$$\bar{X} = \frac{1}{L} \int x\, dL = \frac{1}{L} \sum_{i=1}^{n} x_i L_i = \frac{0 \times 9.87 + 3 \times 10}{19.87} = 1.51\,\text{m}$$

$$\bar{Y} = \frac{1}{L} \sum_{i=1}^{n} y_i L_i = \frac{\pi \times 9.87 + 4 \times 10}{19.87} = 3.57\,\text{m}$$

The center of gravity of a semicircular disk, measured from the diameter, is

$$\frac{4R}{3\pi} = \frac{4\pi}{3\pi} = 1.33\,\text{m}$$

$$\bar{Z} = \frac{1}{L} \sum_{i=1}^{n} z_i L_i = \frac{1.33 \times 9.87 + 0 \times 10}{19.87} = 0.66\,\text{m}$$

The moment of inertia (or second moment) of an area, with respect to some axis $r = 0$, is

$$I = \int_A r^2\, dA \qquad (3\text{-}6)$$

In this expression r is the distance from the axis $r = 0$ to the centroid of the area element dA. Usually r is replaced by either x or y so that the moment of inertia is found with respect to the x or y coordinate axis. The mass moment of inertia is similarly defined as

$$I_m = \int_\Psi r^2\, dm = \int_\Psi r^2 \rho\, d\Psi \qquad (3\text{-}7)$$

If the moment of inertia of a body around its centroidal axis I_0 is known, the moment of inertia around any axis parallel to this centroidal axis may be found from the parallel-axis theorem

$$I = I_0 + Ad^2 \tag{3-8}$$

where A is the area and d is the distance between the two parallel axes. For mass moments of inertia, A is replaced by the mass M in Eq. (3-8). Appendix B gives the moment of inertia I_0 for some common areas.

Example 5

Find the moment of inertia of a rectangle of base b and height h (Fig. 3-6)
   (a) about the centroidal axis.
   (b) about the base of the rectangle.

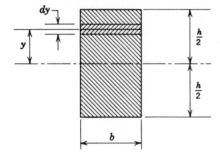

Figure 3-6

Solution.   (a) We select a differential area of width b and height dy that has a distance y to its centroid, or $dA = b\,dy$. Applying Eq. (3-6) with $y = 0$ at the centroidal axis,

$$I_0 = \int_{-h/2}^{h/2} y^2 (b\,dy)$$

Since b is constant,

$$I_0 = b \int_{-h/2}^{h/2} y^2\,dy = b[\tfrac{1}{3}y^3]_{-h/2}^{h/2}$$

$$= \frac{b}{3}\left(\frac{h^3}{8} + \frac{h^3}{8}\right) = \frac{b}{3}\left(\frac{h^3}{4}\right) = \frac{bh^3}{12}$$

   (b) Applying Eq. (3-8) gives

$$I = I_0 + Ad^2 = \frac{bh^3}{12} + (bh)\left(\frac{h}{2}\right)^2$$

$$= \frac{bh^3}{3}$$

PROBLEM 3-1

For the system shown (Fig. 3-7), choose the one true statement concerning the bearing reactions at A and B if the system is in equilibrium.

    (A) Both reactions are vertical.

    (B) Neither reaction is vertical.

    (C) The reaction at A is vertical.

    (D) The reaction at B is vertical.

    (E) Since there are two unknown components at A and B, respectively, the system is statically indeterminate and the reactions cannot be described using methods of statics.

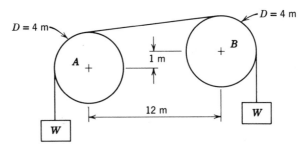

Figure 3-7

Solution.   Draw the free-body diagram:

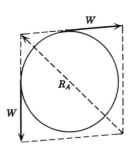 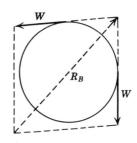

Figure 3-8

Hence neither reaction A nor B is vertical.

Answer is (B)

PROBLEM 3-2

The value of reaction R (Fig. 3-9) is

    (A) 9.6 newtons
    (B) 20.0 newtons
    (C) 26.4 newtons
    (D) 36.0 newtons
    (E) 44.0 newtons

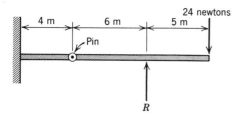

Figure 3-9

Solution.

$$\sum M_{\text{pin}} = 0$$

$$-24N \times 11 + 6R = 0 \qquad R = \frac{24 \times 11}{6} = 44\ \text{N}$$

Answer is (E)

PROBLEM 3-3

The moment at reaction R (Fig. 3-10) is

    (A) Unknown; the structure is statically indeterminate
    (B) 0
    (C) Pa
    (D) $Pa\left(\dfrac{b+c}{a+b+c}\right)$
    (E) $P\left(\dfrac{bc}{a+b}\right)$

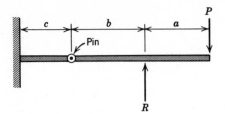

R **Figure 3-10**

Solution.   The moment at any point on a beam may be determined by calculating the moments on either side of the point—hence $M_R = Pa$.

Answer is (C)

PROBLEM 3-4

For the beam loaded as shown in Fig. 3-11, reaction R is

(A) $\dfrac{Pa}{L}$

(B) $\dfrac{Pb}{L}$

(C) $\dfrac{PL}{a}$

(D) $\dfrac{PL}{b}$

(E) $\dfrac{PL}{ab}$

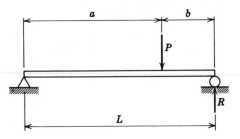

Figure 3-11

Solution.

$$\Sigma M_{\text{left reaction}} = 0$$

$$+RL - Pa = 0 \qquad R = \frac{Pa}{L}$$

Answer is (A)

PROBLEM 3-5

Determine the magnitude of the beam reaction marked R in Fig. 3-12. The value is nearest to

(A) 0
(B) 0.5P
(C) 1.0P
(D) 1.5P
(E) 2.0P

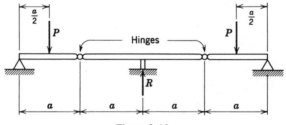

Figure 3-12

Solution.

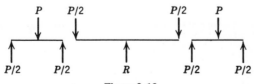

Figure 3-13

We can see that $R = 2(P/2) = P$.

Answer is (C)

PROBLEM 3-6

Determine the magnitude of the reaction marked R in Fig. 3-14. The value is nearest to

    (A) 0
    (B) P
    (C) $2P$
    (D) $3P$
    (E) $4P$

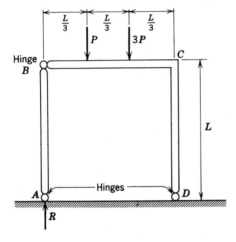

Figure 3-14

Solution. Since joint C in Fig. 3-14 is rigid, we have a stable rigid frame.

$$\sum M_D = 0$$

$$+3P\left(\frac{L}{3}\right)+P\left(\frac{2L}{3}\right)-RL = 0 \qquad PL+\tfrac{2}{3}PL-RL = 0$$

$$R = 1\tfrac{2}{3}P$$

Answer is (C)

PROBLEM 3-7

For the structure loaded as shown in Fig. 3-15, reaction R is

(A) $\dfrac{PL}{H}$

(B) $\dfrac{2PL}{3H}$

(C) $\dfrac{PL}{3H}$

(D) $\dfrac{3PH}{L}$

(E) $\dfrac{3PH}{2L}$

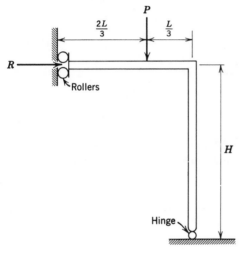

Figure 3-15

Solution.

$\sum M_{\text{base hinge}} = 0$

$$-RH + P\frac{L}{3} = 0 \qquad R = \frac{PL}{3H}$$

Answer is (C)

PROBLEM 3-8

In the vector diagram (Fig. 3-16) $\bar{R}$ represents

(A) $\bar{A} + \bar{B}$
(B) $\bar{A} - \bar{B}$
(C) $\bar{B} - \bar{A}$
(D) $\bar{A} \cdot \bar{B}$
(E) $\sqrt{\bar{A}^2 + \bar{B}^2}$

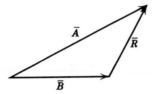

Figure 3-16

Solution.

$$\bar{B} + \bar{R} = \bar{A} \qquad \bar{R} = \bar{A} - \bar{B}$$

Answer is (B)

PROBLEM 3-9

Given a 50-newton pulley, supported as shown (Fig. 3-17) and carrying a cable supporting an additional 50-newton load. The force the beam exerts on the pulley is

(A) 50 N up
(B) 100 N up
(C) 100 N down
(D) 150 N up
(E) 150 N down

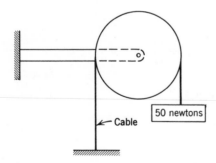

Figure 3-17

Solution.

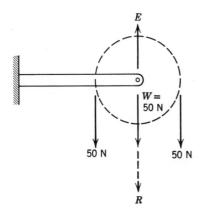

$$R$$ **Figure 3-18**

The resultant of three forces $= 50 + 50 + 50 = 150$ N down.
$$\Sigma F_y = 0$$

$$+E - 50 - 50 - 50 = 0$$

$$E = +150 \text{ N equilibrant}$$

Therefore the force the beam exerts on the pulley is 150 N up.

Answer is (D)

PROBLEM 3-10

A 5000-newton sphere rests against a smooth plane inclined at 45° to the horizontal and against a smooth wall as shown in Fig. 3-19. The magnitude of the reaction R_B is nearest to

(A) 0 newtons
(B) 2500 newtons
(C) 3500 newtons
(D) 5000 newtons
(E) 7000 newtons

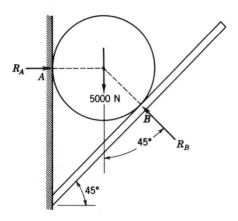

Figure 3-19

Solution.

$$\sum F_V = 0$$

$$-5000 + R_B \cos 45° = 0 \qquad R_B = \frac{5000}{0.707} = 7071 \text{ N}$$

Answer is (E)

PROBLEM 3-11

In Fig. 3-19 a 5000-N sphere rests against a smooth plane inclined at 45° to the horizontal and against a smooth vertical wall. What is the magnitude of the reaction R_A?

(A) 0 newtons
(B) 2500 newtons
(C) 3535 newtons
(D) 5000 newtons
(E) 7071 newtons

Solution.   In Problem 3-10 we found that $R_B = 7071 \text{ N}$

$$\sum F_H = 0$$

$$R_A - R_B \sin 45° = 0 \qquad R_A = 7071(0.707) = 5000 \text{ N}$$

Answer is (D)

PROBLEM 3-12

A 1200-lb solid steel triangular prism is supported by three vertical cables as shown in Fig. 3-20. One cable is at the middle of one edge, with the other two at the corners of the opposite edge. All three cables are the same length and are of the same size. Which one of the following statements is correct concerning the tensile forces in the cables?

(A) The tensile force in cable C_1 is twice that in cable C_2.

(B) The tensile force in cable C_1 is half that in cable C_2.

(C) The tensile force in cable C_3 is equal to the sum of the tensile forces in cables C_1 and C_2.

(D) All three cables have the same tensile force.

(E) None of the statements (A) through (D) is correct.

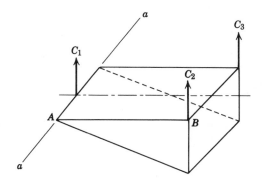

Figure 3-20

Solution. Owing to symmetry, $C_2 = C_3$.

$$\Sigma M_a = 0$$

$$(C_2 + C_3)L - \tfrac{2}{3}L(1200) = 0 \qquad C_2 + C_3 = 800$$

Therefore

$$C_2 = C_3 = 400 \text{ lb}$$

$$\Sigma F_V = 0$$

$$400 + 400 + C_1 - 1200 = 0$$

Thus

$$C_1 = 400 \text{ lb}$$

Therefore all cable tensions are equal and have a value of 400 lb.

Answer is (D)

PROBLEM 3-13

The tension in the cable supporting the beam loaded as shown in Fig. 3-21 is approximately

    (A) 5 kips
    (B) 10 kips
    (C) 15 kips
    (D) 20 kips
    (E) 25 kips

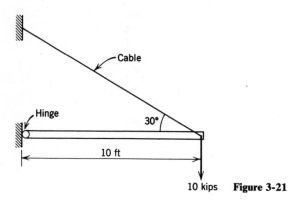

10 kips   **Figure 3-21**

Solution.

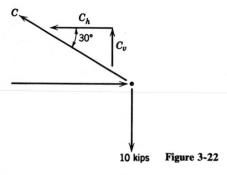

10 kips   **Figure 3-22**

From the free-body diagram, Fig. 3-22, we can see that for $\sum F_y = 0$

$$C_v - 10 \text{ kips} = 0 \qquad C_v = 10 \text{ kips}$$

$$\sin 30° = \frac{C_v}{C} = \frac{1}{2} \qquad C = 2C_v = 2 \times 10 \text{ kips} = 20 \text{ kips}$$

Answer is (D)

PROBLEM 3-14

Two weights are suspended on a cord as shown in Fig. 3-23. The angle θ at equilibrium is nearest to

(A) 35°
(B) 45°
(C) 55°
(D) 65°
(E) 75°

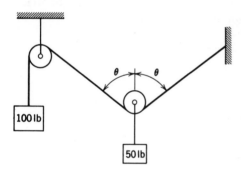

Figure 3-23

Solution. From the diagram we can see that the tension in the cord is 100 lb. We can then draw the free-body diagram and the force triangle.

Figure 3-24

$$\theta = \cos^{-1} \tfrac{25}{100} = \cos^{-1} 0.250 = 75.5°$$

Answer is (E)

PROBLEM 3-15

The center of gravity of a log 10 ft long and weighing 100 lb is 4 ft from one end of the log. It is to be carried by two men. If one is at the heavy end, how far from the other end does the second man have to hold the log if each is to carry 50 lb?

(A) at the end
(B) 2 ft
(C) 4 ft
(D) 5 ft
(E) 6 ft

Solution.   This problem can be treated as a beam with a concentrated load of 100 lb 4 ft from one end. For both men to carry an equal weight of 50 lb, they must be the same distance from the concentrated load.

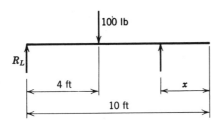

Figure 3-25

The second man is $10 - 4 - 4 = 2$ ft from the other end.

Alternative solution.

$$\Sigma M_{R_L} = 0$$

$$+4 \times 100 - 50(10 - x) = 0$$

$$400 - 500 + 50x = 0$$

$$50x = 100 \quad x = 2 \text{ ft}$$

Answer is (B)

PROBLEM 3-16

The theoretical mechanical advantage of the system shown in the figure is nearest to

(A) 5
(B) 7
(C) 12
(D) 15
(E) 20

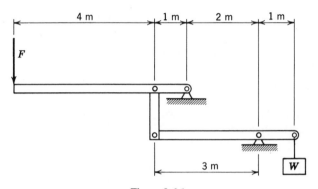

Figure 3-26

Solution. The problem can be solved by considering it as two separate problems:

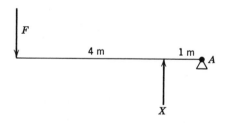

Figure 3-27

$$\sum M_A = 0$$

$$X = \frac{5F}{1} = 5F$$

Thus the mechanical advantage for this section is 5.

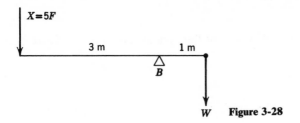

$$\sum M_B = 0$$

$$(5F)(3) = W(1)$$

$$\frac{W}{F} = 15 = \text{mechanical advantage}$$

Answer is (D)

PROBLEM 3-17

The beam ABC is loaded as shown. The equilibrium is maintained by a 4000-lb weight suspended from bar DE. Neglect the weight of the members. The required length L of bar DE is nearest to

(A) 2 ft
(B) 4 ft
(C) 6 ft
(D) 8 ft
(E) 10 ft

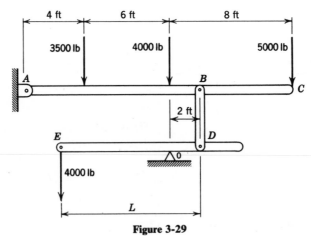

Figure 3-29

Solution.

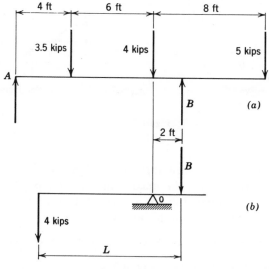

Figure 3-30

Applying $\sum M_A = 0$ to Fig. 3-30a

$$\frac{(4 \times 3500) + (10 \times 4000) + (18 \times 5000)}{12} = \text{force at } B = 12{,}000 \text{ lb}$$

From Fig. 3-30b $\quad \sum M_0 = 0$

$$12{,}000 \times 2 = 4000(L - 2) \qquad L = 8 \text{ ft}$$

Answer is (D)

PROBLEM 3-18

A homogeneous body is composed of a semicylinder and a rectangular parallelepiped as shown in Fig. 3-31. Find the maximum value of h such that the system will be in a *stable* equilibrium on a horizontal plane. Assume no sliding between plane and cylinder. The value of h is nearest to

(A) $0.50R$
(B) $0.70R$
(C) $0.82R$
(D) $1.00R$
(E) $1.25R$

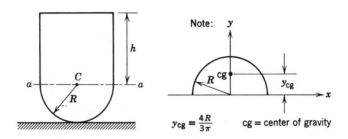

Figure 3-31

Solution. To have stable equilibrium the centroid has to be in the semicylindrical portion; thus the maximum value for h occurs when the centroid is on line a-a. One of the properties of the centroid is that the first moment of area about it is equal to zero. Hence

$$0 = \text{moment of semicylinder} - \text{moment of rectangle}$$

Having the coordinates of the centroid of a semicircle given in the problem, the equation above becomes:

$$\underbrace{\frac{\pi R^2}{2}}_{\text{area}}\underbrace{\left(\frac{4R}{3\pi}\right)}_{\text{arm}} - \underbrace{2Rh}_{\text{area}}\underbrace{\left(\frac{h}{2}\right)}_{\text{arm}} = 0 \qquad \frac{2R^3}{3} = Rh^2$$

$$h^2 = \tfrac{2}{3}R^2 \qquad h = R\sqrt{\tfrac{2}{3}} = 0.82R$$

Answer is (C)

PROBLEM 3-19

The three-hinged arch ABC is loaded as shown. Neglect the weight of the arch. The hinge force at B is nearest to

(A) 1.0 kips
(B) 1.2 kips
(C) 1.4 kips
(D) 1.6 kips
(E) 1.8 kips

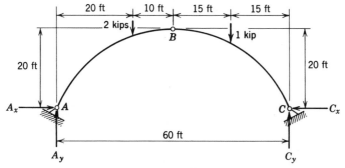

Figure 3-32

Solution.

$$\sum M_A = 0$$

$$-2 \times 20 - 1 \times 45 + 60C_y = 0$$

$$C_y = \tfrac{85}{60} = 1.42 \text{ kips}$$

$$\sum F_y = 0$$

$$A_y - 2 - 1 + 1.42 = 0$$

$$A_y = 1.58 \text{ kips}$$

$$\sum M_B = 0 \text{ (left side)}$$

$$2 \times 10 + 20A_x - 30(1.58) = 0$$

$$A_x = \frac{47.4 - 20}{20} = 1.37 \text{ kips}$$

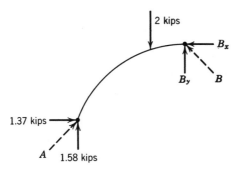

Figure 3-33

Shear at hinge:

$$1.58-2+B_y = 0 \qquad B_y = 0.42 \text{ kips}$$
$$B_x = A_x = 1.37 \text{ kips}$$

Reaction at hinge:

$$B = (1.37^2 + 0.42^2)^{1/2} = 1.43 \text{ kips}$$

Answer is (C)

PROBLEM 3-20

A solid steel bar leans against a smooth vertical wall, with its lower end on a smooth, level floor. A stop on the floor prevents the bar from slipping. The bar is of uniform cross section and weighs 800 newtons. Assume the load acts at the center of gravity of the bar. The value of B_h is nearest to

(A) 0 newtons
(B) 200 newtons
(C) 400 newtons
(D) 600 newtons
(E) 800 newtons

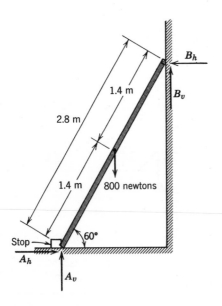

Figure 3-34

Solution. The wall is said to be smooth, hence there is no friction, so $B_v = 0$

$$\sum M_A = 0$$

$$-800(1.4 \cos 60°) + B_h(2.8 \sin 60°) = 0$$

$$B_h = \frac{560}{2.42} = 231 \text{ N}$$

Answer is (B)

PROBLEM 3-21

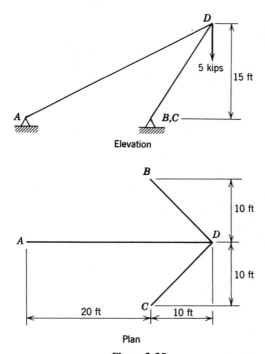

Figure 3-35

The force in member AD of the space frame shown in the plan and elevation views of Fig. 3-35 is nearest to

    (A) 4.5 kips
    (B) 5.0 kips
    (C) 5.6 kips
    (D) 6.5 kips
    (E) 7.6 kips

Solution. The easiest method of solution would be to determine the force in the frame assuming members BD and CD are replaced by a single member ED.

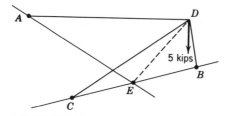

Figure 3-36

Length of members:

$$DE = (15^2 + 10^2)^{1/2} = (325)^{1/2} = 18.0 \text{ ft}$$
$$AD = (15^2 + 30^2)^{1/2} = (1125)^{1/2} = 33.5 \text{ ft}$$

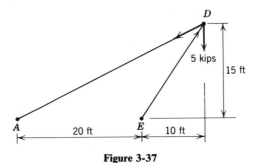

Figure 3-37

$\sum F_y = 0$ (assume DE in compression and AD in tension)

$$-\frac{15}{33.5}AD + \frac{15}{18}DE - 5 = 0$$

$$-0.448AD + 0.833DE - 5 = 0$$

$$-AD + 1.86DE - 11.2 = 0 \qquad (1)$$

$\sum F_x = 0$

$$-\frac{30}{33.5}AD + \frac{10}{18}DE = 0$$

$$-0.896AD + 0.556DE = 0$$

$$-AD + 0.62DE = 0 \qquad (2)$$

Subtracting (2) from (1):

$$-AD + 1.86DE - 11.2 = 0$$

$$\underline{AD - 0.62DE \qquad\quad = 0}$$

$$1.24DE - 11.2 = 0$$

$$DE = \frac{11.2}{1.24} = 9.03 \text{ kips}$$

(The positive result tells us our assumption of compression was correct.) Substituting back in the lower equation,

$$AD - 0.62(9.03) = 0 \qquad AD = 5.60 \text{ kips (tension)}$$

Answer is (C)

PROBLEM 3-22

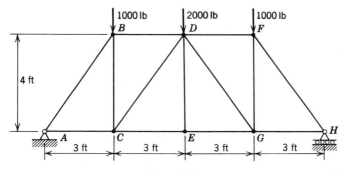

Figure 3-38

Given the pin-jointed structure shown. The force in member CE is nearest to

(A) 0 kips
(B) 0.75 kips
(C) 1.50 kips
(D) 2.25 kips
(E) 3.00 kips

Solution. The structure and the loading pattern are symmetrical.

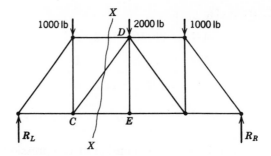

Figure 3-39

Therefore

$$R_L = R_R = \tfrac{4000}{2} = 2000 \text{ lb}$$

To find the force in member CE, take a section $X - X$ and analyze either side.

$$\Sigma M_D = 0$$

$$2 \text{ kips} \times 6 \text{ ft} - 1 \text{ kip} \times 3 \text{ ft} - F_{CE} \times 4 \text{ ft} = 0$$

$$F_{CE} = \frac{12-3}{4} = 2.25 \text{ kips (tension)}$$

Answer is (D)

PROBLEM 3-23

The force in member BC of the truss shown is nearest to

    (A) 0 newtons
    (B) 250 newtons
    (C) 350 newtons
    (D) 500 newtons
    (E) 700 newtons

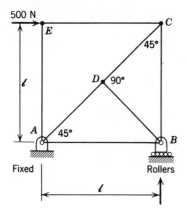

Figure 3-40

Solution. $\sum M_A = 0$ $B_v \times \ell = 500 \times \ell$ $B_v = 500 \text{ N} \uparrow$

$\sum V = 0$ $A_v = 500 \text{ N} \downarrow$

$\sum H = 0$ $A_h = 500 \text{ N} \leftarrow$

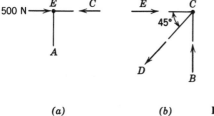

(*a*)   (*b*)   **Figure 3-41**

By the method of joints:

(*a*) Force in $EC = 500$ N compression.
(*b*) Horizontal component of $CD = 500$ N.
   Force in $CD = 707$ N tension.
   Force in $BC =$ Vertical component of $CD = 500$ N compression.

Answer is (D)

PROBLEM 3-24

A pin-connected truss has a horizontal load of 2000 lb and a vertical load of

1200 lb as shown. The total axial force in member CD is closest to

    (A) 1200 lb
    (B) 1950 lb
    (C) 2600 lb
    (D) 3350 lb
    (E) 4000 lb

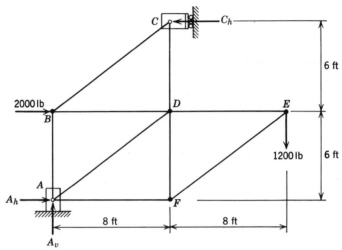

Figure 3-42

Solution.

$$\sum F_y = 0$$

$$A_v - 1200 = 0 \qquad A_v = 1200 \text{ lb}$$

$$\sum M_A = 0$$

$$-1200 \times 16 + C_h \times 12 - 2000 \times 6 = 0$$

$$C_h = \frac{19{,}200 + 12{,}000}{12} = 2600 \text{ lb}$$

Note that all triangles have a $3:4:5$ relationship.

$$\sum F_y = 0$$

$$CD = \tfrac{3}{5}BC \qquad BC = \tfrac{5}{3}CD$$

$$\sum F_x = 0$$

$$\tfrac{4}{5}BC = C_h \qquad \tfrac{4}{5}(\tfrac{5}{3}CD) = 2600$$

$$CD = 1950 \text{ lb tension}$$

$$\text{Answer is (B)}$$

PROBLEM 3-25

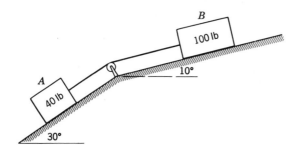

Figure 3-43

Two blocks, A and B, are connected by a cord passing over a smooth pulley. The coefficient of friction under block A is 0.25 and that under block B is 0.40. Which one of the following statements is correct?

(A) Block A will move, but not block B.

(B) Block B will move, but not block A.

(C) Block A and block B will both move.

(D) Neither block A nor block B will move.

(E) None of the above statements is correct.

Solution.

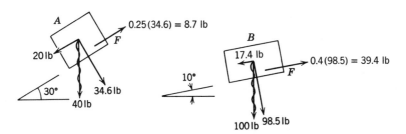

Figure 3-44

	Block A		Block B		Both blocks
Forces down planes	20.0	+	17.4	=	37.4
Friction forces	8.7	+	39.4	=	48.1

Block *B* will not move. Block *A* would move but is restrained from moving by the cord. Thus neither block *A* nor block *B* will move.

Answer is (D)

PROBLEM 3-26

A hangar door weighs 1200 lb and is supported on two rollers. The rollers have rusted, causing the door to slide along the track when moved. Assume the coefficient of friction is 0.40. If the door must be opened in an emergency by a materials-handling machine pushing horizontally at point *P*, the maximum distance *d* that will not cause one roller to leave the track is nearest to

    (A) 0 ft
    (B) 5.0 ft
    (C) 7.5 ft
    (D) 10.0 ft
    (E) 12.5 ft

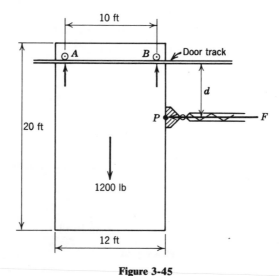

Figure 3-45

Solution.   To slide the door along the track, a force $F = \mu N$ must be applied

$$F = 0.40 \times 1200 = 480 \text{ lb}$$

Taking moments about roller B, we see that roller A will lift off the track only when the moment Fd is greater than $1200\,\text{lb} \times 5\,\text{ft}$.

$$d_{\max} = \tfrac{6000}{480} = 12.5\,\text{ft}$$

Answer is (E)

PROBLEM 3-27

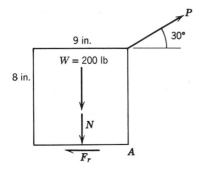

Figure 3-46

A solid block 9 in. × 9 in. × 8 in. high weighs 200 lb. If μ equals 0.25, the force P required to cause the block to slide is closest to

(A) 50 lb

(B) 60 lb

(C) 70 lb

(D) 80 lb

(E) block will overturn before it slides

Solution.

$$\Sigma F_H = 0 \qquad F_r = P \cos 30° = 0.866P$$

$$\Sigma F_V = 0 \qquad N = 200 - P \sin 30° = 200 - 0.5P$$

$$\text{Friction relation } F_r = \mu N$$

$$0.866P = 0.25(200 - 0.5P)$$

$$= 50 - 0.125P$$

$$P = 50.5\,\text{lb}$$

Check to ensure that the block does not overturn around point A when $P = 50.5$ lb.

$$\text{Overturning moment} = 50.5 \cos 30° \times 8 \text{ in.} = 350 \text{ in.-lb}$$

$$\text{Righting moment} = 200 \times 4\tfrac{1}{2} \text{ in.} = 900 \text{ in.-lb}$$

The block will not overturn; it will slide.

Answer in (A)

PROBLEM 3-28

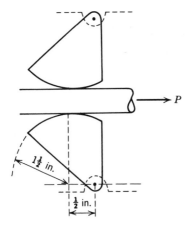

Figure 3-47

The cam arrangement shown was designed to develop large friction forces on a cable that is subjected to a tension force. The coefficient of friction is 0.3. The maximum value of P at which the cable will not slip is nearest to

(A) 0 lb

(B) 200 lb

(C) 400 lb

(D) 600 lb

(E) 800 lb

Solution.

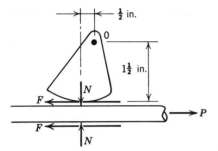

Figure 3-48

The tension in the cable equals P. Therefore F equals $0.5P$. From the cam,
$\Sigma M_0 = 0$:

$$\tfrac{1}{2}''N = 1\tfrac{1}{2}''F$$

$$N = 3F = 3(0.5P) = 1.5P$$

The maximum friction possible $F = \mu N = 0.3(1.5P) = 0.45P$

$$2F = 0.90P$$

The tension in the cable is P; the resisting force is $0.9P$. This indicates that
the cable will slip for all values of P.

$$\text{Answer is (A)}$$

PROBLEM 3-29

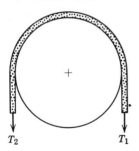

$T_2 \qquad\qquad T_1$ **Figure 3-49**

For a belt passing over a pulley, the ratio of the forces is given as

$$\frac{T_2}{T_1} = e^{\mu\beta}$$

where μ is the coefficient of friction and β is the angle of contact. For the pulley shown and $\mu = 0.30$, the ratio T_2/T_1 is nearest to

(A) 1.0
(B) 2.5
(C) 3.1
(D) 4.5
(E) 6.2

Solution. The angle of contact β is expressed in radians in this equation. Since there are 2π rad in 360° and the angle of contact in this case is 180°, $\beta = \pi$ rad.

$$\frac{T_2}{T_1} = e^{0.3\pi} = 2.57$$

Answer is (B)

PROBLEM 3-30

In a problem involving rolling friction, the value of it would be given as

(A) an angle ϕ
(B) μ_s, the tangent F_s/N of an angle ϕ (F_s = static friction)
(C) μ_k, the tangent F_k/N of an angle ϕ (F_k = kinetic friction)
(D) r_f, the radius of the friction circle ($r_f = r \sin \phi$)
(E) b, a deformation, as a linear dimension

Solution. In rolling friction we encounter a situation where the "ideal" situation fails to give us any appreciation of the actual situation. If we were to place a rigid wheel on a rigid smooth surface and set the wheel in motion, the forces would appear as in Fig. 3-50.

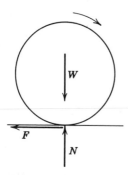

Figure 3-50

In the absence of retarding forces, the wheel would theoretically roll forever. In the actual case there is deformation of the surface and the wheel.

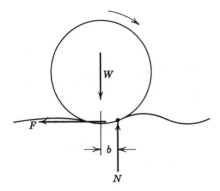

Figure 3-51

The result is that the normal force N acts ahead of the line of action of the weight W. This small distance (b) is called the coefficient of rolling friction and is a linear dimension.

Answer is (E)

PROBLEM 3-31

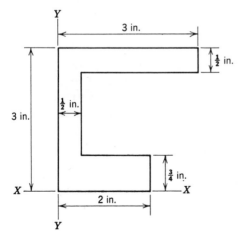

Figure 3-52

The centroid of the figure shown has a value $\bar{Y}$ nearest to

    (A) 1.00 in.
    (B) 1.25 in.
    (C) 1.50 in.
    (D) 1.75 in.
    (E) 2.00 in.

Solution. Although only $\bar{Y}$ is required in the problem, both $\bar{X}$ and $\bar{Y}$ will be calculated to show the method of computation.

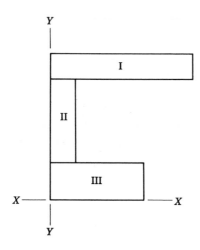

Figure 3-53

Section	b (in.)	h (in.)	A (in.^{2})	X (in.)	Y (in.)	AX (in.^{3})	AY (in.^{3})
I	3.0	0.5	1.5	1.5	2.75	2.25	4.13
II	0.5	1.75	0.875	0.25	1.625	0.22	1.42
III	2.0	0.75	1.5	1.0	0.375	1.5	0.56
			3.875			3.97	6.11

$$\bar{X} = \frac{\Sigma AX}{\Sigma A} = \frac{3.97}{3.875} = 1.02 \text{ in.}$$

$$\bar{Y} = \frac{\Sigma AY}{\Sigma A} = \frac{6.11}{3.875} = 1.58 \text{ in.}$$

Answer is (C)

PROBLEM 3-32

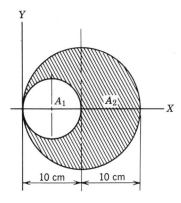

Figure 3-54

A circular disk of uniform density has a hole cut out of it, as shown. $\bar{X}$ of the center of mass is nearest to

(A) 8 cm
(B) 10 cm
(C) 12 cm
(D) 14 cm
(E) 16 cm

Solution. The center of mass, or center of gravity as it probably is more frequently called, is the point through which the resultant of the total weight of the object will pass regardless of the orientation of the object.

Since the disk has an axis of symmetry, the center of mass is on that axis and $\bar{Y}=0$.

$$\bar{X}=\frac{\int X\,dA}{\int dA}=\frac{\Sigma AX}{\Sigma A}$$

$$A_1=\frac{\pi}{4}10^2=25\pi$$

$$A_2(\text{total area of disk incl. } A_1) = \frac{\pi}{4}20^2 = 100\pi$$

$$\bar{X} = \frac{A_2X_2 - A_1X_1}{A_2 - A_1} = \frac{100\pi \times 10 - 25\pi \times 5}{100\pi - 25\pi}$$

$$= \frac{1000\pi - 125\pi}{75\pi} = \frac{875\pi}{75\pi} = 11\tfrac{2}{3}\text{ cm}$$

$$\bar{X} = 11\tfrac{2}{3}\text{ cm}$$

Answer is (C)

PROBLEM 3-33

The formula $I = \int y^2\, dA$ represents the

    (A) product of inertia
    (B) section modulus
    (C) area of cross section
    (D) moment of inertia
    (E) modulus of elasticity

Solution. The formula represents the moment of inertia.

Answer is (D)

PROBLEM 3-34

The term $I = bh^3/12$ refers to the

    (A) radius of gyration
    (B) section modulus
    (C) instantaneous center
    (D) moment of inertia
    (E) product of inertia

Solution. $I = bh^3/12$ is the equation for the moment of inertia for a rectangular cross section.

Answer is (D)

PROBLEM 3-35

The moment of inertia of the area shown in the figure about the x-x axis is

(A) 28 in.^{4}
(B) 40 in.^{4}
(C) 64 in.^{4}
(D) 110 in.^{4}
(E) 256 in.^{4}

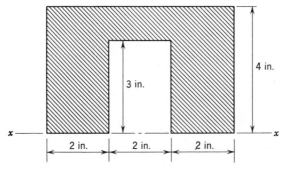

Figure 3-55

Solution. The moment of inertia $I_x = \int y^2 \, dA$. Thus the moment of inertia of a cross-sectional area is equal to the sum of the differential areas dA multiplied by the square of their moment arms about the reference axis.

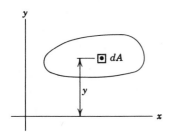

Figure 3-56

This problem can be solved in two ways:

1. Integration of the differential areas.

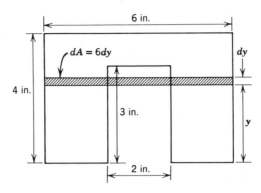

Figure 3-57

$$I_x = \int y^2 \, dA$$

$$= \int_0^4 y^2 6 \, dy - \int_0^3 y^2 2 \, dy$$

$$= 6\left[\frac{y^3}{3}\right]_0^4 - 2\left[\frac{y^3}{3}\right]_0^3$$

$$= 6(\tfrac{64}{3}) - 2(\tfrac{27}{3}) = 128 - 18$$

$$= 110 \text{ in.}^4$$

2. Transfer of the moment of inertia, with respect to the centroid, to a parallel axis. In this method three facts are utilized:

   (*a*) The moment of inertia of an object is the sum of the moments of inertia of its individual parts (all referred to the same axis).

   (*b*) The moment of inertia of a rectangle with respect to its centroid is $I_{x_0} = bh^3/12$, where b is the width of the rectangle parallel to the centroidal axis and h is its depth.

   (*c*) The transfer formula for obtaining the moment of inertia with respect to an axis parallel to the centroidal axis is $I_x = I_{x_0} + Ad^2$, where A is the cross-sectional area and d is the distance between the axes.

For a rectangle

$$I_x = \frac{bh^3}{12} + bh\left(\frac{h}{2}\right)^2 = \frac{bh^3}{3}$$

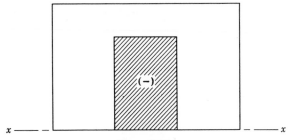

Figure 3-58

Large rectangle $\dfrac{bh^3}{3} = \dfrac{6(4)^3}{3} = 128 \text{ in.}^4$

minus

Small rectangle $\dfrac{bh^3}{3} = \dfrac{2(3)^3}{3} = \underline{\quad 18 \text{ in.}^4}$

$ I_x = 110 \text{ in.}^4$

Answer is (D)

PROBLEM 3-36

The moment of inertia of a rectangle with respect to an axis passing through its base is

(A) $\dfrac{bh^3}{3}$

(B) $\dfrac{bh^3}{12}$

(C) $\dfrac{bh^2}{3}$

(D) $\dfrac{bh}{2}$

(E) none of the above

Solution. Using the transfer formula for the moment of inertia $I_x = I_0 + Ad^2$, where $I_0 = $ moment of inertia about the centroid, $A = $ area $= bh$, and $d = h/2$, we have

$$I_x = \frac{bh^3}{12} + bh\left(\frac{h}{2}\right)^2 = \frac{bh^3}{12} + \frac{bh^3}{4} = \frac{bh^3}{3}$$

Solution using calculus: The general equation for the moment of inertia is $I = \int y^2 \, dA$.

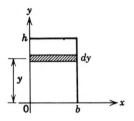

Figure 3-59

$$dA = b \, dy$$

$$dI_x = y^2(b \, dy) = by^2 \, dy$$

$$I_x = b \int_0^h y^2 \, dy = b\left[\frac{y^3}{3}\right]_0^h = \frac{bh^3}{3}$$

Answer is (A)

PROBLEM 3-37

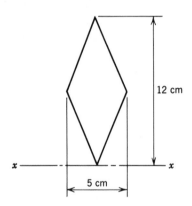

12 cm

5 cm

Figure 3-60

The moment of inertia about the x-x axis is

   (A) 420 cm^4
   (B) 1230 cm^4
   (C) 1260 cm^4
   (D) 1380 cm^4
   (E) 2460 cm^4

Solution.

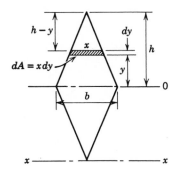

Figure 3-61

From similar triangles in Fig. 3-61,

$$\frac{b}{h} = \frac{x}{h-y} \qquad x = \frac{b}{h}(h-y)$$

$$I_0 = \int y^2 \, dA = 2\int_0^h y^2 x \, dy = 2\int_0^h y^2 \frac{b}{h}(h-y) \, dy$$

$$= \frac{2b}{h}\left(\int_0^h hy^2 \, dy - \int_0^h y^3 \, dy\right)$$

$$= \frac{2b}{h}\left[\frac{hy^3}{3} - \frac{y^4}{4}\right]_0^h = \frac{2b}{h}\left(\frac{h^4}{3} - \frac{h^4}{4}\right) = \frac{2bh^4}{12h}$$

$$I_0 = \frac{bh^3}{6}$$

But we want I_x.

Using the transfer formula $I_x = I_0 + Ad^2$, where $A = 2(b/2)h = bh$ and $d = h$, we have

$$I_x = \frac{bh^3}{6} + bh(h^2) = \frac{7bh^3}{6}$$

In our case $h = 6$ and $b = 5$. Therefore

$$I_x = \frac{7 \times 5 \times 6^3}{6} = 1260 \text{ cm}^4$$

Answer is (C)

PROBLEM 3-38

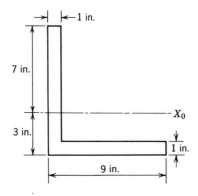

Figure 3-62

The moment of inertia of the angle section shown about the X_0 axis is nearest to

(A) 60 in.^{4}
(B) 90 in.^{4}
(C) 120 in.^{4}
(D) 150 in.^{4}
(E) 180 in.^{4}

Solution.

$$I_x = \sum (I_o + Ad^2) = \frac{1(10)^3}{12} + 1(10)(2)^2 + \frac{8(1)^3}{12} + 8(1)(2.5)^2$$
$$= 83.3 + 40 + 0.7 + 50 = 174 \text{ in.}^4$$

Answer is (E)

PROBLEM 3-39

This is a problem set containing six questions. Given the pin-connected frame shown in Fig. 3-63 with a 10-kip load at point D and a 5-kip load at point F.

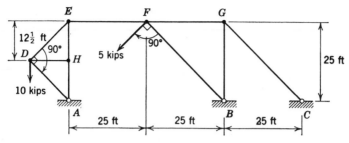

Figure 3-63

Question 1. The magnitude of reaction A is closest to

(A) 10 kips
(B) 11 kips
(C) 12 kips
(D) 13 kips
(E) 14 kips

Solution.

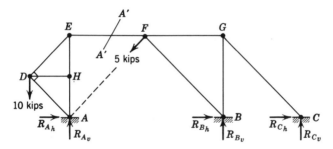

Figure 3-64

Taking the left side of Section A'–A':

$\sum M_E = 0 \qquad 12.5(10) + R_{A_h}(25) = 0$

$R_{A_h} = -5 \text{ kips} \qquad \text{or} \qquad 5 \text{ kips} \leftarrow$

$\sum F_v = 0 \qquad R_{A_v} - 10 \text{ kips} = 0$

$R_{A_v} = 10 \text{ kips} \uparrow$

Reaction $A = (5^2 + 10^2)^{1/2} = 11.2 \text{ kips}$

Answer is (B)

Question 2.   The angular direction of reaction A, measured clockwise from the horizontal, is closest to

(A) 50°
(B) 60°
(C) 70°
(D) 80°
(E) 90°

Solution.

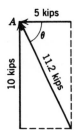

Figure 3-65   Reaction A

$$\theta_A = \tan^{-1} \tfrac{10}{5} = 63.5°$$

Answer is (B)

Question 3.   The force in bar EF is nearest to

(A) 0 kips
(B) 5 kips
(C) 10 kips
(D) 15 kips
(E) 20 kips

Solution.   Taking the left side of Section $A' - A'$:

$$\sum M_A = 0 \qquad 10(12.5) - 25 F_{EF} = 0$$

$$F_{EF} = 5 \text{ kips (tension)}$$

Answer is (B)

Question 4. The magnitude of reaction C is closest to

(A) 9 kips
(B) 11 kips
(C) 13 kips
(D) 15 kips
(E) 17 kips

Solution. Right side of Section $A' - A'$:

$$\Sigma M_B = 0$$

$$EF \times 25' + AF \times (25^2 + 25^2)^{1/2} + R_{C_v} \times 25' = 0$$

$$R_{C_v} = \frac{-5 \times 35.36' - 5 \times 25'}{25'} = -12.07 \text{ kips} = 12.07 \text{ kips} \downarrow$$

$$\Sigma F_v = 0 \qquad -12.07 \text{ kips} + R_{B_v} - \frac{5}{\sqrt{2}} = 0$$

$$R_{B_v} = 15.61 \text{ kips} \uparrow$$

At joint C:

$$\Sigma F_h = 0 \qquad -\frac{F_{GC}}{\sqrt{2}} + R_{C_h} = 0$$

$$\Sigma F_v = 0 \qquad \frac{F_{GC}}{\sqrt{2}} = 12.07 \text{ kips}$$

$$R_{C_h} = 12.07 \text{ kips} \rightarrow$$

$$\text{Reaction } C = (12.07^2 + 12.07^2)^{1/2} = 17.07 \text{ kips}$$

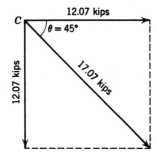

Figure 3-66 Reaction C

Answer is (E)

Question 5.   The magnitude of reaction B is closest to

    (A) 10 kips
    (B) 11 kips
    (C) 12 kips
    (D) 13 kips
    (E) 14 kips

Solution.   Total structure:

$$\sum F_h = 0 \qquad -5 + R_{B_h} + 12.07 \text{ kips} - \frac{5}{\sqrt{2}} = 0$$

$$R_{B_h} = -3.53 \text{ kips} = 3.53 \text{ kips} \leftarrow$$

$$\text{Reaction } B = (15.61^2 + 3.53^2)^{1/2} = 16 \text{ kips}$$

$$\text{Answer is (E)}$$

Question 6.   The angular direction of reaction B, measured clockwise from the horizontal, is closest to

    (A) 30°
    (B) 45°
    (C) 60°
    (D) 75°
    (E) 90°

Solution.

Figure 3-67   Reaction B

$$\theta_B = \tan^{-1}\frac{15.61}{3.53} = 77.3°$$

Answer is (D)

4 Dynamics

Dynamics is the study of the motion of nondeformable bodies. This study is normally divided into *kinematics*, the study of acceleration-velocity-displacement relations, and *kinetics*, which relates these motions and the forces causing, and caused by, them. Statics may be regarded as merely a special (but important) subdivision of dynamics where forces are in equilibrium and no accelerations are present.

KINEMATICS

Kinematics ignores the forces causing motion and considers only the motion itself. The acceleration a, velocity v, and displacement x of a body are related by the very basic expressions

$$a = \frac{dv}{dt} \qquad v = \frac{dx}{dt} \qquad\qquad (4\text{-}1)$$

where t is time. When the acceleration is a known function of time, velocity, or displacement, Eq. (4-1) may be integrated to give direct velocity-time or displacement-time relations. For the important case of rectilinear motion beginning at a point x_0 with a constant acceleration a_0 and initial velocity v_0, the displacement at any later time, by integration, is

$$x(t) = x_0 + v_0 t + \tfrac{1}{2} a_0 t^2 \qquad\qquad (4\text{-}2)$$

Since we also have

$$a_0 = \frac{dv}{dt} = \frac{dv}{dx}\frac{dx}{dt} = v\frac{dv}{dx} = \frac{d}{dx}\left(\frac{v^2}{2}\right)$$

we also find by integration that the velocity and displacement are related by

$$\tfrac{1}{2}(v^2 - v_0^2) = a_0(x - x_0) \qquad (4\text{-}3)$$

Example 1

A driver sees a stoplight when his car is traveling at 55 miles/hr. If it takes him 0.6 sec to apply the brakes, and the brakes produce a deceleration of 15 ft/sec^2 in the car, how many feet does the car travel before coming to a stop?

Solution. Using the conversion factor 88 ft/sec = 60 miles/hr, we find the initial speed $v_0 = 55(88/60) = 80.7$ ft/sec. During the 0.6-sec reaction period the car moves a distance $s_1 = v_0 t = (80.7)(0.6) = 48.4$ ft.

During the deceleration period $a_0 = -15$ ft/sec^2 while the velocity decreases from v_0 to zero. From Eq. (4-3) the distance traveled $s_2 = x - x_0$ is

$$s_2 = \frac{1}{2a_0}(v^2 - v_0^2) = \frac{1}{2(-15)}[0 - (80.7)^2] = 217 \text{ ft}$$

Thus the total distance traveled is

$$s = s_1 + s_2$$

$$s = 48.4 + 217 = 265 \text{ ft}$$

In addition to rectilinear or straight-line motions there are curvilinear motions. Equations (4-1) apply in a vectorial sense. For two- or three-dimensional motion, the trajectory and velocity of the body are often expressed by a set of parametric equations, usually with time t as the parameter. The trajectory of a body in a gravity field is a good example. Choosing x and y to be horizontal and vertical displacements, respectively,

$$\begin{aligned} x &= x_0 + v_0 \cos \theta_0 \, t \\ y &= y_0 + v_0 \sin \theta_0 \, t - \tfrac{1}{2}gt^2 \end{aligned} \qquad (4\text{-}4)$$

when the body is released at $t = 0$ from the point (x_0, y_0) with the initial velocity v_0 and orientation θ_0 from the horizon. Here the location of the body is defined in terms of the time parameter.

The accelerations of a body experiencing curvilinear motion are often split into tangential and normal components. For the case of circular motion of radius r at velocity v, the tangential acceleration a_t and normal acceleration a_n are

$$a_t = \frac{dv}{dt} \qquad a_n = \frac{v^2}{r} \qquad (4\text{-}5)$$

If the speed of the body is unchanging, $a_t = 0$ but a normal acceleration still exists. In terms of the angular velocity ω, $v = \omega r$ and $a_n = \omega^2 r$.

KINETICS

Here we consider the kinetics of bodies whose dynamic behavior is adequately described by reference to only one body property, the mass, which is assumed to be concentrated at a point. Mass is the proportionality constant that relates the acceleration of a body to the net force that acts on it. The Newtonian law expresses this as

$$\sum \mathbf{F} = m\mathbf{a} \tag{4-6}$$

for a body of constant mass m. This vector equation is usually written in component scalar form for use in obtaining numerical answers, however.

Example 2

The blocks A and B, of weights $W_A = 500\,\text{lb}$ and $W_B = 100\,\text{lb}$, are connected by a rope that passes over a small pulley (Fig. 4-1). No friction is present. When the system is released from rest, what is the acceleration a of the bodies and the tension T in the rope?

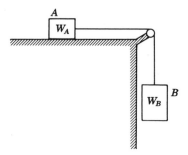

Figure 4-1

Solution. Drawing a diagram of each block (Fig. 4-2) shows that the blocks are clearly not in equilibrium but will accelerate to the right (block A) and

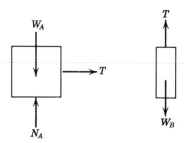

Figure 4-2

downward (block B) at the same rate a. Applying the Newtonian equation, we have for block A

$$\Sigma F_x = ma$$

$$T = \frac{W_A}{g} a$$

and for block B

$$\Sigma F_y = ma$$

$$W_B - T = \frac{W_B}{g} a$$

Hence

$$W_B - \frac{W_A}{g} a = \frac{W_B}{g} a$$

$$a = g \frac{W_B}{W_A + W_B}$$

$$= 32.2 \frac{100}{500 + 100} = 5.37 \text{ ft/sec}^2$$

$$T = \frac{W_A}{g} a = \frac{500}{32.2}(5.37) = 83.3 \text{ lb}$$

The normal and tangential components of Eq. (4-6) are useful in situations involving curvilinear motion. The components are

$$\Sigma F_t = m \frac{dv}{dt} \qquad \Sigma F_n = m \frac{v^2}{r} \qquad (4\text{-}7)$$

The normal force component is also called the centrifugal force.

Example 3

A segment of a flywheel weighs 1300 lb, and its center of gravity is 6 ft from the center of the shaft. The wheel rotates at 125 rpm. What is the pull on the arm supporting the segment?

Solution. The angular velocity $\omega = 125(2\pi/60) = 13.1$ rad/sec. The speed of the segment is $v = \omega r$, so the normal force—Eq. (4-7)—is

$$F_n = m \frac{v^2}{r} = mr\omega^2 = \left(\frac{1300}{32.2}\right)(6)(13.1)^2 = 41,500 \text{ lb}$$

ENERGY CONSERVATION

If Newton's law of motion is rearranged and integrated with respect to distance between two points, we find that the work done in moving a body from one point to the other is equal to the change in kinetic energy KE of that body. If the amount of work done depends only on the end points of the path, that is, if a conservative force produces the work, then the work can be expressed as a change in potential energy PE. Common conservative forces include the weight of a body and the force due to an elastic spring. For conservative forces we may then say that the total mechanical energy of a system is conserved, or

$$PE + KE = \text{constant} \qquad (4\text{-}8)$$

for a process. This may be restated in terms of changes of kinetic and potential energy as

$$\Delta(KE) = -\Delta(PE) \qquad (4\text{-}9)$$

For *non*conservative forces, such as those due to friction, total mechanical energy is not conserved, and Eqs. (4-8) and (4-9) should not be used then.

Example 4

A block of mass m is released from rest on an inclined frictionless plane, as in Fig. 4-3. What is the block's speed when it has dropped a vertical distance of 4 ft?

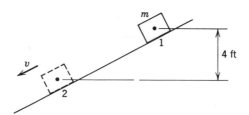

Figure 4-3

Solution. Here we apply Eq. (4-9):

$$\Delta(KE) = -\Delta(PE)$$

$$KE_2 - KE_1 = -(PE_2 - PE_1)$$

$$\tfrac{1}{2}mv^2 - 0 = -(0 - mgh)$$

$$v^2 = 2gh$$

$$v^2 = 2(32.2)(4)$$

$$v = 16.05 \text{ ft/sec}$$

Example 5

The spring of a spring gun has an uncompressed length of 8 in. The modulus of the spring $k = 1$ lb/in. The spring is compressed to a length of 4 in., and a ball weighing 1 oz is put in the barrel against the compressed spring, as in Fig. 4-4. If the spring is then released, find the velocity with which the ball leaves the gun. Neglect friction.

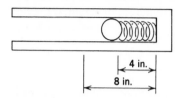

4 in.

8 in.

Figure 4-4

Solution. The force involved in the compression of an elastic spring is a conservative force whose magnitude is proportional to the distance x that the spring is compressed, that is, $F = kx$, where k is the spring constant. The work done on the spring is

$$W = \int_0^x F\,dx = \int_0^x kx\,dx = \tfrac{1}{2}kx^2$$

In our case

$$W = \tfrac{1}{2}(1 \text{ lb/in.})(4 \text{ in.})^2 = 8 \text{ in.-lb} = \tfrac{2}{3} \text{ ft-lb}$$

This work represents stored or potential energy. When the spring is released, this is converted into the kinetic energy of the ball, which was initially at rest.

$$\Delta(KE) = -\Delta(PE)$$

$$\tfrac{1}{2}mv^2 = W$$

$$v = \left(\frac{2W}{m}\right)^{1/2}$$

$$v = \left[\frac{2(\tfrac{2}{3})}{(\tfrac{1}{16})/32.2}\right]^{1/2} = 26.2 \text{ ft/sec}$$

MOMENTUM CONSERVATION

Newton's law of motion, when viewed another way, yields a useful principle that relates impulse and momentum. Expressing Eq. (4-6) as

$$\sum \mathbf{F} = \frac{d}{dt}(m\mathbf{v}) \tag{4-10}$$

we may integrate with respect to time to obtain

$$\sum \int \mathbf{F} \, dt = (m\mathbf{v})_2 - (m\mathbf{v})_1 \qquad (4\text{-}11)$$

The left term represents an external impulse that changes the momentum $m\mathbf{v}$ of a body from that of state 1 to that of state 2. In the absence of impulsive external forces, we note that momentum is conserved for the body. Extended to a system of several bodies, we may state again, if the net external impulsive force on the system is zero, that system momentum is conserved, or

$$(\sum m\mathbf{v})_1 = (\sum m\mathbf{v})_2 \qquad (4\text{-}12)$$

This is true even though individual bodies within a system may impact with one another; the reason is that the impulsive forces are internal, not external, to the system.

The nature of the impact that occurs between two bodies is important even when no external impulsive forces act. An index of the kind of impact that occurs between two bodies is the coefficient of restitution e, which is the ratio of the relative velocity between the two bodies after impact to the relative velocity before impact. Two cases are of particular interest:

1. The case $e = 1$ represents elastic impact where the relative velocities before and after impact are equal. This is the only impact situation where energy is conserved; all other impact cases involve a change in total mechanical energy.

2. The other extreme $e = 0$ represents inelastic or plastic impact. After impact the two bodies move together with a common velocity. Energy is not conserved.

Different parts of a single problem commonly use the principles of momentum and energy conservation and the equation of motion together to achieve a solution.

Example 6

A bullet of mass m, traveling at velocity v_1, impacts inelastically with a simple pendulum composed of a mass M on the end of a flexible cord. If the impact occurs a distance L below the pendulum's suspension point, as illustrated in Fig. 4-5, through what angle θ will the pendulum move? (Assume the impact occurs at a right angle to the vertical.)

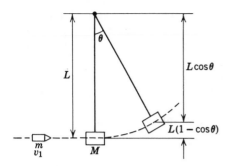

Figure 4-5

Solution. By the principle of conservation of momentum for an inelastic impact, the sum of the momenta of the bullet and the pendulum before impact equals the momentum of the combination after impact, or

$$mv_1 + 0 = (m + M)V_c$$

where V_c is the velocity of the combination immediately after impact. Thus

$$V_c = \frac{mv_1}{m + M}$$

Using the energy conservation principle for the motion after the impact, we write

$$Wh = \tfrac{1}{2}mV^2$$

which in this case gives

$$g(m + M)L(1 - \cos \theta) = \tfrac{1}{2}(m + M)V_c^2 = \tfrac{1}{2}(m + M)\left(\frac{mv_1}{m + M}\right)^2$$

Solving for θ,

$$\theta = \cos^{-1}\left[1 - \frac{1}{2gL}\left(\frac{mv_1}{m + M}\right)^2\right]$$

RIGID BODY DYNAMICS

The motion of some bodies cannot be properly analyzed by assuming the mass of the body is concentrated at a point and analyzing it as a particle. In particular this is true when bodies execute rotational motion. We consider briefly the analysis of such motions now.

A general plane motion can always be regarded as the sum of a translational motion and a rotational motion. We have already examined translational motion. For rotational motion the velocity v of a point which is a distance r from the center of rotation on a body rotating with an angular velocity ω is $v = \omega r$, and the normal and tangential acceleration components are $a_n = \omega^2 r$, $a_t = \alpha r$, where $\alpha = d\omega/dt$ is the angular acceleration of the body. These quantities are often easier to calculate when one first finds the instantaneous center of rotation, the point around which the body appears to rotate at a given instant.

Example 7

A 5-in.-radius cylinder rolls to the right at a constant rate of 10 in./sec. For the instant shown in Fig. 4-6, what is the
    (a) velocity at point A?
    (b) acceleration of point A?

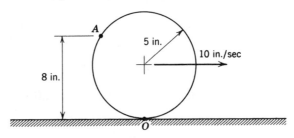

Figure 4-6

Solution. We first note that the point on the cylinder that is in contact with the plane, point O, is the instantaneous center of rotation. At a given instant it is stationary, and all other points are executing a purely rotational motion about point O.

From Fig. 4-7 the angular velocity ω of the body is

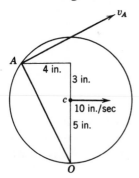

Figure 4-7

$$\omega = \frac{v}{r} = \frac{10}{5} = 2 \text{ rad/sec}$$

Also the distance $OA = d = (8^2 + 4^2)^{1/2} = 8.94$ in.

(a) The velocity $v_A = \omega d = (2)(8.94) = 17.88$ in./sec.

(b) Since ω is constant, the angular acceleration $\alpha = d\omega/dt = 0$, and the tangential acceleration $a_t = 0$ also. The normal acceleration

$$a_n = \omega^2 d = (2)^2(8.94) = 35.8 \text{ in./sec}^2$$

The kinetics of the plane motion of rigid bodies is quite similar to the kinetics of particles. The differences arise because we are now analyzing the motion of distributed masses rather than point masses.

To study the plane motion of a rigid body directly, we must supplement the equation of motion, Eq. (4-6), with an equation that says the net external moment on a body is equal to the product of the moment of inertia I and the angular acceleration α of the body, or

$$\sum M_0 = I_0 \alpha \tag{4-13}$$

where the subscript denotes some common reference axis.

For plane motion Eq. (4-9), which says that energy is conserved, remains valid so long as we now express the kinetic energy as the sum of the translational kinetic energy $\frac{1}{2}mV^2$ and the rotational kinetic energy $\frac{1}{2}I_0\omega^2$.

The momentum of a body may be treated as the sum of linear momentum and angular momentum just as a general motion is the sum of a translational and a rotational motion. The treatment of linear momentum conservation is unchanged from the presentation given in Eqs. (4-10)–(4-12). Conservation of angular momentum may be expressed as

$$I_0\omega_1 + \int M_0 \, dt = I_0\omega_2 \tag{4-14}$$

In this equation the $I_0\omega$ terms are the angular momentum of the body before and after an angular impulse, given by the middle term, is applied to the body. M_0 is the net moment, computed with respect to axis O, of the forces acting on the body at an instant. One then integrates over the time period of interest to find the total angular impulse. In the absence of an applied angular impulse, the angular momentum $I_0\omega$ is constant.

PROBLEM 4-1

A pebble is dropped in a well, and it is found that 4.25 sec elapse after release of the pebble before the splash is heard. If the velocity of sound in the well is 1030 ft/sec, the depth to the water surface is most nearly

(A) 230 ft
(B) 260 ft
(C) 290 ft
(D) 450 ft
(E) 550 ft

Solution. The pebble is initially at rest ($V_0 = 0$) and thereafter falls with a constant acceleration due to gravity g so that

$$V(t) = \frac{dy}{dt} = gt$$

The fall distance is then $y = \frac{1}{2}gt_1^2$ by integration, where t_1 is the time for the pebble to reach the water. The sound wave will return the same distance y at a constant velocity so that $y = 1030t_2$, where t_2 is the time for the sound to travel from the base to the top of the well. By the problem statement,

$$t_1 + t_2 = 4.25 \text{ sec}$$

and we also have

$$\tfrac{1}{2}gt_1^2 = 1030t_2$$

Solving these equations simultaneously,

$$\tfrac{1}{2}(32.2)t_1^2 - 1030t_2 = 0$$

$$\underline{1030t_1 + 1030t_2 = 4.25(1030)}$$

$$16.1t_1^2 + 1030t_1 - 4380 \quad = 0$$

or

$$t_1^2 + 64.0t_1 - 272 = 0$$

$$t_2 = \tfrac{1}{2}[-64.0 \pm \sqrt{64.0^2 + 4(272)}] = \tfrac{1}{2}(-64.0 \pm 72.0)$$

Since t_1 must be positive, $t_1 = 4.0$ sec, and

$$y = \tfrac{1}{2}gt_1^2 = \tfrac{1}{2}(32.2)(4.0)^2 = 258 \text{ ft}$$

Answer is (B)

PROBLEM 4-2

A body moves so that the x component of acceleration is given by the equation $6 - t$ and the y component of acceleration is given by $6 + t$. If the initial x and y components of velocity are both 2, the speed of the body at the end of 2 sec is closest to

(A) 12
(B) 14
(C) 16
(D) 18
(E) 20

Solution.

$$a_x = \frac{d^2x}{dt^2} = 6 - t \qquad a_y = \frac{d^2y}{dt^2} = 6 + t$$

Integrating these expressions once,

$$V_x = \frac{dx}{dt} = 6t - \frac{1}{2}t^2 + C_1 \qquad V_y = \frac{dy}{dt} = 6t + \frac{1}{2}t^2 + C_2$$

At $t = 0$, $V_x = 2$ and $V_y = 2$ so $C_1 = C_2 = 2$ by substitution into the velocity expressions. At the end of 2 sec we have

$$V_x = 6(2) - \tfrac{1}{2}(2)^2 + 2 \qquad V_y = 6(2) + \tfrac{1}{2}(2)^2 + 2$$
$$V_x = 12 \qquad\qquad V_y = 16$$

The speed is

$$V = (V_x^2 + V_y^2)^{1/2} = (12^2 + 16^2)^{1/2}$$
$$V = 20$$

Answer is (E)

PROBLEM 4-3

The driver of a car traveling 30 miles/hr suddenly applies the brakes and skids 60 ft before coming to a stop.

(1) If the car weighs 1600 lb and a constant rate of deceleration is assumed, the rate of deceleration is nearest to

(A) -8.0 ft/sec^2
(B) -16.1 ft/sec^2
(C) -24.1 ft/sec^2
(D) -32.2 ft/sec^2
(E) -40.2 ft/sec^2

Solution. The general displacement equation for rectilinear motion can be obtained by integrating the expression $d^2x/dt^2 = a$ twice. The result is

$$x = x_0 + V_0 t + \tfrac{1}{2}at^2$$

where x_0 and V_0 are the initial displacement and velocity, respectively, and a is the (constant) acceleration. If t is measured from the moment the brakes are applied and the initial location of the car is $x_0 = 0$, then $V_0 = 30$ miles/hr $= 44$ ft/sec. When the car stops at time t later, $x = 60$ ft, so

$$60 = 44t + \tfrac{1}{2}at^2$$

or, solving for a,

$$a = \frac{120}{t^2} - \frac{88}{t}$$

Since the deceleration is to be constant,

$$\frac{da}{dt} = 0 = \frac{120(-2)}{t^3} - \frac{88(-1)}{t^2}$$

yielding $t = 240/88 = 2.73$ sec as the time to stop. Hence

$$a = \frac{120}{(2.73)^2} - \frac{88}{2.73} = -16.1 \text{ ft/sec}^2$$

(The minus sign denotes deceleration.)

Answer is (B)

(2) If the car weighs 1600 lb and a constant rate of deceleration is assumed, the average coefficient of sliding friction is closest to

(A) 0.3
(B) 0.4
(C) 0.5
(D) 0.6
(E) 0.7

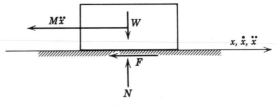

Figure 4-8

Solution. For sliding,

$$F = -\mu N = -1600\mu$$

Also

$$F = M\ddot{x} = Ma$$

Equating these expressions,

$$-1600\mu = \left(\frac{1600}{32.2}\right)(-16.1) \qquad \text{and} \qquad \mu = 0.5$$

Answer is (C)

PROBLEM 4-4

An aircraft begins its take-off run with an acceleration of 4 m/sec^2, which then decreases uniformly to zero in 15 sec, at which time the craft becomes airborne.

(1) The take-off speed, in meters per second, is closest to

(A) 60
(B) 50
(C) 40
(D) 30
(E) 20

Solution. The aircraft's acceleration, as a function of time, is

$$a = \frac{d^2x}{dt^2} = 4 - \frac{4}{15}t = 4\left(1 - \frac{t}{15}\right)$$

At the initial instant when $t = 0$, $V = 0$ and $x = 0$. By integration, the velocity is

$$V = \frac{dx}{dt} = 4\left(t - \frac{t^2}{30}\right) + C_1$$

At $t = 0$, $V = 0$, so $C_1 = 0$. At $t = 15$ sec the velocity is then

$$V = 4\left[15 - \frac{(15)^2}{30}\right] = 4[15 - 7.5]$$

$$V = 30 \text{ m/sec}$$

Answer is (D)

(2) The length (in meters) of the take-off run is nearest to

(A) 150
(B) 225
(C) 300
(D) 375
(E) 450

Solution. Integrating once more, the position $x(t)$ is

$$x = 4\left(\frac{t^2}{2} - \frac{t^3}{90}\right) + C_2$$

At $t = 0$, $x = 0$, hence $C_2 = 0$. At $t = 15$ sec the length of the take-off run is

$$x = 4\left[\frac{(15)^2}{2} - \frac{(15)^3}{90}\right] = 4[112.5 - 37.5]$$

$$x = 300 \text{ m}$$

Answer is (C)

PROBLEM 4-5

A body weighing 322 lb is subjected to an acceleration (in the positive x direction) which is a linearly decreasing function of the velocity. The body is stationary at $x = 0$ when $t = 0$, and the force acting on the body at this instant is 100 lb. The acceleration is zero when the velocity reaches 100 ft/sec.

(1) The differential equation that expresses this situation mathematically is

(A) $\ddot{x} + 0.1\dot{x} - 10 = 0$
(B) $\ddot{x} + 0.01\dot{x} - 100 = 0$
(C) $\ddot{x} - \dot{x} = 0$
(D) $\ddot{x} + 0.1\dot{x} + 10 = 0$
(E) $\ddot{x} + \dot{x} - 100 = 0$

Solution. Using $F = Ma$, the initial acceleration is

$$a(0) = \ddot{x} = \frac{F}{M} = \frac{F}{W/g} = \frac{100(32.2)}{322} = 10 \text{ ft/sec}^2$$

The other initial conditions are $\dot{x}(0)=0$, $x(0)=0$. The acceleration as a function of time is of the form

$$\ddot{x} = -K\dot{x} + C \qquad K > 0$$

At $t = 0$, $\ddot{x} = 10$ when $\dot{x} = 0$, so $C = 10$. Also when $\dot{x} = 100$, $\ddot{x} = 0$, or

$$0 = -100K + 10 \qquad K = 0.1$$

Hence

$$\ddot{x} + 0.1\dot{x} - 10 = 0$$

Answer is (A)

(2) The displacement $x(t)$ is

(A) $100t$
(B) $100(t + 10e^{-t})$
(C) $1000(e^{-t/10} - 1)$
(D) $-100(1 - e^{t/10})$
(E) $1000(e^{-0.1t} - 1) + 100t$

Solution. We seek the general solution to the differential equation for x, which is the sum of a particular solution and the complementary solution to

$$\ddot{x} + 0.1\dot{x} = 10$$

By inspection, a particular solution is $x = 100t$. To find the complementary solution, that is, the solution to $\ddot{x} + 0.1\dot{x} = 0$, assume a solution of the form

$$x = Ae^{\alpha t} + B$$

Then

$$\dot{x} = A\alpha e^{\alpha t}$$

and

$$\ddot{x} = A\alpha^2 e^{\alpha t}$$

Substituting into the homogeneous differential equation, we obtain

$$A\alpha^2 e^{\alpha t} + 0.1A\alpha e^{\alpha t} = 0$$

or

$$\alpha(\alpha + 0.1) = 0$$

The nontrivial solution is $\alpha = -0.1$. The general solution is then

$$x = Ae^{-0.1t} + B + 100t$$

subject to the initial conditions $\dot{x} = x = 0$ at $t = 0$.

$$x = 0 = A + B$$
$$\dot{x} = 0 = -0.1A + 100$$

The solution is $A = 1000$, $B = -1000$. Hence

$$x(t) = 1000(e^{-0.1t} - 1) + 100t$$

Answer is (E)

PROBLEM 4-6

A weight is attached to one end of a 53-ft rope passing over a small pulley 29 ft above the ground. A man whose hand is 5 ft above the ground grasps the other end of the rope and walks away at the rate of 5 ft/sec. When the man is 7 ft from a point directly under the pulley, the rate at which the weight is rising is closest to

(A) 1.0 ft/sec
(B) 1.4 ft/sec
(C) 2.5 ft/sec
(D) 3.6 ft/sec
(E) 5.0 ft/sec

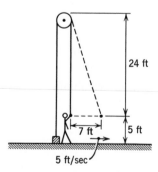

24 ft

7 ft   5 ft

5 ft/sec   **Figure 4-9**

Solution.

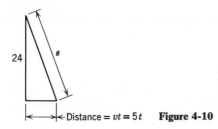

Distance $= vt = 5t$ **Figure 4-10**

At $t = 0$ we note that the rope is taut since 29 ft + 24 ft = 53 ft. Let s be the distance between the pulley and the man's hand:

$$s = [24^2 + (5t)^2]^{1/2} = (576 + 25t^2)^{1/2}$$

$$v = \frac{ds}{dt} = \tfrac{1}{2}(576 + 25t^2)^{-1/2} \times 50t$$

When the horizontal distance $5t = 7$, $t = 1.4$ sec, and

$$v = \tfrac{1}{2}[576 + 25(1.4)^2]^{-1/2} \times 50(1.4) = 1.4 \text{ ft/sec}$$

The rope is therefore moving at 1.4 ft/sec, and consequently the weight is rising at this same velocity.

Answer is (B)

PROBLEM 4-7

A river flows north with a speed of 3 miles/hr. A man rows a boat across the river. His speed relative to the water is 4 miles/hr. What is his velocity relative to the earth?

(A) 4 miles/hr

(B) 3 miles/hr

(C) 7 miles/hr

(D) 1 miles/hr

(E) 5 miles/hr

Solution.

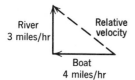

River
3 miles/hr

Relative velocity

Boat
4 miles/hr   **Figure 4-11**

$$V = (3^2 + 4^2)^{1/2} = 5 \text{ miles/hr}$$

Answer is (E)

PROBLEM 4-8

A car jumps across a 10-ft-wide ditch with a constant velocity V. The ditch is 6 in. lower on the far side. The minimum velocity (in miles per hour) that will keep the car from falling into the ditch is nearest to

(A) 35
(B) 40
(C) 45
(D) 50
(E) 55

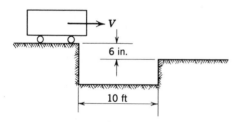

Figure 4-12

Solution.   The vertical drop $y = \frac{1}{2}gt^2 = 0.5$ ft. Hence

$$t = \left(\frac{2y}{g}\right)^{1/2} = \left[\frac{2(0.5)}{32.2}\right]^{1/2} = 0.176 \text{ sec}$$

The horizontal motion is not accelerated and is

$$x = Vt = 10 \text{ ft}$$

$$V = \frac{10}{t} = \frac{10}{0.176} = 56.8 \text{ ft/sec}$$

Since 60 miles/hr = 88 ft/sec,

$$V = \left(\frac{60}{88}\right)(56.8) = 38.8 \text{ miles/hr}$$

Answer is (B)

PROBLEM 4-9

A batted baseball leaves the bat at an angle of 30° above the horizontal and is caught by an outfielder 400 ft from home plate. Assume the ball's height when hit is the same as when caught and that air resistance is negligible. The initial velocity of the ball is closest to

(A) 200 ft/sec
(B) 175 ft/sec
(C) 150 ft/sec
(D) 125 ft/sec
(E) 100 ft/sec

Solution.

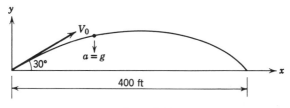

Figure 4-13

Since we neglect air resistance, the only force acting on the ball is due to its weight. Hence the acceleration of the ball at all times is g directed downward.

The horizontal distance traversed is $x = V_x t$, where $V_x = V_0 \cos 30°$ and $x = 400$ ft.

$$400 = V_0 \frac{\sqrt{3}}{2} t \quad \text{or} \quad t = \frac{800}{V_0 \sqrt{3}} \text{ sec}$$

The displacement vertically is

$$y = V_y t - \tfrac{1}{2}gt^2$$

or

$$0 = V_0 \sin 30° t - \tfrac{1}{2}gt^2$$

Substituting for t,

$$0 = \frac{800}{\sqrt{3}} \sin 30° - \frac{1}{2} g \left(\frac{800}{V_0 \sqrt{3}} \right)^2$$

$$V_0^2 = 32.2 \left(\frac{800}{\sqrt{3}} \right) \qquad V_0 = 122 \text{ ft/sec}$$

Answer is (D)

PROBLEM 4-10

A satellite travels in a perfectly circular orbit around the earth at an altitude of 1000 miles. Assume the earth is a perfect sphere with a radius of 4000 miles and that the force of the earth's gravity g at the height of the satellite is 20.6 ft/sec^2. The speed of the satellite is nearest to

(A) 10,000 miles/hr
(B) 12,000 miles/hr
(C) 14,000 miles/hr
(D) 16,000 miles/hr
(E) 18,000 miles/hr

Solution.

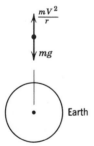

Figure 4-14

The two forces acting on the satellite are the attractive force of the earth mg, and the centrifugal force mV^2/r. If the satellite is to remain a constant distance from earth, these two forces must be in equilibrium. Hence

$$\frac{mV^2}{r} = mg$$

$V = (rg)^{1/2}$

$\quad = [(4000 + 1000) \text{ miles} \times (5280 \text{ ft/mile})(20.6 \text{ ft/sec}^2)]^{1/2}$

$\quad = 2.33 \times 10^4 \text{ ft/sec} = 15{,}900 \text{ miles/hr}$

Answer is (D)

PROBLEM 4-11

The distance of the planet Neptune from the sun is 30 times that of the earth from the sun. The force of attraction between two bodies is directly proportional to the product of their masses and inversely proportional to the square of the distance between them. The approximate period of Neptune's revolution about the sun is nearest to

(A) 900 yr
(B) 160 yr
(C) 90 yr
(D) 30 yr
(E) 5 yr

Solution.   Mathematically, the attractive force F is

$$F = k\frac{MM'}{R^2}$$

Between the sun and earth it is

$$F = k\frac{M_S M_E}{R^2}$$

and between the sun and Neptune

$$F = k\frac{M_S M_N}{(30R)^2}$$

For uniform circular motion the force F causes a centripetal acceleration $a = V^2/R$, where $V = 2\pi R/T$ and T is the period of revolution. Hence

$$a_E = \frac{4\pi^2 R}{T_E^2} \quad \text{and} \quad a_N = \frac{4\pi^2(30R)}{T_N^2}$$

Using $F = ma$,

$$k\frac{M_S M_E}{R^2} = \frac{M_E 4\pi^2 R}{T_E^2} \quad \text{and} \quad k\frac{M_S M_N}{(30R)^2} = \frac{M_N 4\pi^2(30R)}{T_N^2}$$

or

$$\frac{R^3}{T_E^2} = \frac{kM_S}{4\pi^2} = \frac{(30R)^3}{T_N^2}$$

The period of the earth's revolution about the sun T_E is one year, so the period for Neptune's revolution is

$$T_N = (30)^{3/2} \text{ yr} = 164.3 \text{ yr}$$

(Using Kepler's third law of planetary motion would yield the same result.)

Answer is (B)

PROBLEM 4-12

A traveling crane lifts a 1000-lb load on a 20-ft hoisting cable. The maximum horizontal acceleration that the crane may have without producing a deviation of the cable of more than 30° from the vertical is closest to

(A) 19 ft/sec²
(B) 25 ft/sec²
(C) 32 ft/sec²
(D) 43 ft/sec²
(E) 55 ft/sec²

Solution.

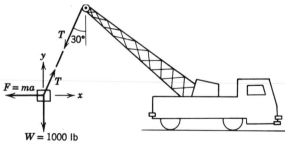

Figure 4-15

Writing Newton's equation of motion in the x and y directions for the weight

$$\Sigma F_y = ma_y \qquad T \cos 30° - W = 0$$

$$\Sigma F_x = ma_x \qquad T \sin 30° = \frac{W}{g} a$$

Eliminating W from these equations and simplifying gives

$$a = g \tan 30° = (32.2)/\sqrt{3} = 18.6 \text{ ft/sec}^2$$

Answer is (A)

PROBLEM 4–13

An elevator, which with its load weighs 70,000 N (newtons), is descending at a speed of 4.5 m/sec. If the load on the cable must not exceed 125,000 N, the shortest distance in which the elevator should be stopped is most nearly

(A) 0.6 m
(B) 1.0 m
(C) 1.3 m
(D) 2.0 m
(E) 2.6 m

Solution. The maximum allowable inertia force is the difference between the maximum cable load and the elevator weight; in this case the maximum inertia force to stop the elevator is $125,000 - 70,000 = 55,000$ N. Writing the inertia force as

$$F = -ma = -\frac{W}{g} a$$

$$55,000 \text{ N} = -\frac{70,000 \text{ N}}{9.81 \text{ m/sec}^2} a$$

and the maximum allowable deceleration is $a = -7.71 \text{ m/sec}^2$. Since

$$a = \frac{dv}{dt} = \frac{dv}{ds}\frac{ds}{dt} = \frac{d}{ds}\left(\frac{v^2}{2}\right)$$

$$as = \tfrac{1}{2}(V_2^2 - V_1^2)$$

The initial elevator velocity $V_1 = 4.5$ m/sec, and the final velocity $V_2 = 0$.

Hence the minimum stopping distance s is

$$s = \frac{1}{2(-7.71)}(-4.5)^2 = 1.31 \text{ m}$$

Answer is (C)

PROBLEM 4-14

The system of pulleys carries two loads as shown in Fig. 4-16. The acceleration of the 8-lb weight is closest to

(A) 38.6 ft/sec^2
(B) 32.2 ft/sec^2
(C) 17.6 ft/sec^2
(D) 9.2 ft/sec^2
(E) 7.4 ft/sec^2

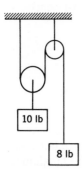

Figure 4-16

Solution. The pulleys are assumed to be weightless and frictionless. We will select the downward direction to be positive in each free-body diagram. For the movable pulley,

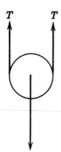

$Mg = 10$ lb   **Figure 4-17**

$$\sum F = -2T + Mg = Ma_1 \tag{1}$$

and for the 8-lb weight

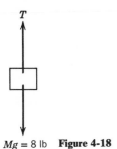

$Mg = 8$ lb   **Figure 4-18**

$$\sum F = -T + mg = ma_2 \qquad (2)$$

The accelerations a_1 and a_2 are related by the pulley geometry since the 8-lb weight will rise twice the distance the 10-lb weight will fall; hence

$$a_1 = \frac{-a_2}{2} \qquad (3)$$

Equations (1), (2), and (3) are three equations for the unknowns T, a_1, and a_2. First eliminate a_1:

$$2T - Mg = \frac{Ma_2}{2}$$

$$-2T + 2mg = 2ma_2$$

Now add:

$$a_2 \left(2m + \frac{M}{2} \right) = 2mg - Mg$$

$$a_2 = \left(\frac{2mg - Mg}{2mg + Mg/2} \right) g$$

$$a_2 = \left[\frac{2(8) - 10}{2(8) + 10/2} \right] (32.2)$$

Downward acceleration $= a_2 = 9.20$ ft/sec^2

Answer is (D)

PROBLEM 4-15

A 16-lb weight and an 8-lb weight resting on a horizontal frictionless surface are connected by a cord A and are pulled along the surface with a uniform acceleration of 4 ft/sec^2 by a second cord attached to the 16-lb weight. The tension in cord A is closest to

(A) 4 lb
(B) 1 lb
(C) 2 lb
(D) 3 lb
(E) 32 lb

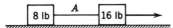

Figure 4-19

Solution. The presence of the 16-lb weight does not affect the tension in A. The tensile force is that needed to accelerate the 8-lb weight.

$$F = ma = \frac{W}{g}a = \left(\frac{8}{32.2}\right)(4) \approx 1 \text{ lb}$$

Answer is (B)

PROBLEM 4-16

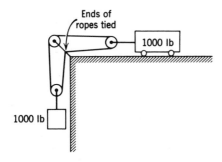

Figure 4-20

The tension in the rope for the arrangement shown in Fig. 4-20 (assume no friction and weightless pulleys) is nearest to

(A) 1000 lb
(B) 500 lb
(C) 333 lb
(D) 250 lb
(E) 200 lb

Solution.

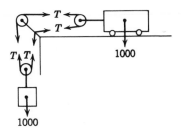

1000

Figure 4-21

For the suspended weight, $\sum F_y = ma$

$$1000 - 2T = \frac{1000}{32.2} a \qquad (1)$$

For the cart, $\sum F_x = ma$

$$2T = \frac{1000}{32.2} a \qquad (2)$$

Solving Eqs. (1) and (2) for T,

$$1000 - 2T = 2T \qquad T = 250 \text{ lb}$$

Answer is (D)

PROBLEM 4-17

Two 3-N weights are connected by a massless string hanging over a smooth frictionless peg. If a third weight of 3 N is added to one of the weights and the system is released, the amount of the increased force on the peg is most nearly

(A) 0 N
(B) 1.5 N
(C) 2 N
(D) 3 N
(E) 6 N

Solution.

The initial system is in static equilibrium with $T = 3$ N, $2T = 6$ N. The additional weight is then added. Applying $\sum F = ma$ successively to the 6-

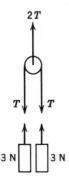

Figure 4-22

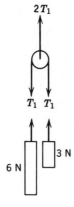

Figure 4-23

and 3-N weights, we obtain

$$6 - T_1 = \frac{6}{g}a \qquad (1)$$

$$T_1 - 3 = \frac{3}{g}a \qquad (2)$$

The increase in force on the peg is represented by the quantity $2T_1 - 2T$, so $2T_1$ must be found. Doubling Eq. (2) and comparing it with Eq. (1),

$$6 - T_1 = 2(T_1 - 3)$$

$$3T_1 = 12 \qquad \text{and} \qquad 2T_1 = 8$$

Hence the increase in force on the peg $= 8 - 6 = 2$ N.

Answer is (C)

PROBLEM 4-18

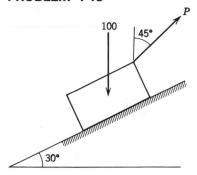

Figure 4-24

This system is initially at rest. The force P that will be required to give the 100-lb block a velocity V of 10 ft/sec up the plane in a time interval of 5 sec is most nearly

(A) 360 lb
(B) 310 lb
(C) 116 lb
(D) 58 lb
(E) 45 lb

Solution.

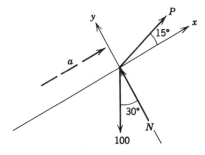

Figure 4-25

Since no coefficient of friction is mentioned, let us assume the plane to be smooth. Considering the block to be a particle, we choose the coordinate system to be as shown and write the equation of motion

$$\sum F_x = \frac{W}{g} a_x$$

$$P \cos 15° - 100 \sin 30° = \frac{100}{32.2} a_x$$

To determine a_x, $V = V_0 + at$ with $V_0 = 0$. At $t = 5$ sec,

$$10 \text{ ft/sec} = a(5 \text{ sec})$$

$$a = 2 \text{ ft/sec}^2$$

Hence

$$P = \left[\frac{100}{32.2}(2) + 100(0.500) \right] \Big/ 0.966$$

from the earlier equation, or $P = 58.2$ lb.

Answer is (D)

PROBLEM 4-19

A locomotive weighing 120 tons is coupled to, and pulls, a car weighing 40 tons. The resistances to motion on a level track are $\frac{1}{100}$th of its weight for the locomotive and $\frac{1}{160}$th of its weight for the car. The tractive force exerted by the locomotive is 8000 lb. The tension T in the coupling is nearest to

(A) 1775 lb
(B) 2225 lb
(C) 4500 lb
(D) 5100 lb
(E) 5775 lb

Solution.

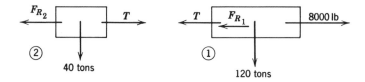

Figure 4-26

The diagram on the right represents the locomotive; the car is to the left. The forces resisting motion are

$$F_{R_1} = \frac{W_1}{100} = \frac{120(2000)}{100} = 2400 \text{ lb}$$

$$F_{R_2} = \frac{W_2}{160} = \frac{40(2000)}{160} = 500 \text{ lb}$$

Successively applying $\sum F_x = ma_x$ to the two bodies,

$$8000 - 2400 - T = \frac{120(2000)}{g} a \tag{1}$$

$$T - 500 = \frac{40(2000)}{g} a \tag{2}$$

We wish to solve for the tension T. Since the common factor $2000a/g$ appears on the right side of Eqs. (1) and (2), we can immediately write

$$\frac{5600 - T}{120} = \frac{T - 500}{40}$$

$$T = 1775 \text{ lb}$$

Answer is (A)

PROBLEM 4-20

The coefficient of friction between the 50-lb weight and the 300-lb weight is 0.5. The 300-lb weight is free to roll, and the weight and friction for the

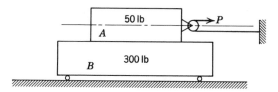

Figure 4-27

pulley are negligible. When $P = 16$ lb, the acceleration of block B is closest to

(A) 1.5 ft/sec²
(B) 2.7 ft/sec²
(C) 2.9 ft/sec²
(D) 3.4 ft/sec²
(E) 5.4 ft/sec²

Solution.

Let us first assume that slipping occurs between the blocks. Then the

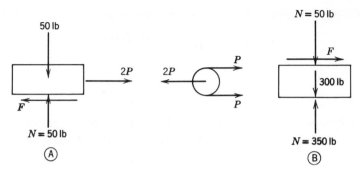

Figure 4-28

frictional force developed is

$$F = \mu N = 0.5(50) = 25 \text{ lb}$$

Now apply $\sum F_x = ma$ to block B,

$$25 = \frac{300}{g} a_B$$

$$a_B = \frac{25}{300} g = 2.68 \text{ ft/sec}^2 \quad \text{to the right}$$

To check our original assumption, assume the blocks move together. For the two blocks, $\sum F_x = ma$

$$32 = 2P = \frac{350}{g} a \qquad a = 2.94 \text{ ft/sec}^2$$

For block B alone, this acceleration requires a frictional force

$$F = \frac{300}{g}(2.94) = 27.4 \text{ lb}$$

However, since this force is greater than the maximum frictional force that can be developed (25 lb), the first assumption was correct. Hence

$$a_B = 2.68 \text{ ft/sec}^2 \quad \text{to the right}$$

Answer is (B)

Problem 4-21

A weight W is suspended by two strings, AB and AC, as shown. The ratio of

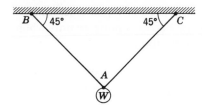

Figure 4-29

the forces in AC (1) just before, to (2) just after the instant that AB is cut is most nearly

(A) 2.0

(B) 1.6

(C) 1.4

(D) 1.0

(E) 0.5

Solution.

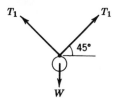

Figure 4-30

Before string AB is cut, the tensile forces in strings AB and AC are equal because of symmetry. Summing forces vertically,

$$2T_1 \sin 45° = W \qquad T_1 = \frac{W}{\sqrt{2}}$$

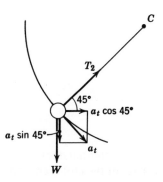

Figure 4-31

At the initial instant after string AB is cut, the body possesses only a tangential acceleration; the radial acceleration becomes nonzero only when the body is moving. Writing the equations of motion,

$$\sum F_y = ma_y$$

$$W - T_2 \sin 45° = ma_t \sin 45°$$

$$\sum F_x = ma_x$$

$$T_2 \cos 45° = ma_t \cos 45°$$

$$T_2 = ma_t$$

Inserting this relation to eliminate the acceleration term in the first equation, we obtain

$$W - T_2 \sin 45° = T_2 \sin 45°$$

$$W = 2T_2 \sin 45°$$

$$T_2 = \frac{W}{\sqrt{2}}$$

Hence $T_1 = T_2$ and the ratio $T_1/T_2 = 1$.

Answer is (D)

PROBLEM 4-22

A 0.25-kg ball is thrown horizontally with a kinetic energy of 50 joules from a vertical cliff 20 m high.

(1) The ball's kinetic energy when it strikes the ground, level with the base of the cliff, is most nearly

(A) 50 J
(B) 55 J
(C) 100 J
(D) 250 J
(E) 5000 J

Solution.

The mechanical energy, which is the sum of kinetic and potential energy, of the ball remains constant while the ball falls. If the base of the cliff is chosen

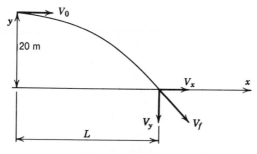

Figure 4-32

as a datum, then relative to this datum the ball will possess only kinetic energy when it hits the ground. Hence

Ground kinetic energy = initial kinetic energy + change in potential energy

$$= 50 + Wh$$

$$= 50 + mgh$$

$$= 50 + (0.25)(9.81)(20) = 99 \text{ J}$$

Answer is (C)

(2) The distance from the foot of the cliff to the point where the ball strikes the ground is most nearly

(A) 20 m
(B) 25 m
(C) 30 m
(D) 35 m
(E) 40 m

Solution. Initially $KE = 50 = \frac{1}{2}mV_0^2$

$$V_0^2 = \frac{100}{m} = \frac{100}{0.25} = 400$$

$$V_0 = V_x = 20 \text{ m/sec}$$

The ball falls a vertical distance of 20 m, so

$$20 = \frac{1}{2}gt^2 \quad \text{or} \quad t = \left(\frac{40}{g}\right)^{1/2} = \left(\frac{40}{9.81}\right)^{1/2} = 2.02 \text{ sec}$$

Finally, $L = V_0 t = 20(2.02) = 40.4$ m.

Answer is (E)

PROBLEM 4-23

A static weight W produces a static deflection of 2 in. in a spring having a spring constant k. The mass of the spring is neglected.

(1) The maximum deflection x when the weight W is dropped on the spring from a point 6 in. above the free position of the spring is most nearly

   (A) 2.0 in.
   (B) 4.6 in.
   (C) 5.3 in.
   (D) 6.6 in.
   (E) 7.3 in.

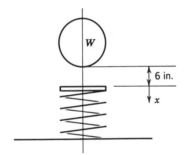

Figure 4-33

Solution. Conservation of the sum of kinetic and potential energy can be expressed here as

$$W(6+x)=\tfrac{1}{2}mV^2+\tfrac{1}{2}kx^2 \qquad (1)$$

The spring constant can be found from the given data for the static case; then $F = kx = W$, and $k = W/x = W/2$.

The maximum deflection occurs when the weight W has compressed the spring and its velocity has become zero. Then

$$W(6+x)=\tfrac{1}{2}kx^2=\frac{1}{2}\left(\frac{W}{2}\right)x^2$$

or

$$x^2 - 4x - 24 = 0$$
$$x = \tfrac{1}{2}[+4 \pm \sqrt{16 + 4(24)}]$$
$$x = 2 \pm \sqrt{28} = 2 \pm 5.3$$

Only the positive root is relevant here: $x = 7.3$ in.

Answer is (E)

(2) The maximum velocity of the falling weight is most nearly

(A) 5.5 ft/sec
(B) 5.7 ft/sec
(C) 5.9 ft/sec
(D) 6.1 ft/sec
(E) 6.3 ft/sec

Solution. The maximum velocity occurs when $dV/dx = 0$ in the motion governed by equation (1). By direct differentiation

$$W = mV\frac{dV}{dx} + kx$$

Thus the maximum occurs at $x = W/k = 2$ in. $= 0.167$ ft, which also is the point at which the weight and spring are in static equilibrium. From Eq. (1),

$$W(0.5 + 0.167) = \frac{1}{2}\frac{W}{g}V^2 + \frac{1}{2}\left[\frac{W}{0.167}\right](0.167)^2$$

$$0.667 = \frac{V^2}{2(32.2)} + \frac{0.167}{2} \qquad V = 6.13 \text{ ft/sec}$$

Answer is (D)

PROBLEM 4-24

A projectile weighing 100 lb strikes the concrete wall of a fort with an impact velocity of 1200 ft/sec. The projectile comes to rest in 0.01 sec, having penetrated the 8-ft thick wall to a distance of 6 ft. The average force exerted on the wall by the projectile is closest to

    (A) 2×10^5 lb
    (B) 3×10^5 lb
    (C) 1×10^7 lb
    (D) 2×10^8 lb
    (E) 1×10^{10} lb

Solution. Knowing that the impulse exerted on the wall is equal to the change of momentum of the projectile, one may write

$$\int F\,dt = m\,\Delta V$$

If F is assumed to be the average force, then it is a constant, and the equation becomes

$$F\,\Delta t = \frac{W}{g}\Delta V$$

$$F(0.01) = \frac{100}{32.2}(1200)$$

$$F = 3.73(10^5)\,\text{lb}$$

Answer is (B)

PROBLEM 4-25

A 10-lb block which is suspended by a long cord is at rest when a 0.05-lb bullet traveling horizontally to the left strikes and is embedded in it. The impact causes the block to swing upward 0.5 ft measured vertically from its lowest position.

(1) The velocity of the bullet just before it strikes the block is most nearly

    (A) 80 ft/sec
    (B) 200 ft/sec
    (C) 1135 ft/sec
    (D) 1140 ft/sec
    (E) 1600 ft/sec

Solution. Let m = mass of bullet, M = mass of block, V_b = velocity of bullet before impact, and V = system velocity immediately after impact. Energy is conserved for the motion of the system after impact so that

$$\Delta(KE) = -\Delta(PE)$$
$$\tfrac{1}{2}(M+m)V^2 = (M+m)gh$$
$$V^2 = 2gh = 2g(0.5)$$
$$V = g^{1/2} \text{ ft/sec}$$

Also momentum is conserved during impact.

$$mV_b = (M+m)V$$

$$V_b = \left(\frac{M}{m}+1\right)V$$

$$V_b = \left(\frac{10}{0.05}+1\right)(32.2)^{1/2} = 1140 \text{ ft/sec}$$

Answer is (D)

(2) The loss of kinetic energy of the system during impact is most nearly

(A) 5 ft-lb
(B) 30 ft-lb
(C) 1005 ft-lb
(D) 1010 ft-lb
(E) 1980 ft-lb

Solution. The initial kinetic energy of the bullet is

$$\frac{1}{2}mV_b^2 = \frac{1}{2}\left(\frac{0.05}{32.2}\right)(1140)^2 = 1010 \text{ ft-lb}$$

The kinetic energy immediately after impact is

$$\tfrac{1}{2}(M+m)V^2 = \tfrac{1}{2}(M+m)g$$
$$= \tfrac{1}{2}(10+0.05) = 5.025 \text{ ft-lb}$$

The system loss of kinetic energy is the difference between these two values, or

$$1010 - 5.025 = 1005 \text{ ft-lb}$$

Answer is (C)

PROBLEM 4-26

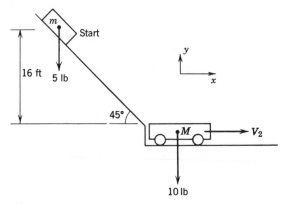

Figure 4-34

A 5-lb block of wood slides down a frictionless inclined plane at an angle of 45° with the horizontal and lands on a 10-lb cart with frictionless wheels. The slant length of the plane is $16\sqrt{2}$ ft. If the block sticks to the cart, the cart and the block will move away from the bottom of the inclined plane at a speed most nearly equal to

(A) 7.6 ft/sec
(B) 10.7 ft/sec
(C) 18.5 ft/sec
(D) 22.7 ft/sec
(E) 32.1 ft/sec

Solution. Since the plane is frictionless, the fall of the block is unimpeded. Equating the changes in kinetic and potential energy between the top and bottom of the plane,

$$\tfrac{1}{2}mV^2 = mgh$$
$$V^2 = 2gh = 2(32.2)(16)$$

Because of the 45° slope, V_x and V_y are equal, or

$$V_x^2 = V_y^2 = \tfrac{1}{2}V^2 = (32.2)(16)$$
$$V_x = 22.7 \text{ ft/sec}$$

Momentum in the x-direction is conserved when the block and cart collide:

$$mV_x = (m+M)V_2$$

$$V_2 = \frac{m}{m+M}V_x = \left(\frac{1}{1+M/m}\right)V_x$$

$$= \left(\frac{1}{1+2}\right)(22.7)$$

$$V_2 = 7.57 \text{ ft/sec}$$

Answer is (A)

PROBLEM 4-27

A 150-lb man stands at the rear of a 250-lb boat. The distance from the man to the pier is 30 ft, and the length of the boat is 16 ft. Assume no friction between the boat and the water. The distance of the man from the pier after he walks to the front of the boat at a velocity of 3 miles/hr is most nearly

(A) 14 ft
(B) 16 ft
(C) 20 ft
(D) 24 ft
(E) 30 ft

Solution.

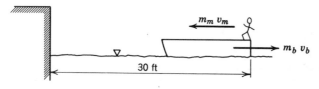

Figure 4-35

So that the momentum is conserved, $m_m v_m = m_b v_b$. Let $v_{m/b}$ be the velocity of the man relative to the boat so that $v_{m/b} = 3$ miles/hr $= 4.4$ ft/sec. Since

$$v_m = v_{m/b} - v_b$$

$$m_m(v_{m/b} - v_b) = m_b v_b$$

or, by rearrangement,

$$v_b = \frac{m_m}{m_m + m_b} v_{m/b} = \frac{150}{400}(4.4) = 1.65 \text{ ft/sec}$$

The man walks to the front of the boat in

$$t = \frac{s_m}{v_m} = \frac{16.0}{4.4} = 3.64 \text{ sec}$$

In this time the boat will move away from the pier a distance

$$s_b = v_b t = (1.65)(3.64) = 6.0 \text{ ft}$$

The rear of the boat is now $30 + 6 = 36$ ft from the pier, and the man is

$$36 - 16 = 20 \text{ ft from the pier}$$

Answer is (C)

PROBLEM 4-28

A pile weighing 2000 lb is driven vertically into the ground by a "monkey" (pile hammer) weighing 6000 lb which falls freely from rest through a height of 16 ft to the head of the pile. The impact of the monkey on the pile is assumed to be inelastic. The resistance of the ground can be assumed uniform and equivalent to a force of 150,000 lb. With each blow the pile moves into the ground a distance that is most nearly

(A) 0.10 ft
(B) 0.25 ft
(C) 0.50 ft
(D) 0.75 ft
(E) 1.00 ft

Solution.   For conservation of momentum,

$$M_h V_1 + M_p V_{1_p} = M_h V_{2_h} + M_p V_{2_p}$$

Here $V_{1_p} = 0$ and $V_{2_h} = V_{2_p} = V_2$ for inelastic impact. The equation is now

$$M_h V_1 = (M_h + M_p)V_2$$

From the work-energy relation for a freely falling body (the hammer),

$$V_1 = (2gh)^{1/2} = [2(32.2)(16)]^{1/2} = 32.1 \text{ ft/sec}$$

Hence

$$V_2 = \frac{M_h}{M_h + M_p} V_1 = \frac{6000}{8000}(32.1) = 24.1 \text{ ft/sec}$$

Since the ground resistance F is uniform, the deceleration of the pile-hammer combination will be constant.

$$+8000 - 150,000 = F = (M_h + M_p)a = \frac{8000}{32.2}a$$

or the deceleration $a = -572$ ft/sec.²

The pile displacement $x = x_0 + V_0 t + \frac{1}{2}at^2$, where $x_0 = 0$, $V_0 = V_2$, and $a = -572$ ft/sec² in this case. The pile stops moving when $dx/dt = 0$.

$$0 = \frac{dx}{dt} = V_2 + at$$

$$t = -\frac{V_2}{a} = -\frac{24.1}{-572} = 0.0421 \text{ sec}$$

With each blow the pile movement or "set" is

$$x = 24.1(0.0421) - \frac{1}{2}(572)(0.0421)^2 = 0.51 \text{ ft}$$

Answer is (C)

PROBLEM 4-29

The crankshaft of an engine is turning at the rate of 20 rev/sec. The

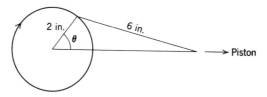

Figure 4-36

connecting rod is 6 in. long, and the radius is 2 in. When the angle θ is 30°, the piston is moving at a rate that is most nearly

(A) 21.0 ft/sec
(B) 10.5 ft/sec
(C) 7.0 ft/sec
(D) 18.2 ft/sec
(E) 13.5 ft/sec

Solution.

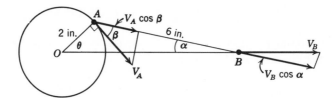

Figure 4-37

The velocity at point A is $V_A = 20(2\pi)2 = 80\pi$ in./sec. The velocity of the piston is V_B. From the diagram $V_B \cos \alpha = V_A \cos \beta$, and β is a function of the given angles α and θ.

Using the laws of sines for triangle OAB,

$$\frac{6}{\sin \theta} = \frac{2}{\sin \alpha} \qquad \sin \alpha = \tfrac{1}{3} \sin \theta$$

Also from triangle OAB,

$$\alpha + \theta + 90° + \beta = 180°$$

For $\theta = 30°$,

$$\sin \alpha = \tfrac{1}{3} \sin 30° \qquad \alpha = 9°36'$$

$$\beta = 180° - 90° - 30° - 9°36' = 50°24'$$

and

$$V_B = V_A \frac{\cos \beta}{\cos \alpha} = 80\pi \frac{\cos 50°24'}{\cos 9°36'} = 80\pi \frac{0.637}{0.986}$$

$$V_B = 162 \text{ in./sec} = 13.5 \text{ ft/sec}$$

Answer is (E)

PROBLEM 4-30

A block weighs 30 N. A pin on the rotating arm OP, which moves in a smooth vertical slot, causes the block to slide on a horizontal plane, as shown in Fig. 4-38. In this 45° position, the 15-cm-long arm OP is rotating clockwise at a constant angular velocity of 5 rad/sec. The coefficient of friction between the block and the plane is 0.2. The torque exerted by the

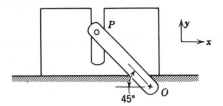

Figure 4-38

arm OP is most nearly

    (A) 0.9 N-m
    (B) 1.5 N-m
    (C) 1.9 N-m
    (D) 2.1 N-m
    (E) 2.6 N-m

Solution. Consider first the rotating arm. The normal acceleration of point P is

$$a_n = r\omega^2 = (0.15)(5)^2 = 3.75 \text{ m/sec}^2$$

For the position shown, the component in the x direction is $(a_n)_x = 3.75 \sin 45° = 2.65 \text{ m/sec}^2$. The block slides along the plane, creating a frictional force

$$F_r = \mu N = \mu W = (0.2)(30) = 6 \text{ N}$$

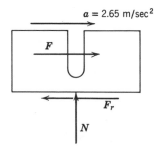

Figure 4-39

Writing the equation of motion in the x direction, $\sum F_x = ma_x$ or

$$F - F_r = ma$$

$$F - 6 = \frac{30}{9.81}(2.65) = 8.10$$

$$F = 14.10 \text{ N}$$

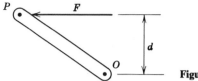

Figure 4-40

The torque on the arm T is

$$T = Fd = (14.10)(0.15 \sin 45°) = 1.50 \text{ N-m}$$

The torque exerted by the arm is equal and opposite to this torque.

Torque by arm $= 1.50$ N-m $\checkmark$

Answer is (B)

PROBLEM 4-31

A homogeneous cylinder begins to roll without slipping down a plane that is inclined at an angle θ. The moment of inertia of the cylinder about the axis through O is $I_O = \frac{1}{2}mr^2$.

(1) The acceleration of the center O as a function of θ is

   (A) $2g \sin \theta$
   (B) $\frac{1}{2}g \tan \theta$
   (C) $\frac{2}{3}g \tan \theta$
   (D) $\frac{2}{3}g \sin \theta$
   (E) $\frac{1}{2}g \sin \theta$

Solution. The cylinder rotates about the contact point A between the plane and the cylinder in a case of noncentroidal rotation. The governing

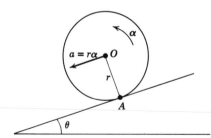

Figure 4-41

equation is $\sum M_A = I_A \alpha$. The only force that can produce a nonzero moment about A is the weight W of the cylinder. This moment is equal to rH, where

Figure 4-42

H is the component of W parallel to the inclined plane. Note that $H = W \sin \theta = mg \sin \theta$. The moment of inertia I_A, by the parallel-axis theorem, is

$$I_A = I_O + md^2$$
$$= \tfrac{1}{2}mr^2 + mr^2$$
$$= \tfrac{3}{2}mr^2$$

We now have $rmg \sin \theta = \tfrac{3}{2}mr^2\alpha$. Using $a = r\alpha$ and simplifying,

$$a = \tfrac{2}{3}g \sin \theta$$

Answer is (D)

(2) The minimum coefficient of friction μ to prevent the cylinder from slipping is

(A) $\tfrac{1}{3}$

(B) $\tfrac{1}{2}$

(C) $\tfrac{1}{2}\cos \theta$

(D) $\tfrac{1}{2}\tan \theta$

(E) $\tfrac{1}{3}\tan \theta$

Solution.

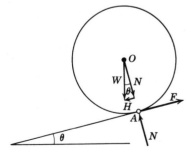

Figure 4-43

Parallel to the inclined plane,

$$\Sigma F = ma$$

$$H - F = ma$$

$$W \sin \theta - F = \frac{W}{g}\left(\frac{2}{3}g \sin \theta\right)$$

$$F = \frac{W}{3} \sin \theta$$

For impending slipping $F = \mu N = \mu W \cos \theta$. Hence

$$\mu = \frac{F}{W \cos \theta} = \frac{W \sin \theta}{3W \cos \theta} \qquad \mu = \frac{1}{3}\tan \theta$$

Answer is (E)

PROBLEM 4-32

A homogeneous solid cylinder of mass m and radius R has a string wound around it. One end of the string is fastened to a fixed point, and the cylinder is allowed to fall as shown.

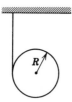

Figure 4-44

The moment of inertia of a cylinder about its axis is $\frac{1}{2}mR^2$. The tension in the string is

(A) $\frac{1}{3}mg$

(B) $\frac{1}{2}mg$

(C) $\frac{3}{5}mg$

(D) $\frac{2}{3}mg$

(E) mg

Solution.

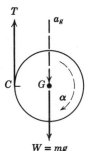

$W = mg$ Figure 4-45

For dynamic equilibrium vertically, $\sum F_y = ma_y$ or

$$mg - T = ma_g$$

Since the contact point C has no vertical acceleration, $a_g = R\alpha$ and

$$mg - T = mR\alpha \qquad (1)$$

Also, about point G, $\sum M_G = I_G\alpha$ so that

$$TR = \tfrac{1}{2}mR^2\alpha \qquad \text{or} \qquad 2T = mR\alpha \qquad (2)$$

Substituting Eq. (2) for $mR\alpha$ into Eq. (1),

$$T = \tfrac{1}{3}mg$$

Answer is (A)

PROBLEM 4-33

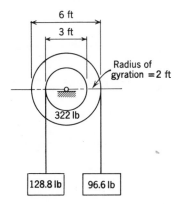

Figure 4-46

If the system shown in Fig. 4-46 is released from rest, the angular acceleration of the circular drum is most nearly

(A) 4.4 rad/sec²
(B) 1.3 rad/sec²
(C) 8.3 rad/sec²
(D) 1.7 rad/sec²
(E) 24.2 rad/sec²

Solution.

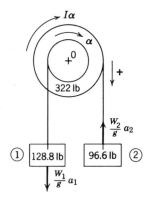

Figure 4-47

For dynamic equilibrium $\sum M = I\alpha$. The moment of inertia of the drum is
$I = (W/g)k^2$, where k is the radius of gyration of the drum.

$$I = \frac{322}{32.2}(2)^2 = 40 \text{ lb-ft-sec}^2$$

The acceleration of each weight is $a = r\alpha$. Therefore $a_1 = 1.5\alpha$ and $a_2 = 3\alpha$.
Now summing moments about 0 gives

$$3\left(W_2 - \frac{W_2}{g}a_2\right) - 1.5\left(W_1 + \frac{W_1}{g}a_1\right) = I\alpha$$

$$3\left[96.6 - \frac{96.6}{32.2}(3\alpha)\right] - 1.5\left[128.8 + \frac{128.8}{32.2}(1.5\alpha)\right] = 40\alpha$$

$$289.8 - 27\alpha - 193.2 - 9\alpha = 40\alpha$$

$$\alpha = \frac{96.6}{76} = 1.27 \text{ rad/sec}^2$$

Answer is (B)

PROBLEM 4-34

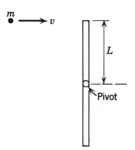

Figure 4-48

A putty ball of mass m, moving with a velocity v, as shown, makes inelastic impact with the end of a long thin bar of length $2L$ that has a moment of inertia I_0 about its center. The bar is frictionlessly pivoted at its center. The expression for the angular velocity of the bar just after the impact occurs is

(A) v/L

(B) mvL/I_0

(C) $mvL/(mL^2 + I_0)$

(D) $[mv^2/(I_0 + mL^2)]^{1/2}$

(E) $v(m/I_0)^{1/2}$

Solution. In this problem we employ the principle of conservation of angular momentum (moment of momentum). This may be expressed as $H_{p1} = H_{p2}$, where H_p is the moment of momentum at any given instant about the pivot.

Before impact, the moment of momentum of the putty ball is $H_{p1} = mvL$. After impact, the moment of momentum of the ball-bar combination is $H_{p2} = mv_2L + I_0\omega$, where $v_2 = L\omega$. Hence

$$mvL = mL^2\omega + I_0\omega$$

$$\omega = \frac{mvL}{mL^2 + I_0}$$

Answer is (C)

PROBLEM 4-35

A homogeneous cylindrical wheel, of 3-ft radius, weighing 400 lb carries two symmetrically placed weights, each weighing 64.4 lb, attached 2 ft from its center. If each weight moves radially outward 1 ft when the wheel is

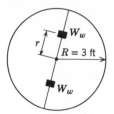

Figure 4-49

rotating at 2000 rpm, the change, in radians per second, which will occur in the angular velocity of the wheel, is most nearly

- (A) 13 rad/sec
- (B) 28 rad/sec
- (C) 46 rad/sec
- (D) 82 rad/sec
- (E) 290 rad/sec

Solution. Angular momentum is conserved for the wheel-weight combination. Expressed mathematically, $I_1\omega_1 = I_2\omega_2$. The initial angular velocity ω_1 is

$$\omega_1 = (2000 \text{ rpm})(\tfrac{1}{60} \text{ min/sec})(2\pi \text{ rad/rev}) = 210 \text{ rad/sec}$$

The polar moment of inertia of the wheel is

$$I_0 = \tfrac{1}{2}mR^2 = \frac{1}{2}\frac{W}{g}R^2 = \frac{1}{2}\left(\frac{400}{32.2}\right)(3)^2 = 56.0 \text{ lb-ft-sec}^2$$

The moment of inertia of the combination is then

$$I = I_0 + 2\frac{W_w}{g}r^2$$

For $r = 2$ ft,

$$I_1 = 56.0 + 2\left(\frac{64.4}{32.2}\right)(2)^2 = 72.0 \text{ lb-ft-sec}^2$$

For $r = 3$ ft,

$$I_2 = 56.0 + 2\left(\frac{64.4}{32.2}\right)(3)^2 = 92.0 \text{ lb-ft-sec}^2$$

For conservation of angular momentum,

$$I_1\omega_1 = (72.0)(210) = 92.0\omega_2$$

$$\omega_2 = 164 \text{ rad/sec}$$

Therefore the *decrease* in angular velocity is $210 - 164 = 46 \text{ rad/sec}$

$$\text{Answer is (C)}$$

PROBLEM 4-36

An unbalanced flywheel has its center of mass 4.00 in. from the axis of rotation. The radius of gyration of the flywheel with respect to an axis through the center of mass parallel to the axis of rotation is 16.00 in. The flywheel, which weighs 145.0 lb, is rotating clockwise about its axis at an angular speed of 3600 rpm when a counterclockwise torque $T = 18.0t^2$ is applied, where T is in pound-feet and t is in seconds. Neglecting friction, the angular speed in revolutions per minute of the flywheel when t is 10.00 sec is closest to

(A) 1500 rpm clockwise
(B) 2900 rpm clockwise
(C) 3600 rpm counterclockwise
(D) 3100 rpm counterclockwise
(E) 16,600 rpm counterclockwise

Solution.

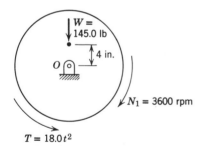

$T = 18.0t^2$
Figure 4-50

By the parallel-axis theorem, the moment of inertia about axis O is

$$I_0 = \bar{I} + Md^2 = Mk^2 + Md^2 = M(k^2 + d^2)$$

where $M = $ flywheel mass $= 145/32.2 = 4.50$ slugs, $k = $ radius of gyration $= 16$ in., and $d = $ distance from O to center of mass $= 4$ in. Thus

$$I_0 = 4.5[(\tfrac{16}{12})^2 + (\tfrac{4}{12})^2] = 8.50 \text{ lb-ft-sec}^2$$

In this problem angular momentum is not constant, but instead the change in angular momentum is equal to the angular impulse caused by the torque T.

$$\text{Angular impulse} = \int T\,dt = \int_0^{10} 18.0t^2\,dt = 6.0t^3\Big|_0^{10} = 6000 \text{ lb-ft-sec}$$

This is now equated to the change in angular momentum, so

$$6000 = I_0(\omega_2 - \omega_1) = 8.50(\omega_2 - \omega_1)$$

$$\omega_2 = \omega_1 + \frac{6000}{8.50} = \omega_1 + 706$$

Hence

$$N_2 = N_1 + \frac{60}{2\pi}(706)$$

$$= -3600 + 6740 = 3140 \text{ rpm counterclockwise}$$

Answer is (D)

PROBLEM 4-37

A simple spring-mass system possesses a certain natural frequency. If the mass is quadrupled in value, the ratio of the new period of oscillation to the original value is closest to which of the following?

(A) 4
(B) $\frac{1}{4}$
(C) 2
(D) $\frac{1}{2}$
(E) 16

Solution.   The period T of a spring may be written as

$$T = 2\pi\left(\frac{m}{k}\right)^{1/2}$$

where m is the mass and k is the spring constant. Hence the ratio

$$\frac{\text{New } T}{\text{Old } T} = \left(\frac{4m}{m}\right)^{1/2} = 2$$

Answer is (C)

5 Mechanics of Materials

This field of mechanics considers the equilibrium behavior of *deformable* solid bodies. It differs from statics in that statics is primarily concerned with the action of external forces on *rigid* bodies, whereas mechanics of materials is primarily concerned with the stress-deformation behavior of nonrigid solid bodies subjected to these external force systems.

In this chapter we consider some fundamental relations that describe the behavior of a solid body subjected to axial tension or compression, or to twisting or torsion, or to transverse bending or flexure. These forces or force combinations create internal stresses in the body and cause related material displacements and deformations. A knowledge of these relations is valuable in the analysis and design of structures and machinery.

AXIAL STRESS AND STRAIN

The axial stress σ is the axial force per unit area acting on a member. If P is the magnitude of the force and A is the cross-sectional area, then the axial stress acting on that area is

$$\sigma = \frac{P}{A} \tag{5-1}$$

When the axial force P acts on a deformable solid bar, the bar changes length. The elongation (or shortening) per unit length is the strain ε, or

$$\varepsilon = \frac{\delta}{L} \tag{5-2}$$

163

where L is the initial length of the bar and δ is the change in that length. For elastic materials, that is, materials that completely return to their initially undeformed state when an applied stress is removed, it is found that stress is proportional to strain

$$\sigma = E\varepsilon \tag{5-3}$$

This is Hooke's law; within limits (below the yield point of the material) many common engineering materials follow this law. E is called the modulus of elasticity.

Application of the foregoing equations to an axially loaded bar segment of differential length dx that elongates a distance $d\delta$ shows

$$\delta = \int_0^\delta d\delta = \int_0^L \frac{P}{EA}\, dx \tag{5-4}$$

when integrated over the entire bar length L. If P, E, and A are constant for the bar, Eq. (5-4) is easily evaluated to give $\delta = PL/EA$.

When a bar is stretched axially in tension, it is found that the cross-sectional area of the bar is decreased by a small amount. A measure of the behavior is Poisson's ratio μ, sometimes called ν, which is formed by dividing the lateral strain by the axial strain. The range of Poisson's ratio is $0 < \mu < 0.5$, although 0.25–0.30 is typical of most metals.

These very basic stress-deformation relations can be used to solve many statically indeterminate problems which could not be solved by statics alone. Here we use the observation that all body members will deform when stressed and require that the deformations be *geometrically consistent*.

Example 1

A steel rod containing a turnbuckle has its ends attached to rigid walls and is tightened by the turnbuckle in summer when the temperature is 90°F to give a stress of 2000 psi. What is the stress in the rod in winter when its temperature is −20°F? (For steel, $E = 30 \times 10^6$ psi and $\alpha = 6.5 \times 10^{-6}$.)

Solution.   The strain ε_t induced in an equivalent unrestrained rod by the temperature change is

$$\varepsilon_t = \alpha\, \Delta T = 6.5(10^{-6})[90 - (-20)] = 7.15 \times 10^{-4}$$

This shortening, however, cannot occur, and instead the stress in the rod increases. The thermally induced additional tensile stress is

$$\sigma_t = \varepsilon_t E = (7.15 \times 10^{-4})(30 \times 10^6) = 21{,}450 \text{ psi}$$

The total stress σ in the rod is

$$\sigma = 2000 + 21,450 = 23,450 \text{ psi tension}$$

Example 2

A short column is made of a $\frac{1}{2}$-in.-thick pipe (8-in. ID) filled with concrete and capped with a rigid plate in contact with both the steel and the concrete. What load P can be carried by the column if the maximum allowable compressive stress is 15,000 psi in the steel and 900 psi in the concrete? Assume $E_s = 30(10^6)$ psi for steel and $E_c = 2(10^6)$ psi for concrete.

Solution.

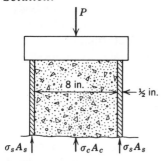

$\sigma_s A_s$ $\sigma_c A_c$ $\sigma_s A_s$ **Figure 5-1**

Figure 5-1 shows a free-body diagram of the column. Summing forces vertically,

$$P = \sigma_c A_c + \sigma_s A_s$$

In addition, each material must compress or shorten by the same amount since the rigid plate is in contact with both materials. Thus our consistent deformation requirement requires equal strains $\varepsilon_c = \varepsilon_s$. In terms of stress

$$\frac{\sigma_s}{E_s} = \frac{\sigma_c}{E_c}$$

or

$$\sigma_s = \sigma_c \left(\frac{E_s}{E_c} \right) = \sigma_c \left[\frac{30(10^6)}{2(10^6)} \right] = 15\sigma_c$$

If the maximum allowable steel stress of 15,000 psi is attained, the resulting concrete stress would be

$$\frac{15,000}{15} = 1000 \text{ psi}$$

which is greater than the allowable stress. Hence the concrete stress governs, and

$$P = \sigma_c A_c + 15\sigma_c A_s$$

$$P = (900)\frac{\pi}{4}(8)^2 + 15(900)\frac{\pi}{4}(9^2 - 8^2)$$

$$P = 45,200 + 180,300 = 225,500 \text{ lb}$$

TORSION

When a circular shaft or other body is twisted, as in cases where shafts transmit power from one point to another by rotational motion, the body is in a state of torsion; shear stresses τ and angular rotations ϕ are the result. A moment or torque T causes these stresses and deformations. When a circular shaft is in a state of pure torsion (when no axial or bending stresses are present), the shear stress at any point in the shaft is

$$\tau = \frac{Tr}{J} \tag{5-5}$$

Thus the stress increases in direct proportion to the distance r from the shaft axis. For a hollow circular shaft of outer diameter d_o and inner diameter d_i, the polar moment of inertia J of the cross section is

$$J = \frac{\pi}{32}(d_o^4 - d_i^4) \tag{5-6}$$

For solid shafts we set $d_i = 0$.

The effect of this torque and resulting shear stress is to cause a relative rotation between different shaft sections. For a shaft section of length L the total angle of twist is

$$\phi = \frac{TL}{GJ} \tag{5-7}$$

where G is the shearing modulus of the material. In these relations the shear equivalent of Hooke's law $\tau = G\gamma$ has been used, where γ is the shear strain.

Example 3

A solid steel shaft 8 ft long is to transmit a torque $T = 20,000$ ft-lb. The shear modulus of the material is $G = 12 \times 10^6$ psi, and the allowable shearing

stress is 10,000 psi. Compute
  (a) the required shaft diameter
  (b) the angle of twist between the two ends of the shaft

Solution. (a) Since the shear stress is largest when $r = R$, the radius of the shaft, we have

$$\tau_{max} = \frac{TR}{J}$$

For a solid shaft,

$$J = \frac{\pi}{32} d_o^4 = \frac{\pi}{2} R^4$$

and

$$R^3 = \frac{2T}{\pi\tau} = \frac{2(20,000)}{\pi(10,000)(144)} = 0.00884 \text{ ft}^3$$

$$R = 0.207 \text{ ft} = 2.48 \text{ in.}$$

The required shaft diameter is $2R = 4.96$ in. or, practically, 5 in.
  (b) The twist is

$$\phi = \frac{TL}{GJ}$$

$$\phi = \frac{TL}{G\left(\frac{\pi}{2}R^4\right)}$$

$$\phi = \frac{(20,000)(8)}{12(10^6)(144)\frac{\pi}{2}(0.207)^4}$$

$$\phi = 0.0321 \text{ rad} = 0.0321\left(\frac{180}{\pi}\right) = 1.84°$$

BEAM EQUILIBRIUM

The principles of statics are adequate to analyze the external equilibrium of any statically determinate beam for any combination of concentrated or distributed loads. Here, however, we wish to determine the shear force V and bending moment M which exist at a point *in* the beam so that the beam is internally in equilibrium. With the aid of a systematic sign convention, the same statics principles may still be directly used. This information is commonly displayed in diagrams which give V and M for every cross section of the beam. This information can then be used to determine stresses and deformations.

The shear force V, bending moment M, and distributed beam loading w (directed downward) are related by the expressions

$$\frac{dV}{dx} = -w \qquad \frac{dM}{dx} = V \tag{5-8}$$

where w, V, and M are all functions of the distance along the beam x. Here we have adopted the sign convention shown in Fig. 5-2 for positive shear and moment.

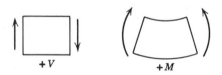

Figure 5-2   Sign convention for positive shear and moment.

$+V$ $+M$

The shear diagram can be constructed a bit more easily when one proceeds from left to right along the beam, for then the loads on the beam act in the same direction as the change in the shear ordinate. From Eqs. (5-8) the slope in the shear diagram is equal to minus the loading intensity at that point. Wherever concentrated loads are applied, the shear ordinate abruptly changes by that amount. Also from Eqs. (5-8) a positive shear gives a positive slope to the moment diagram, and a zero shear occurs where the moment M is a maximum or minimum, since the ordinate of the shear diagram equals the slope of the moment diagram. Finally, if Eqs. (5-8) are integrated, we find that (a) minus the area under the load curve between two points equals the change in shear between these points, and (b) the area under the shear curve between two points equals the change in the bending moment between these same two points.

Example 4

The loading diagram for a beam is shown in Fig. 5-3. The beam is 6 in. × 12 in. and is placed on edge with respect to the loads. Determine the reactions R_1 and R_2 and construct the shear and bending moment diagrams.

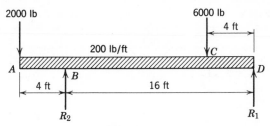

Figure 5-3

Solution. To find R_2 we sum moments around point D.

$$\sum M_D = 0 = 2000(20) + 200(20)(10) + 6000(4) - 16R_2$$

$$R_2 = \tfrac{1}{16}(40,000 + 40,000 + 24,000)$$

$$R_2 = 6500 \text{ lb}$$

Summing forces vertically,

$$\sum F_v = 0$$

$$R_1 = 2000 + 6000 + 200(20) - 6500$$

$$R_1 = 5500 \text{ lb}$$

Diagrams for the shear force V and bending moment M are given in Fig. 5-4.

The shear curve drops continuously down to the right at a slope of 200 lb/ft, the magnitude of the distributed loading. At points A, B, and C the shear ordinate jumps by the amount, and in the direction, of the concentrated loads at those points. At D the 5500-lb reaction returns the shear ordinate to zero.

The moment diagram M can be constructed directly from the shear diagram since no concentrated couples act on the beam. The change in the moment ordinate equals the area under the shear curve. From A to B we find

$$M_B = -\tfrac{1}{2}(4)(2000 + 2800) = -9600 \text{ ft-lb}$$

since $M_A = 0$. From B to C the shear is positive, giving

$$M_C = M_B + \tfrac{1}{2}(12)(3700 + 1300)$$

$$= -9600 + 30,000 = 20,400 \text{ ft-lb}$$

From C to D the shear is negative, which causes the moment to decrease to zero at D. The slope of the moment diagram is steepest where the ordinate of

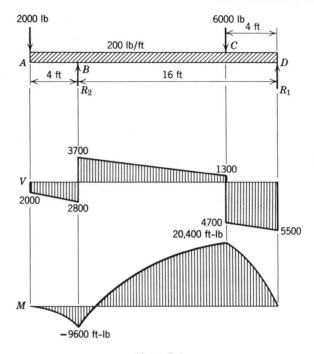

Figure 5-4

the shear diagram is largest. Finally, we note that the maximum positive moment is at C where the shear curve passes through zero at the support. [As an alternative to this procedure, we could mathematically integrate Eqs. (5-8) and plot them.]

BEAM STRESSES AND DEFLECTIONS

The existence of a shearing force or bending moment at an internal beam cross section creates axial and shearing stresses at the section.

The bending stress σ at a point is given by the flexure formula

$$\sigma = \frac{My}{I} \tag{5-9}$$

where M is the bending moment acting on a beam cross section, I is the moment of inertia around the neutral axis of the section, and y is the distance from the neutral axis to the point. The largest bending stress thus occurs in the beam fibers most distant from the neutral axis.

A shear force V acting on a section causes an internal longitudinal (horizontal) shear stress distribution to be set up across the section. The magnitude of this shear stress τ at some distance y from the neutral axis is

$$\tau = \frac{VQ}{Ib} \tag{5-10}$$

Here I is again the section moment of inertia and b is the section width at height y. The quantity Q is the first moment of the area of a portion of the beam cross section. The area is that portion of the section which lies outside the distance y from the neutral axis. If c denotes the extreme fiber in the section, then Q is

$$Q = \int_y^c y\, dA \tag{5-11}$$

This equation shows immediately that the extreme fibers in a beam carry zero shear stress since $Q = 0$ when $y = c$.

Example 5

For the beam in example 4 determine

    (a) the location and magnitude of the maximum bending stress
    (b) the location and magnitude of the maximum shear stress

Solution. (a) The maximum bending moment occurs at point C and is $M_C = 20{,}400$ ft-lb. The moment of inertia of the 6 in.$\times$12 in. section is $I = bh^3/12 = (6)(12)^3/12 = 864$ in.^{4}. The maximum bending stress occurs farthest from the neutral axis where $y = h/2 = 6$ in. Equation (5-9) now gives the maximum stress as

$$\sigma = \frac{My}{I} = \frac{(20{,}400)(12)(6)}{864} = 1700 \text{ psi}$$

(b) Since cross-sectional properties are constant throughout the length of the beam, the maximum shear stress will occur at some point in the section where the shear force V is a maximum. From the shear diagram in example 4 the shear is a maximum at the right support where $V = 5500$ lb. (Notice here that the maximum shear ordinate does not occur at the point of application of the largest concentrated force.) The maximum stress occurs at the point in this section where Q is maximized. First we evaluate Eq. (5-11) for any point y on the rectangular beam section of height h and width b:

$$Q = \int_y^{h/2} y(b\, dy) = \frac{b}{2}\left(\frac{h^2}{4} - y^2\right)$$

Thus Q, and consequently τ, is a maximum at the neutral axis $y = 0$.

$$\tau = \frac{VQ}{Ib} = \frac{V(bh^2/8)}{(bh^3/12)b} = \frac{3V}{2bh} = \frac{3V}{2A}$$

is the maximum shear stress in a rectangular beam. Hence

$$\tau = \frac{3(5500)}{2(6)(12)} = 114.6 \text{ psi}$$

For small deflections of beams the basic relation between the beam deflection y, curvature $1/\rho$, and moment M at a point is

$$\frac{d^2y}{dx^2} = \frac{1}{\rho} = \frac{M}{EI} \tag{5-12}$$

Here we assume that y is positive upward when M is positive according to the earlier sign convention for moment diagrams. If M is written as a function of x, Eq. (5-12) may be integrated twice. The constants of integration are evaluated by noting the slope and deflection constraints for the particular beam.

The moment-area method is a rapid, efficient alternative method of interpreting and using the basic deflection equation. Integrated once between points A and B on the beam, Eq. (5-12) gives

$$\theta_B - \theta_A = \int_A^B \frac{M}{EI} dx \tag{5-13}$$

which shows that the change in slope between A and B equals the area of the M/EI diagram between A and B. It can also be shown that the deflection δ of point A from the tangent to point B is

$$\delta = \int_A^B \frac{M}{EI} x \, dx \tag{5-14}$$

The integral is the first moment of the M/EI diagram about point A. In applying the moment-area principles, one should treat positive and negative sections of the moment diagram separately.

Since Eq. (5-12) is a linear equation, the principle of linear superposition is valid for finding slopes and deflections for a beam which is simultaneously acted upon by several loads. This principle is therefore a useful tool in the solution of statically indeterminate beam problems.

Example 6

A propped cantilever beam 12 ft long carries a uniformly distributed load of 2000 lb/ft (2 kips/ft), as shown in Fig. 5-5. Calculate the reactions.

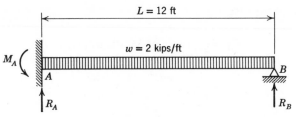

Figure 5-5

Solution. The beam is statically indeterminate to the first degree. We consider the problem as the sum of two statically determinate problems and require the net deflection of the propped end (point B) to be zero. Shown beneath each statically determinate cantilever in Fig. 5-6 is its associated

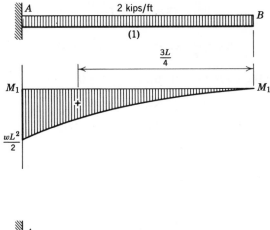

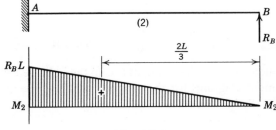

Figure 5-6

bending moment diagram; also shown is the distance to the centroid of the moment diagram from the free end of the beam. For this problem we note that EI is constant throughout the beam. According to Eq. (5-14), the deflection at B for case 1 is

$$\delta_{B_1} = \frac{1}{EI}(\text{area of } M_1 \text{ diagram}) \, \bar{x}_1$$

$$\delta_{B_1} = \frac{1}{EI}\left(\frac{-wL^3}{6}\right)\frac{3L}{4} = \frac{-wL^4}{8EI}$$

For case 2,

$$\delta_{B_2} = \frac{1}{EI}\left(\frac{+R_B L^2}{2}\right)\frac{2L}{3} = \frac{+R_B L^3}{3EI}$$

Since point B cannot deflect,

$$\delta_B = \delta_{B_1} + \delta_{B_2} = 0$$

$$\frac{R_B L^3}{3EI} - \frac{wL^4}{8EI} = 0$$

$$R_B = \frac{+3}{8}wL = \frac{+3}{8}(2)(12) = +9 \text{ kips}$$

Now, by using the equations of statics, $\sum F_v = 0$

$$R_A + R_B = wL$$

$$R_A = \tfrac{5}{8}wL = \tfrac{5}{8}(2)(12) \ = +15 \text{ kips}$$

At point A, $\quad \sum M_A = 0$

$$M_A + R_B L - \frac{wL^2}{2} = 0$$

$$M_A = \frac{wL^2}{2} - \frac{3}{8}wL^2$$

$$M_A = \frac{wL^2}{8} = \frac{2(12)^2}{8} = +36 \text{ kip-ft}$$

The plus sign indicates that the moment acts in the indicated direction. The shear and moment diagrams for the propped cantilever beam can now be constructed directly, if desired.

EULER COLUMN BUCKLING

Columns are slender members which carry primarily axial loads. The critical load P_{cr} which will cause lateral buckling, an instability phenomenon, in long slender columns is given by the Euler column formula, which may be expressed as

$$P_{cr} = \frac{\pi^2 EI}{L_1^2}$$ (5-15)

where L_1 is related to the length of the column. The moment of inertia may be written $I = Ar^2$, r being the radius of gyration. Euler's formula is applicable only when the modified slenderness ratio L_1/r is greater than 100, approximately. Shorter columns do not fail by buckling which is described by the Euler equation. The relation between L_1 and the column length L depends on the end conditions of the column. For both ends fixed $L_1/L = \frac{1}{2}$; for one end fixed and one end pinned $L_1/L = 0.7$; for both ends pinned $L_1/L = 1$; and for one end fixed and one end free $L_1/L = 2$.

Example 7

A $\frac{3}{4}$-in.-diameter solid steel rod ($E = 30 \times 10^6$ psi) is 5 ft long and is pin-connected at its ends. How large a compressive load can be applied before buckling occurs?

Solution.   The moment of inertia for a circular rod of diameter d is

$$I = \frac{\pi}{64} d^4 = Ar^2$$

The radius of gyration is

$$r = \left(\frac{\pi}{64} d^4 \Big/ \frac{\pi}{4} d^2 \right)^{1/2} = \frac{d}{4} = \frac{3}{16} \text{ in.}$$

For pinned ends $L_1 = L$ so

$$\frac{L_1}{r} = \frac{5(12)}{\frac{3}{16}} = 320 > 100$$

and the Euler formula applies. The critical load is

$$P_{cr} = \frac{\pi^2 EI}{L_1^2}$$

$$P_{cr} = \frac{\pi^2 E}{L^2} \left(\frac{\pi}{64} d^4 \right)$$

$$P_{cr} = \frac{\pi^3 (30)(10^6)(0.75)^4}{64[5(12)]^2}$$

$$P_{cr} = 1277 \text{ lb}$$

PROBLEM 5-1

The only strength property of stress grade lumber that can be precisely determined without breaking the piece is

(A) rigidity
(B) factor of safety
(C) modulus of rupture
(D) stiffness
(E) horizontal shear

Solution. Stiffness is a measure of the wood's ability to resist deformation or bending. It is expressed in terms of the *modulus of elasticity* and applies only within the proportional limit. It can therefore be determined without breaking the piece.

Answer is (D)

PROBLEM 5-2

The chief difference between exterior grade plywood and interior grade plywood is

(A) in the number of laminations used
(B) in the weight per square foot
(C) in the size of knotholes permitted
(D) that only selected heartwoods are used for the exterior grade
(E) that different types of glues are used

Solution. The distinguishing feature between exterior and interior grade plywood is the difference in the durability of the adhesives employed.

Answer is (E)

PROBLEM 5-3

Warp, wane, and checks are all factors that help to determine the quality of

(A) brick masonry
(B) ceramic veneer
(C) structural glass
(D) concrete
(E) stress grade lumber

Solution. Warp (any variation from a true or plane surface), wane (bark or lack of wood on a corner or edge of a piece), and checks (a lengthwise separation of the wood) are irregularities or defects in wood.

Answer is (E)

PROBLEM 5-4

The terms igneous, sedimentary, and metamorphic are commonly applied in

(A) geodesy
(B) matrix analysis
(C) physics
(D) chemistry
(E) geology

Solution. The terms igneous (congealed from a molten mass), sedimentary (originated by sedimentation or settling of particles), and metamorphic (preexisting rock masses in which new minerals or structures are formed at high temperatures and pressures) are geological terms.

Answer is (E)

PROBLEM 5-5

When earth is deposited as an embankment, it will assume a natural slope. The angle formed between this natural slope line and a horizontal plane is called the

(A) angle of incidence
(B) angle of refraction
(C) dip
(D) spectral angle
(E) angle of repose

Solution. By definition this is the angle of repose. It is somewhat similar to the angle of internal friction ϕ, but the value may vary greatly from ϕ.

Answer is (E)

PROBLEM 5-6

The Brinell number of a material is a measure of

(A) specific gravity
(B) weight
(C) specific heat
(D) density
(E) hardness

Solution.   The Brinell test is a static indentation hardness test. The Brinell hardness number is found by dividing the applied load by the contact area indented by the hardened steel ball that applies the load.

Answer is (E)

PROBLEM 5-7

A major classification group of stainless steels contains 6% to 22% nickel in addition to 16% to 26% chromium. This group differs from the other two major groupings in that its steels

(A) have a coefficient of linear expansion almost zero
(B) have a coefficient of linear expansion small enough to make it useful in the manufacture of precision measuring devices
(C) are nonmagnetic
(D) have their iron in the ferritic form
(E) have their iron in the martensitic form

Solution.   The classification includes 18-8 stainless (18% Cr, 8% Ni). These stainless steels have sufficient chromium to make them austenitic and nonmagnetic.

Answer is (C)

PROBLEM 5-8

The ultimate strength divided by the allowable stress is the

(A) yield point
(B) percentage of elongation

(C) percentage of reduction in area

(D) working stress

(E) factor of safety

Solution.   The ultimate strength divided by the allowable stress is the factor of safety.

Answer is (E)

PROBLEM 5-9

If an engineering structure is to be designed against rupture of any of its members under steady load, the factor of safety is a ratio based on working stress and

(A) elastic strength

(B) ultimate strength

(C) toughness

(D) resilience

(E) endurance limit

Solution.   Where a structure is designed against rupture with a steady load, the governing property is ultimate strength. Therefore the factor of safety is the ratio of ultimate strength to working stress.

Answer is (B)

PROBLEM 5-10

A body having the same elastic properties in all directions is

(A) isotropic

(B) homogeneous

(C) isothermal

(D) orthotropic

(E) inhomogeneous

Solution.   A body is said to be isotropic if its elastic properties are the same in all directions.

Answer is (A)

PROBLEM 5-11

The coefficient of thermal expansion of steel is approximately what percent of the coefficient of thermal expansion of concrete?

    (A) 50%
    (B) 75%
    (C) 100%
    (D) 125%
    (E) 150%

Solution.

Coefficient of thermal expansion of steel = 0.0000065 in./in./°F
Coefficient of thermal expansion of concrete = 0.000006 in./in./°F

$$\frac{6.5 \times 10^{-6}}{6.0 \times 10^{-6}} = 1.08 = 108\%$$

Answer is (C)

PROBLEM 5-12

Split rings are commonly used in the construction of

    (A) bearing piles
    (B) concrete pipes
    (C) steel joists
    (D) timber trusses
    (E) earth fill dams

Solution. Split rings are timber connectors used to transmit forces between wood members or between wood and metal members.

Answer is (D)

PROBLEM 5-13

The unit lateral deformation of a body under stress divided by the unit longitudinal deformation is known as

(A) Poisson's ratio

(B) Hooke's law

(C) Polar moment of inertia

(D) Euler's ratio

(E) Mohr's modulus

Solution.

$$\text{Poisson's ratio } \mu = \frac{\text{unit lateral deformation (lateral strain)}}{\text{unit longitudinal deformation (axial strain)}}$$

Answer is (A)

PROBLEM 5-14

A thin wire is subject to a tensile stress. If the temperature is constant, the electric resistance of the stressed wire with respect to the electric resistance of the unstressed wire will

(A) increase

(B) decrease

(C) remain the same

(D) become negative as the elastic limit of the material is exceeded

(E) any of the above, depending on the material

Solution. The resistance of wire varies inversely with the area. Since the stressed wire has a reduced cross-sectional area and increased length, its resistance is increased compared to the unstressed wire. This property is utilized in strain gages.

Answer is (A)

PROBLEM 5-15

The offset method is used to find a property of metals that do not have a well-defined stress-strain curve, such as steel. This property is called the

(A) yield point

(B) modulus of elasticity

(C) section modulus

(D) proportional limit

(E) moment of inertia

Solution. The offset method is used to determine the yield point.

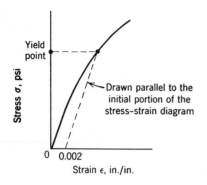

Figure 5-7

Answer is (A)

PROBLEM 5-16

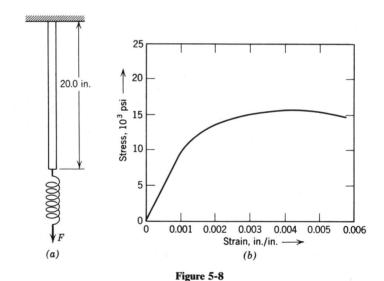

Figure 5-8

The rod in Fig. 5-8*a* of a material having the stress-strain curve shown in Fig. 5-8*b* has a spring attached at one end. The spring constant of this spring is 20,000 lb/in. The rod has a cross-sectional area of 1.00 in.^{2} and is 20.00 in. long. The load F is increased until the spring has elongated 0.75 in. and then decreased to zero. The length of the rod after the load is removed is closest to

(A) 20.000 in.
(B) 20.003 in.
(C) 20.030 in.
(D) 20.060 in.
(E) rod will break

Solution. The spring tells us the magnitude of the load applied. The load F is increased to $0.75 \times 20,000 = 15,000$ lb, then removed. For a rod of 1.00 in.^{2} the stress applied is

$$\sigma = \frac{P}{A} = \frac{15,000}{1.00} = 15,000 \text{ psi}$$

Up to $10,000$ psi on the stress-strain diagram there is a linear relationship between stress and strain (Hooke's law applies). For stresses above the elastic limit ($10,000$ psi) there is a permanent deformation. As the load is increased, the stress-strain relation is as represented by the curve. Beyond the elastic limit the stress-strain curve is not retraced as the load is removed

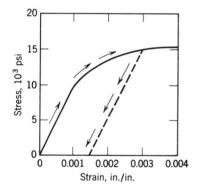

Figure 5-9

but instead decreases with a slope equal to that below the elastic limit. The result is a permanent elongation of the rod.

$$\Delta L = \varepsilon L = 0.0015 \times 20.00 = 0.03 \text{ in.}$$

Thus the new rod length will be

$$20.00 + 0.03 = 20.03 \text{ in.}$$

Answer is (C)

PROBLEM 5-17

The rails of a tramway are welded together at +50°F. Assume the coefficient of linear expansion of rails is 70×10^{-7} in./in.-°F and the modulus of elasticity is 30×10^6 lb/in.^{2}. The stress produced in the rails when heated by the sun to +100°F is nearest to

    (A) 0 psi
    (B) 10,500 psi
    (C) 32,000 psi
    (D) 35,000 psi
    (E) 52,000 psi

Solution.

$$\text{Stress } \sigma = \text{unit strain } (\varepsilon) \times \text{modulus of elasticity } (E)$$

If the rails were free to expand, then their increase per unit length caused by temperature would be

$$\Delta L = \text{coefficient of expansion } (\alpha) \times \text{length } (L) \times \text{change of temperature } (\Delta T)$$

$$= 70 \times 10^{-7} \times 1 \times (100 - 50) = 350 \times 10^{-6} \text{ in./in.}$$

Since ΔL is found per unit length of the rail, it is the unit strain, or $\Delta L = \varepsilon$. Thus the stress developed in the rail is given by

$$\sigma = \varepsilon E = 350 \times 10^{-6} \times 30 \times 10^6 = 10,500 \text{ psi}$$

Answer is (B)

PROBLEM 5-18

A $\frac{1}{2}$-in. O.D. brass rod is 100 ft 5 in. long and a 4-in. nominal cast-iron pipe (4.80 in. O.D., 0.38 in. wall thickness) is 100 ft $5\frac{1}{2}$ in. long when both are at the same temperature (60°F).

Given: α brass   = 10×10^{-6} units per unit length per degree Fahrenheit
        α cast iron = 6×10^{-6} units per unit length per degree Fahrenheit

The rod and the pipe are the same length at a temperature closest to

    (A) -40°F
    (B) 0°F

(C) 80°F

(D) 120°F

(E) 160°F

Solution. Since brass has a greater coefficient of expansion than cast iron and the brass piece is presently shorter than the cast-iron one, they will be the same length at some elevated temperature.

$$\Delta T(\alpha_{brass})(length_{brass}) = 0.5 \text{ in.} + \Delta T(\alpha_{c.i.})(length_{c.i.})$$

$$\Delta T = \frac{0.5 \text{ in.}}{(\alpha_{brass})(length_{brass}) - (\alpha_{c.i.})(length_{c.i.})}$$

$$\Delta T = \frac{0.5 \text{ in.}}{(10 \times 10^{-6})(1205 \text{ in.}) - (6 \times 10^{-6})(1205.5 \text{ in.})}$$

$$\Delta T = \frac{0.5}{0.01205 - 0.00723} = \frac{0.5}{0.00482} = 103.7°F$$

Therefore the temperature at which both pieces are the same length is $60 + 103.7 = 163.7°F$.

Answer is (E)

PROBLEM 5-19

A $\frac{1}{2}$-in.-diameter steel tie rod 18 ft in length is joined to two rigid walls in such a way that an axial tensile stress of 20,000 psi is induced in the rod. Assume a Poisson's ratio of 0.25 and $E = 30 \times 10^6$ psi.

The change in the diameter of the rod caused by the application of this tensile load is closest to

(A) 6.66×10^{-4}

(B) 1.66×10^{-4}

(C) 8.33×10^{-5}

(D) 6.50×10^{-6}

(E) No change in the diameter

Solution.

$$\text{Unit strain } \varepsilon = \frac{\sigma}{E} = \frac{20,000}{30 \times 10^6} = \frac{2}{3} \times 10^{-3}$$

$$\text{Unit change in diameter} = \mu\varepsilon = \tfrac{1}{4}\varepsilon = \tfrac{1}{6}\times 10^{-3}$$

$$\text{Total change in diameter} = d\mu\varepsilon = \tfrac{1}{2}\times\tfrac{1}{6}\times 10^{-3}$$

$$= 8.33\times 10^{-5}\,\text{in.}$$

Answer is (C)

PROBLEM 5-20

The stress in an elastic material is

(A) inversely proportional to the material's yield strength
(B) inversely proportional to the force acting
(C) proportional to the displacement of the material acted on by the force
(D) inversely proportional to the strain
(E) proportional to the length of the material subject to the force

Solution. According to Hooke's law, stress is directly proportional to strain. Strain is the deformation (or displacement) of the material per unit length. Thus we can say: Stress is proportional to the displacement of the material acted on by the force.

Answer is (C)

PROBLEM 5-21

What load must be applied to a 1-in. round steel bar 8 ft long ($E = 30,000,000$ psi) to stretch the bar 0.05 in.?

(A) 7,200 lb
(B) 9,850 lb
(C) 8,600 lb
(D) 12,250 lb
(E) 15,000 lb

Solution.

$$\delta = \frac{PL}{AE} \qquad P = \frac{AE\delta}{L} = \frac{\pi}{4}(1)^2\frac{30,000,000(0.05)}{96} = 12,250\,\text{lb}$$

Answer is (D)

PROBLEM 5-22

A steel bar having a 1-in.^{2} cross section is 150 in. long when lying on a horizontal surface. Assume $E = 30 \times 10^6$ psi and $W = 0.283$ lb/in.^{3}. The increase in the length of the bar when it is suspended vertically from one end is nearest to

(A) 1×10^{-5} in.
(B) 1×10^{-4} in.
(C) 1×10^{-3} in.
(D) 1×10^{-2} in.
(E) 1×10^{-1} in.

Solution.

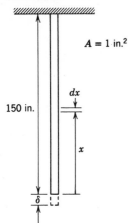

$A = 1$ in.^{2}

dx

150 in.

x

δ

Figure 5-10

$$d\delta = \frac{P(x)\,dx}{AE} \qquad \text{where} \quad P(x) = WAx$$

$$\int_0^\delta d\delta = \int_0^L \frac{WAx\,dx}{AE}$$

$$\int_0^\delta d\delta = \frac{W}{E} \int_0^L x\,dx$$

$$\delta = \frac{W}{E}\frac{L^2}{2}$$

$$\delta = \frac{0.283\ \text{lb/in.}^3}{30 \times 10^6\ \text{lb/in.}^2} \times \frac{150^2\ \text{in.}^2}{2}$$

$$\delta = 1.06 \times 10^{-4}\ \text{in.}$$

Answer is (B)

PROBLEM 5-23

A steel test specimen is $\frac{5}{8}$ in. in diameter at the root of the thread. It is to be stressed to 50,000 psi tension. What load must be applied?

    (A) 12,790 lb
    (B) 15,340 lb
    (C) 16,320 lb
    (D) 25,600 lb
    (E) 31,250 lb

Solution.   Stress $= P/A$. Here $A = (\pi/4)(\frac{5}{8})^2$ and stress $= 50,000$ psi.

$$P = 50,000 \, \frac{\pi}{4}\left(\frac{5}{8}\right)^2 = 15,340 \text{ lb}$$

Answer is (B)

PROBLEM 5-24

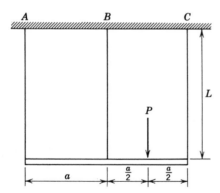

Figure 5-11

Given a rigid bar hanging from three wires of length L, modulus of elasticity E, cross-sectional area A, spaced a distance a apart as shown in the figure. Neglect the weight of the bar. The force exerted by wire A is closest to

    (A) $P/3$
    (B) $P/4$
    (C) $P/5$
    (D) $P/8$
    (E) $P/12$

Solution.   When load P is applied, we know that $P = P_A + P_B + P_C$, and since A, B, and C are wires, none are in compression. Further, we know that each wire will elongate; hence

$$\text{length } A = L + \Delta_A$$

$$\text{length } B = L + \Delta_B$$

$$\text{length } C = L + \Delta_C$$

As the bar is rigid, B will have a deflection intermediate to that of A and C, or

$$\Delta_B = \frac{\Delta_A + \Delta_C}{2}$$

But since all wires are identical, load is proportional to deflection, so

$$P_B = \frac{P_A + P_C}{2}$$

Taking moments about wire A, $\sum M_A = 0$

$$aP_B - 1.5aP + 2aP_C = 0$$

Dividing by a,

$$P_B - 1.5P + 2P_C = 0$$

Thus we have three equations in three unknowns:

$$P = P_A + P_B + P_C \tag{1}$$

$$0 = P_A - 2P_B + P_C \tag{2}$$

$$1.5P = +P_B + 2P_C \tag{3}$$

Solving simultaneously,

(1) $-P = -P_A - P_B - P_C$

(2) $\dfrac{0 = P_A - 2P_B + P_C}{-P = \qquad -3P_B}$ $P_B = \dfrac{P}{3}$

(3) $1.5P = \quad +\dfrac{P}{3} + 2P_C$ $P_C = \dfrac{1.5P - P/3}{2} = \dfrac{7P}{12}$

(1) $P = P_A + \dfrac{P}{3} + \dfrac{7P}{12}$ $P_A = P - \dfrac{4P + 7P}{12} = \dfrac{P}{12}$

Answer is (E)

PROBLEM 5-25

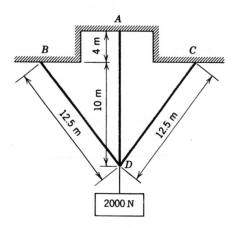

Figure 5-12

Three rods, each 2 cm in diameter, labeled *AD*, *BD*, and *CD*, support the 2000-newton load as shown. The load in rod *A* is nearest to

(A) 580 newtons
(B) 670 newtons
(C) 730 newtons
(D) 820 newtons
(E) 2000 newtons

Solution.

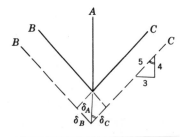

Figure 5-13

By geometry, $\delta_B = \delta_C = \frac{4}{5}\delta_A$. By symmetry, $P_B = P_C$. Using $\delta = PL/EA$,

$$\frac{4}{5}\frac{P_A(14)}{AE} = \frac{P_B(12.5)}{AE} \qquad P_A = 1.12P_B = 1.12P_C$$

$$\sum F_v = 0 \qquad \tfrac{4}{5}P_B + P_A + \tfrac{4}{5}P_C = 2000$$

$$\tfrac{4}{5}P_B + 1.12P_B + \tfrac{4}{5}P_B = 2000$$

$$P_B = P_C = 735 \text{ newtons (tension)}$$

$$P_A = 823 \text{ newtons (tension)}$$

Answer is (D)

PROBLEM 5-26

Normally the diameter of a rivet is $\tfrac{1}{16}$ in. less than that of the hole in which it is to fit. This is to insure easy entrance of a hot rivet. In boiler calculations the rivet is assumed to

(A) remain the same size after driving
(B) shrink an additional $\tfrac{1}{16}$ in. on cooling
(C) shrink on cooling in proportion to its cross-sectional area
(D) attain a size found by tables
(E) fill the hole after driving

Solution. The ASME Boiler and Pressure Vessel Code, among others, requires that all rivets be driven to fill rivet holes completely.

Answer is (E)

PROBLEM 5-27

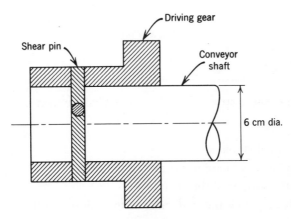

Figure 5-14

A shear pin 1 cm in diameter, as shown in the figure, is used on a screw conveyor to protect the mechanism when it jams. If the screw conveyor revolves with a torque of 519.51 newton-meters, the unit shearing stress on the shear pin is closest to

(A) 2×10^3 pascals
(B) 2×10^4 pascals
(C) 5×10^5 pascals
(D) 1×10^7 pascals
(E) 1×10^8 pascals

Solution. The *newton* is the force that, when applied to a body having a mass of 1 kilogram, gives it an acceleration of 1 meter/sec^2.

The *pascal* is a pressure or stress of 1 newton per square meter.

Torque (newton-meters) = force (newtons) × distance (meters)

$$T = \tau \ (\text{pascals}) \times \text{area (meters}^2)$$

$$\times \text{lever arm (meters)}$$

$$519.51 = \tau \times 2\left(\frac{\pi}{4} \times 0.01^2\right) \times 0.03$$

$$= \tau \times 4.71 \times 10^{-6}$$

$$\tau = \frac{519.51}{4.71 \times 10^{-6}} = 110.3 \times 10^6 \text{ pascals}$$

[Since 1 psi = pascals/6895, $\tau = (110.3 \times 10^6)/6895 = 16,000$ psi]

Answer is (E)

PROBLEM 5-28

The maximum shear stress in a solid round shaft subjected only to torsion occurs

(A) on principal planes
(B) on planes containing the axis of the shaft
(C) on the surface of the shaft
(D) only on planes perpendicular to the axis of the shaft
(E) at the neutral axis

Solution.

$$\tau = \frac{Tc}{J}$$

where

T = torque

c = distance from the center of the shaft

J = polar moment of inertia

Thus the torsional shearing stress at any point is proportional to its distance from the center of the shaft. The maximum shearing stress, therefore, is at the surface of the shaft.

Answer is (C)

PROBLEM 5-29

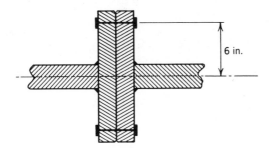

6 in.

Figure 5-15

A shaft coupling is to be designed, using 1-in.-diameter bolts at a distance of 6 in. from the center of the shaft. Allowable shearing stress on the bolts is 15,000 psi. If the shaft is to transmit 5800 hp at a speed of 1200 rpm, how many bolts are needed in the connection?

(A) 2 bolts
(B) 3 bolts
(C) 4 bolts
(D) 5 bolts
(E) 6 or more bolts

Solution.

$$\text{Work/revolution} = \text{force} \times \text{distance} = F(2\pi R) = 2\pi T$$

where T = torque in pound-feet. At 1200 rpm

$$\text{Work/minute} = 1200 \times 2\pi T \text{ lb-ft/min}$$

Since 1 hp = 33,000 ft-lb/min,

$$\text{hp} = \frac{1200 \times 2\pi T}{33,000} = 5800$$

$$\text{Torque} = \frac{33,000 \times 5800}{1200 \times 2\pi} = 25,400 \text{ lb-ft}$$

Assuming that the shearing stress is uniform over the bolt cross section, we can determine the torsional resistance per bolt.

$$\text{Torque} = \text{force} \times \text{lever arm} = \tau A \times \text{lever arm}$$

$$= 15,000 \text{ psi} \times \frac{\pi}{4} \ 1^2 \text{ in.}^2 \times 0.5 \text{ ft}$$

$$= 5890 \text{ lb-ft}$$

$$\text{No. of bolts required} = \frac{25,400}{5890} = 4.31$$

Five bolts are required.

Answer is (D)

PROBLEM 5-30

The intermittent welds generally used to hold material in place temporarily are called

(A) double vee
(B) tack
(C) fillet
(D) groove
(E) butt

Solution. A tack weld is a temporary or auxiliary weld. The other alternatives listed are types of structural welds.

Answer is (B)

PROBLEM 5-31

In an I-beam subjected to simple bending, the maximum bending stress occurs

(A) at the neutral axis
(B) in the web above the neutral axis
(C) in the web below the neutral axis
(D) at the top and bottom surfaces of the beam
(E) where the web joins the lower flange

Solution. In simple bending the stress at each point in the cross section of a beam is directly proportional to that point's distance from the neutral axis. For a symmetrical section, like an I-beam, the maximum bending stress occurs at both the top and bottom surfaces of the beam.

Answer is (D)

PROBLEM 5-32

The bending moment of a beam

(A) depends on the modulus of elasticity of the beam
(B) is minimum where the shear is zero
(C) is maximum at the free end of a cantilever
(D) is plotted as a straight line for a simple beam with a uniformly distributed load
(E) may be determined from the area of the shear diagram

Solution. The shear at any point along the beam is

$$V = \frac{dM}{dx}$$

so

$$M = \int V \, dx$$

or the bending moment is the area of the shear diagram.

Answer is (E)

PROBLEM 5-33

The moment curve for a simple beam with a concentrated load at midspan takes the shape of a

    (A) triangle
    (B) semicircle
    (C) semiellipse
    (D) parabola
    (E) rectangle

Solution.

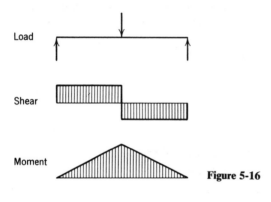

Figure 5-16

Answer is (A)

PROBLEM 5-34

The moment diagram for a beam uniformly loaded with a concentrated load in the center is the sum of

    (A) two triangles
    (B) a rectangle and a triangle
    (C) a parabola and a rectangle
    (D) a parabola and a triangle
    (E) a rectangle and a trapezoid

Solution.

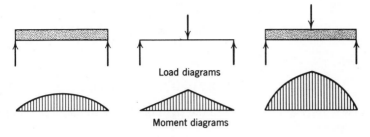

Load diagrams

Moment diagrams

Figure 5-17

The moment diagram for a beam with a uniform load is a parabola, and the moment diagram for a concentrated load is a triangle. By superposition, the combined diagram is the sum of the two individual diagrams.

Answer is (D)

PROBLEM 5-35

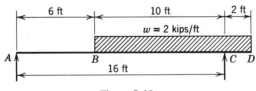

Figure 5-18

The maximum moment in the beam AD is nearest to

(A) 4 kip-ft
(B) 18 kip-ft
(C) 36 kip-ft
(D) 45 kip-ft
(E) 66 kip-ft

Solution.

$$\sum M_A = 0$$

$$+16R_C - (12 \times 2)(12) = 0 \qquad R_C = \frac{24 \times 12}{16} = 18 \text{ kips}$$

$$\sum F_y = 0$$

$$R_A + 18 - (12 \times 2) = 0 \qquad R_A = 6 \text{ kips}$$

The shear is zero at a point 3 ft to the right of *B*. The moment at that point is

$$M_0 = 6 \text{ kips} \times 6 \text{ ft} + (6 \text{ kips} \times 3 \text{ ft})/2 = 45 \text{ kip-ft}$$

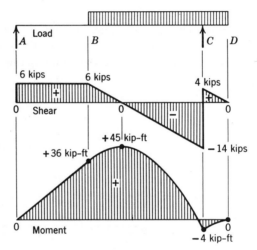

Figure 5-19

Answer is (D)

PROBLEM 5-36

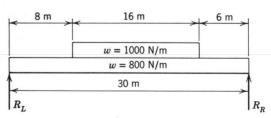

Figure 5-20

A simple beam is loaded as shown. The value of the maximum moment is closest to

    (A) 18,000 newton-meters

    (B) 36,000 newton-meters

    (C) 72,000 newton-meters

    (D) 144,000 newton-meters

    (E) 177,000 newton-meters

Solution. Solve first for the left reaction R_L and the right reaction R_R.

$$\sum M_{R_L} = 0$$

$$+30R_R - 800 \times 30 \times 15 - 1000 \times 16 \times 16 = 0$$

$$+30R_R - 360,000 - 256,000 = 0 \qquad R_R = \frac{616,000}{30} = 20,533 \text{ N}$$

$$\sum F_y = 0$$

$$R_L + R_R - 800 \times 30 - 1000 \times 16 = 0$$

$$R_L + 20,533 - 24,000 - 16,000 = 0 \qquad R_L = 19,467 \text{ N}$$

Then draw the shear diagram:

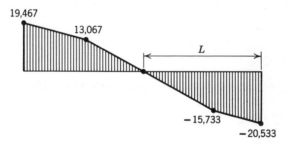

Figure 5-21 Shear diagram

$$L = 6 + \frac{15,733}{1800} = 14.74 \text{ m}$$

The maximum moment occurs at the point where the shear is zero, or 14.74 m from the right end of the beam.

$$M_{max} = +14.74 \times 20,533 - 14.74 \times 800 \times \frac{14.74}{2} - 8.74 \times 1000 \times \frac{8.74}{2}$$

$$= 177,556 \text{ N-m}$$

Answer is (E)

PROBLEM 5-37

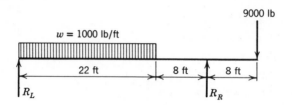

Figure 5-22

For the beam loaded as shown, the value of the maximum moment is nearest to

    (A) 11 kip-ft
    (B) 19 kip-ft
    (C) 66 kip-ft
    (D) 72 kip-ft
    (E) 104 kip-ft

Solution.

$$\Sigma M_{R_L} = 0$$

$$22 \times 1 \text{ kip} \times 11 + 9 \text{ kips} \times 38 - 30 R_R = 0$$

$$+242 + 342 - 30 R_R = 0 \qquad R_R = \frac{242 + 342}{30} = 19.47 \text{ kips}$$

$$\Sigma F_y = 0$$

$$R_L + 19.47 - 22 \times 1 \text{ kip} - 9 \text{ kips} = 0 \qquad R_L = 11.53 \text{ kips}$$

The moment is a maximum at the point where the shear is zero. Expressing the shear as a function of x from the left end, $V(x) = 11.53 \text{ kips} - x \text{ kips} = 0$ when $x = x_0$. $x_0 = 11.53$ ft.

$$M(x_0) = M_{x_0} = x_0 R_L - \frac{wx_0^2}{2} = 11.53(11.53) - \frac{1(11.53)^2}{2}$$

$$= \frac{11.53^2}{2} = 66.5 \text{ kip-ft}$$

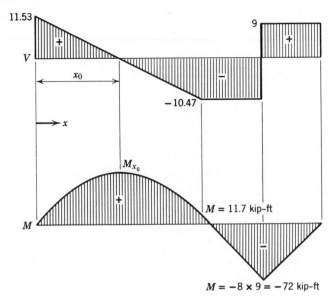

Figure 5-23

Thus the maximum moment is a negative moment of 72 kip-ft, located at the right support, 8 ft to the left of the right end of the beam.

Answer is (D)

PROBLEM 5-38

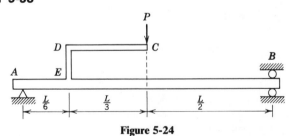

Figure 5-24

For beam AB loaded as shown the value of the right reaction R_B is

(A) $\frac{1}{6}P$

(B) $\frac{1}{3}P$

(C) $\frac{1}{2}P$

(D) $\frac{2}{3}P$

(E) $\frac{5}{6}P$

Solution.

$$\Sigma M_{R_A} = 0$$

$$-P\left(\frac{L}{6}+\frac{L}{3}\right)+LR_B = 0 \qquad R_B = \frac{0.5LP}{L} = 0.5P$$

Answer is (C)

PROBLEM 5-39

Which is the proper shape of the moment diagram for the cantilever beam *AB* loaded as shown?

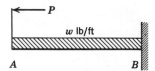

Figure 5-25

Solution.

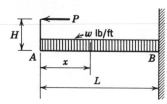

Figure 5-26

The force P contributes a constant moment $-PH$ from A to B. The uniform load w, at any point x, produces a moment $-wx^2/2$. The total moment diagram is the sum of the moment diagrams of the component loads.

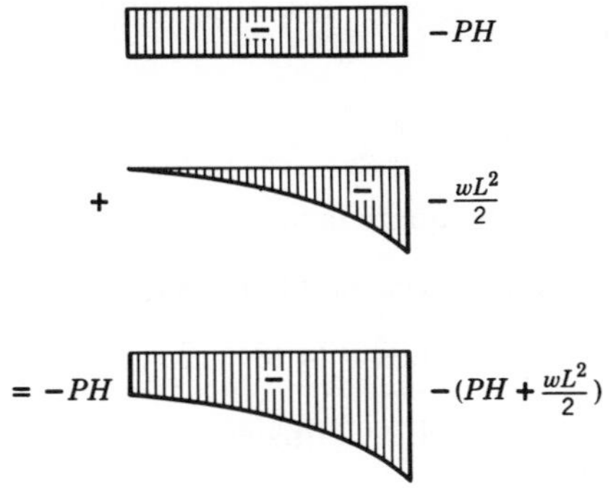

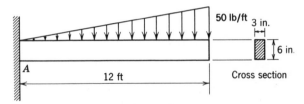

$-PH$

$+$ $\quad -\dfrac{wL^2}{2}$

$= -PH$ $\quad -(PH + \dfrac{wL^2}{2})$

Figure 5-27

Answer is (C)

PROBLEM 5-40

50 lb/ft   3 in.

6 in.

A

12 ft

Cross section

Figure 5-28

A cantilever beam 3 in. × 6 in. supports a uniformly varying load as shown. Neglect the weight of the beam. The maximum bending stress in the beam is nearest to

(A) 135 psi
(B) 1350 psi
(C) 1600 psi
(D) 2250 psi
(E) 4500 psi

Solution. The moment at any section is the algebraic sum of the moments of all external forces on one side of the section.

$$M_{\max} = M_A = \text{force} \times \text{lever arm}$$

$$= (\tfrac{1}{2} \times 12 \times 50)(\tfrac{2}{3} \times 12)$$

$$= 300 \times 8 = 2400 \text{ lb-ft}$$

Since the lower fibers of the beam are in compression, the moment is negative.

The maximum bending stress is

$$\sigma = \frac{M_{max}c}{I}$$

where

σ = bending stress in pounds per square inch

c = distance from neutral axis to extreme fiber = 3 in.

M_{max} = maximum moment in inch-pounds = 2400×12

I = moment of inertia ($=bh^3/12$ for rectangular beams)

$$\sigma = \frac{2400 \times 12 \times 3}{(3 \times 6^3)/12}$$

$$= \frac{2400 \times 12^2}{6^3} = 1600 \text{ psi}$$

Answer is (C)

PROBLEM 5-41

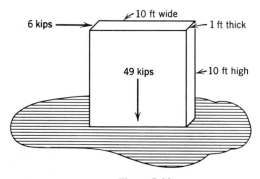

Figure 5-29

A solid steel block, which weighs 49 kips, rests on a level concrete slab. If a horizontal force of 6 kips is applied to the top, the minimum pressure under the base of the block is equal to

(A) 0 lb/ft^2

(B) 1300 lb/ft^2

(C) $3600 \, lb/ft^2$

(D) $4900 \, lb/ft^2$

(E) $8500 \, lb/ft^2$

Solution.

$$\sigma_{max} = \frac{P}{A} + \frac{My}{I} = \frac{49,000}{10 \times 1} + \frac{6000 \times 10 \times 5}{\frac{1}{12} \times 1 \times 10^3} = 4900 + 3600 = 8500 \, lb/ft^2$$

$$\sigma_{min} = \frac{P}{A} - \frac{My}{I} = 4900 - 3600 = 1300 \, lb/ft^2$$

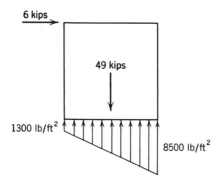

Figure 5-30

Answer is (B)

PROBLEM 5-42

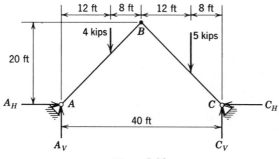

Figure 5-31

Two steel beams are fabricated to form a three-hinged structure as shown. The horizontal component of the reaction at A is to be computed. A_H is nearest to

   (A) 0 kips
   (B) 1.1 kips
   (C) 2.2 kips
   (D) 3.3 kips
   (E) 4.4 kips

Solution.

$$\sum M_A = 0$$

$$-4\times12-5\times32+C_V\times40=0 \qquad C_V=\frac{48+160}{40}=5.2 \text{ kips}$$

$$\sum F_y = 0$$

$$A_V+C_V-4-5=0 \qquad A_V=9-5.2=3.8 \text{ kips}$$

In member AB, $\sum M_B = 0$ since B is a hinge.

$$-A_V\times20+4\times8+A_H\times20=0$$

$$A_H=\frac{76-32}{20}=2.2 \text{ kips}$$

Answer is (C)

PROBLEM 5-43

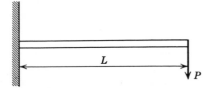

L

P

Figure 5-32

A cantilever beam of length L is loaded at the free end by a force P. Assume the modulus of elasticity $E = 20\times10^6$ psi, the cross-sectional area of the beam $A = 6$ in.^{2}, the moment of inertia $I = 2$ in.^{4}, and $L = 30$ in. The force P that produces a 0.3-in. deflection of the beam at the free end is nearest to

   (A) 333 lb
   (B) 667 lb
   (C) 1000 lb
   (D) 1333 lb
   (E) 1667 lb

Solution. To solve this problem, we must first derive the equation relating P and beam deflection. The second moment-area proposition says that the vertical displacement Δ of point A from the tangent to the elastic curve at B equals the moment (with respect to A) of the area of the bending moment diagram between A and B divided by EI.

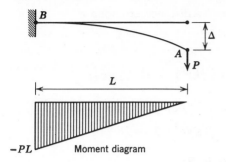

Moment diagram   **Figure 5-33**

$$\Delta = \frac{\text{area of } M \text{ diagram} \times \text{lever arm}}{EI}$$

$$= \frac{-PL(L/2)\times\frac{2}{3}L}{EI} = \frac{-PL^3}{3EI}$$

For $\Delta = 0.3$,

$$\frac{P\times 30^3}{3(20\times 10^6)\times 2} = 0.3$$

$$P = \frac{0.3\times 2\times 3\times 20\times 10^6}{30^3} = 1333 \text{ lb}$$

Answer is (D)

PROBLEM 5-44

A helical spring has a natural length of 6 in. It requires a force of 20 lb to hold it extended to a length of 12 in. Assume the spring does not exceed its elastic limit. The work done in inch-pounds to stretch the spring from a total length of 9 in. to 11 in. is nearest to

(A) 5 in.-lb
(B) 10 in.-lb
(C) 15 in.-lb
(D) 20 in.-lb
(E) 25 in.-lb

Solution.

$$\text{Spring constant } k = \frac{\text{force}}{\text{deflection}} = \frac{20 \text{ lb}}{(12-6) \text{ in.}} = \frac{20 \text{ lb}}{6 \text{ in.}}$$

The energy required to extend a spring a distance dx is $F\,dx$.

Force $F =$ spring constant $k \times$ distance extended x

For a displacement $S_2 - S_1$,

$$\text{Total work} = \int_{S_1}^{S_2} kx\,dx = \tfrac{1}{2}kx^2 \Big|_{S_1}^{S_2} = \tfrac{1}{2}k(S_2^2 - S_1^2)$$

In this case $S_1 = 9$ in. $- 6$ in. $= 3$ in. and $S_2 = 11$ in. $- 6$ in. $= 5$ in.

$$\text{Total work} = \tfrac{1}{2}k(S_2^2 - S_1^2) = \tfrac{1}{2}(\tfrac{20}{6})(5^2 - 3^2) = 26.7 \text{ in.-lb}$$

Answer is (E)

PROBLEM 5-45

A 3-in.-diameter solid steel shaft 10 ft long is subjected to a constant torque of 100,000 in.-lb at each end, together with an axial tensile load of 70,000 lb also applied at each end. For a circular cross section the polar moment of inertia $J = \pi d^4 / 32$. The maximum compressive stress in the shaft under this loading is nearest to

(A) 0 psi
(B) 10,000 psi
(C) 15,000 psi
(D) 18,800 psi
(E) 19,500 psi

Solution.

$$\sigma_x = \sigma_{\text{tension}} = \frac{P}{A} = \frac{70,000}{\dfrac{\pi}{4}3^2} = 9900 \text{ psi}$$

$$\sigma_y = 0$$

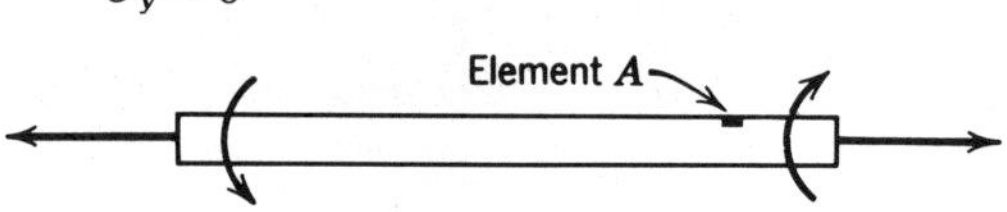

Figure 5-34

Torsion

$$\tau_{xy} = \frac{Tr}{J} = \frac{T(d/2)}{\pi(d^4/32)} = \frac{16T}{\pi d^3} = \frac{16(100,000)}{\pi(27)} = 18,860 \text{ psi}$$

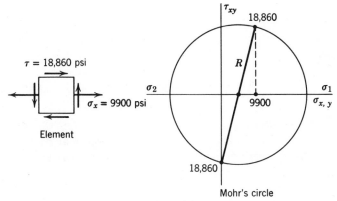

Figure 5-35

$$R = (18,860^2 + 4950^2)^{1/2} = 19,500$$

$$(\tau_{xy})_{max} = R = 19,500 \text{ psi}$$

$$\sigma_1 = 4950 + R = 24,450 \text{ psi tension}$$

$$\sigma_2 = 4950 - R = -14,550 \text{ psi compression}$$

Answer is (C)

PROBLEM 5-46

The *least radius of gyration* is required in the design of

    (A) shaft couplings
    (B) columns
    (C) helical springs
    (D) cantilevered beams
    (E) riveted joints

Solution. The ratio L/r of the column length to the least radius of gyration is the column slenderness ratio.

Answer is (B)

PROBLEM 5-47

A rectangular shape has a cross section 30 cm wide and 60 cm in height. The least radius of gyration for this shape is closest to

(A) 10 cm
(B) 15 cm
(C) 20 cm
(D) 25 cm
(E) 30 cm

Solution. The radius of gyration $r = (I/A)^{1/2}$ where I is the moment of inertia about a centroidal axis and A is the cross-sectional area.

When the least moment of inertia is used, the computation gives the least radius of gyration. For a rectangular cross section,

$$I = \frac{bh^3}{12} = \frac{60 \times 30^3}{12} = 135{,}000 \text{ cm}^4$$

$$A = 30 \times 60 = 1800 \text{ cm}^2$$

$$\text{Least } r = \left(\frac{135{,}000}{1800}\right)^{1/2} = 8.66 \text{ cm}$$

Answer is (A)

PROBLEM 5-48

A beam has a cross-sectional area of 72 cm^2 and a radius of gyration of 3 cm. The corresponding value of the moment of inertia for the beam is nearest to

(A) 24
(B) 216
(C) 650
(D) 1300
(E) 15,000

Solution. Radius of gyration $r = (I/A)^{1/2}$. Here $I = r^2 A = 3^2 \times 72 = 648 \text{ cm}^4$.

Answer is (C)

PROBLEM 5-49

The *slenderness ratio* of a column is generally defined as the ratio of its

(A) length to its minimum width
(B) unsupported length to its maximum radius of gyration
(C) length to its moment of inertia
(D) unsupported length to its least radius of gyration
(E) unsupported length to its minimum cross-sectional area

Solution.   Euler's formula

$$P_{cr} = \frac{\pi^2 EI}{L^2}$$

may be rewritten to compute the critical stress.

$$I = r^2 A$$

$$\sigma_{cr} = \frac{P_{cr}}{A} = \frac{\pi^2 E r^2 A}{AL^2} = \frac{\pi^2 E}{(L/r)^2}$$

where L is the unsupported length and r is the least radius of gyration, the ratio L/r is called the column *slenderness ratio*.

Answer is (D)

PROBLEM 5-50

The section modulus about the central axis of a 10-cm × 25-cm beam on edge is closest to

(A) 250 cm^3
(B) 1000 cm^3
(C) 6000 cm^3
(D) 13,000 cm^3
(E) 15,625 cm^3

Solution.

$$\text{Section modulus } Z = \frac{I}{c}$$

where I is the moment of inertia and c is the distance from the neutral axis to the extreme fiber.

For a rectangular cross section,

$$I = \frac{bh^3}{12} \qquad c = \frac{h}{2}$$

$$Z = \frac{I}{c} = \frac{bh^3/12}{h/2} = \frac{bh^2}{6}$$

In this case

$$Z = \frac{10(25)^2}{6} = 1042 \text{ cm}^3$$

Answer is (B)

PROBLEM 5-51

A cantilevered beam 20 in. long and of square cross section, 1 in. on a side, is loaded at its end, through the centroid of the cross section by a vertical force of magnitude 150 lb. The magnitude of the maximum bending stress is nearest to which value?

(A) 3000 psi
(B) 6000 psi
(C) 12,000 psi
(D) 15,000 psi
(E) 18,000 psi

Solution.   The stress $\sigma = Mc/I$ with $I = bh^3/12 = 1(1)^3/12 = \frac{1}{12}$ in.^{4} and $M = (150 \text{ lb})(20 \text{ in.}) = 3000$ in.-lb.

$$\sigma = \frac{(3000)(\frac{1}{2})}{\frac{1}{12}} = 18,000 \text{ psi}$$

Answer is (E)

PROBLEM 5-52

The maximum unit fiber stress at any vertical section in a beam is obtained by dividing the moment at that section by

(A) the section modulus
(B) the cross-sectional area
(C) one-half the distance to the point where the shear is zero
(D) the radius of gyration
(E) the moment of inertia

Solution. The basic formula for maximum fiber stress in a beam is

$$\text{Unit stress} = \frac{Mc}{I}$$

where M = moment at the section

c = distance from the neutral axis to the extreme fiber

I = moment of inertia

The section modulus Z is defined as I/c, so the basic formula reduces to

$$\text{Unit stress} = \frac{M}{Z} = \frac{\text{moment}}{\text{section modulus}}$$

Answer is (A)

PROBLEM 5-53

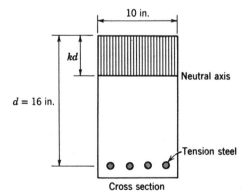

Cross section   **Figure 5-36**

Consider the reinforced concrete beam shown. Assume a balanced design, using straight-line theory.

Data:
  f_c = maximum allowable compressive unit stress = 1350 psi
  f_s = maximum allowable tensile unit stress in the longitudinal
      reinforcement = 20,000 psi
  A_s = area of longitudinal steel reinforcement = 2.18 in.^{2}
  n = ratio of modulus of elasticity of steel to that of concrete = 10

The distance kd is closest to

  (A) 4.3 in.
  (B) 6.5 in.
  (C) 7.1 in.
  (D) 8.0 in.
  (E) 8.6 in.

Solution.

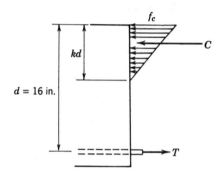

Figure 5-37

Width $b = 10$ in.

$$C = \tfrac{1}{2} f_c k d b = \tfrac{1}{2}(1350)(kd)(10) = 6750 kd$$

$$T = A_s f_s = 2.18(20,000) = 43,600 \text{ lb}$$

In a *balanced design* the maximum allowable concrete stress f_c and allowable steel stress f_s are simultaneously developed.

$$\sum F_x = 0 \qquad T = C$$

$$43,600 = 6750 kd \qquad kd = \frac{43,600}{6750} = 6.46 \text{ in.}$$

Answer is (B)

PROBLEM 5-54

The principal difference between regular concrete and lightweight concrete is in the

    (A) cement used

    (B) water proportions used

    (C) admixtures used

    (D) relative mixing time

    (E) aggregates used

Solution. Regular concrete weighs about 150 lb/ft^3, but by using special lightweight aggregates a concrete weighing about 90 lb/ft^3 can be produced. The difference is in the weight of the aggregates used.

Answer is (E)

PROBLEM 5-55

In the design of a reinforced concrete beam, it is commonly assumed that

    (A) the concrete has one-fourth of the tensile strength of the steel

    (B) the modulus of elasticity for concrete is the same as the modulus of elasticity for steel

    (C) the steel has a tensile stress of 1800 psi induced by shrinkage of the concrete

    (D) the dead load of the beam may be neglected for long spans

    (E) the tensile strength of the concrete is nearly zero

Solution. Conventional concrete design is done by a cracked section analysis; that is, it is assumed that the portion of beam in tension will crack with the result that the effective concrete tensile strength is zero.

Answer is (E)

PROBLEM 5-56

Stirrups in a reinforced concrete beam are designed primarily to resist stresses caused by

(A) diagonal tension

(B) axial tension

(C) horizontal shear

(D) axial compression

(E) web crippling

Solution.   At each point in a loaded concrete beam there are vertical or transverse shearing forces and, for equilibrium, longitudinal shearing forces of equal intensity. These forces tend to distort the beam so that tension exists along a diagonal plane. This diagonal tension acts along a line of action 45° from the beam's axis. Stirrups or other web reinforcement are designed to resist this diagonal tension.

Answer is (A)

PROBLEM 5-57

A concrete mix with an ultimate compressive strength f'_c of 3750 psi would have a n value of

(A) 6

(B) 8

(C) 10

(D) 12

(E) 15

Solution.

f'_c = ultimate compressive strength of concrete, usually at 28 days
n = ratio of modulus of elasticity of steel to that of concrete
E_s = modulus of elasticity of steel = 30×10^6 psi
E_c = modulus of elasticity of concrete $\cong 1000 f'_c$

For a concrete mix with $f'_c = 3750$ psi,

$$n = \frac{E_s}{E_c} = \frac{30 \times 10^6}{1000 \times 3750} = 8$$

Answer is (B)

PROBLEM 5-58

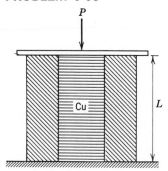

Figure 5-38

Two concentric cylinders of length L are loaded between rigid smooth plates as shown. The inner cylinder is copper and the outer cylinder is steel.

Let

$$E_s = \text{modulus of elasticity of steel}$$
$$E_c = \text{modulus of elasticity of copper}$$
$$A_s = \text{cross-sectional area of steel cylinder}$$
$$A_c = \text{cross-sectional area of copper cylinder}$$
$$P = \text{total applied load } P = P_c + P_s$$

Which one of the following represents the formula for stress in the copper cylinder?

(A) $\sigma_c = \dfrac{P}{A_c}$

(B) $\sigma_c = \dfrac{E_c P}{A_c E_s + A_s E_c}$

(C) $\sigma_c = \dfrac{P_s}{A_c}$

(D) $\sigma_c = \dfrac{E_c P}{A_c E_c + A_s E_s}$

(E) None of the above is correct

Solution.

$$\text{Stress} = \frac{\text{force}}{\text{area}} = \frac{P}{A} \qquad \text{Deflection} = \text{length} \times \text{strain} = L\varepsilon = \delta$$

$$\text{Modulus of elasticity} = \frac{\text{stress}}{\text{strain}} = \frac{\sigma}{\varepsilon} \qquad \delta = L\varepsilon = \frac{L\sigma}{E} = \frac{LP}{AE}$$

But $\delta_s = \delta_c$. Therefore

$$\frac{LP_s}{A_s E_s} = \frac{LP_c}{A_c E_c} \qquad P_s = \frac{P_c A_s E_s}{A_c E_c} \qquad P = P_s + P_c$$

$$P = \frac{P_c A_s E_s}{A_c E_c} + P_c \qquad P = P_c\left(1 + \frac{A_s E_s}{A_c E_c}\right)$$

Therefore

$$P_c = \frac{A_c E_c P}{A_c E_c + A_s E_s}$$

But

$$\sigma_c = \frac{P_c}{A_c} \qquad \sigma_c = \frac{E_c P}{A_c E_c + A_s E_s}$$

Answer is (D)

6 Fluid Mechanics

Fluid mechanics studies fluids at rest and in motion by the application of basic principles of mechanics. The discipline covers a broad field since it endeavors to study both liquids and gases under all conditions. In this chapter we restrict our attention almost entirely to a study of the mechanical behavior of incompressible liquids such as water. Topics related to the internal energy changes and the compressibility and chemical behavior of gases may be found in Chapters 7 and 8.

Paralleling the study of the mechanics of solid bodies, fluid mechanics is normally divided into fluid statics and fluid dynamics. Hydrostatics studies the variation of pressure in a stationary fluid body and leads to an understanding of such subjects as buoyancy and manometry; one also learns how to compute the forces exerted on submerged bodies because of this pressure variation. The fundamental principles underlying fluid dynamics are contained in three conservation laws: those expressing the conservation of mass, momentum, and energy. In connection with these principles the calculation of power, head loss, and flow rate is considered. In almost all the dynamics problems the flow is assumed to be steady and also one-dimensional; that is, the flow variables are only a function of the distance along the conveyance structure and do not vary across the flow cross section.

HYDROSTATICS

The fundamental equation of fluid statics indicates that the rate of change of the pressure p is directly proportional to the rate of change of the depth z, or

$$\frac{dp}{dz} = -\gamma = -\rho g \qquad (6\text{-}1)$$

219

where γ is the unit weight (also called specific weight or weight density) of the fluid (for water $\gamma_w = 62.4 \text{ lb/ft}^3$) and z is positive upward. The fluid density (or mass density) is ρ, and g is the acceleration of gravity. In some cases fluid properties are reported in terms of relative density $S = \rho/\rho_r$, which is the ratio of one fluid density to that of some chosen reference fluid. The ratio could also be written in terms of unit weights as γ/γ_r; if water is chosen as the reference fluid, S is then often called the specific gravity of the fluid.

For incompressible fluids γ is constant and Eq. (6-1) may be integrated to give

$$p_2 = p_1 + \gamma h \tag{6-2}$$

Here point 2 is located a distance h below point 1. In most problems in fluid mechanics one may work in either absolute or gage pressures if proper care is taken to be consistent. Gage pressure is equal to the difference between the absolute pressure and atmospheric pressure; when it is negative, it is often called a vacuum. In English units lb/in.^2 is often written psi, and absolute and gage pressures are distinguished by writing psia and psig.

Manometers are devices that measure pressure differences by a direct application of the hydrostatic principle given in Eq. (6-2). Pressure differences in manometers can easily be computed by systematically applying the equation to each manometer limb.

Example 1

Mercury (Hg) is poured into a U-tube. Then 18-in. of oil of specific gravity $S_0 = 0.90$ is poured into one leg on top of the mercury (Fig. 6-1). What suction, in pounds per square inch, applied to the leg containing only mercury will bring the upper surfaces of the oil and mercury in the two legs to the same level?

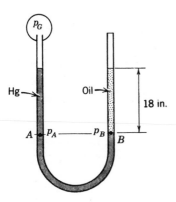

Figure 6-1

Solution. For one fluid, Eq. (6-2) shows that the pressures at two points are equal when the elevation difference between the points is zero so that $p_A = p_B$. But Eq. (6-2) also shows that

$$p_A = p_G + \gamma_M h$$

and

$$p_B = \gamma_0 h$$

Hence

$$p_G = h(\gamma_0 - \gamma_M) = h\gamma_w(S_0 - S_M)$$

where the specific gravity of mercury $S_M = 13.55$.

$$p_G = \tfrac{18}{12}(62.4)(0.90 - 13.55) = -1184 \text{ psfg}$$

$$p_G = \frac{-1184}{144} = -8.22 \text{ psig} = 8.22 \text{ psi vacuum}$$

The buoyant force F_B exerted upward on a floating or submerged object is equal to the weight of the fluid displaced by that object, as can be shown by properly integrating Eq. (6-2) over the surface of the body. Restated,

$$F_B = \gamma_f(\text{volume}) \tag{6-3}$$

where γ_f is the unit weight of the fluid, and the volume is the amount of fluid displaced.

Example 2

A piece of lead (specific gravity $S_L = 11.3$) is tied to 8 in.^{3} of cork whose specific gravity is $S_C = 0.25$. They float submerged in water (Fig. 6-2). What is the weight of the lead?

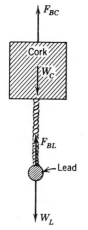

Figure 6-2

Solution. The net upward force on the cork is $F_{BC} - W_C$, which is just balanced by the net downward force on the lead $W_L - F_{BL}$, or

$$F_{BC} - W_C = W_L - F_{BL}$$

Thus

$$(1 - S_C)\gamma_w \left(\frac{8}{12^3}\right) = (S_L - 1)\gamma_w V_L$$

where V_L is the unknown volume of the lead. Solving first for V_L and then for the weight W_L, we have

$$W_L = S_L \gamma_w V_L = \gamma_w \left(\frac{S_L}{S_L - 1}\right)(1 - S_C)\frac{8}{12^3}$$

$$W_L = 62.4\left(\frac{11.3}{10.3}\right)(0.75)\frac{8}{12^3}$$

$$W_L = 0.238 \text{ lb lead}$$

Distributed fluid pressures cause hydrostatic forces to act on submerged surfaces. It is important to be able to compute the direction, magnitude, and location of these resultant forces. The force on any submerged surface is determinable if one knows how to compute the force acting on a submerged plane area and the centroid or center of gravity of areas and volumes.

The magnitude of the force F on a submerged plane surface is

$$F = \int_A p\, dA = p_C A = \gamma h_C A \qquad (6\text{-}4)$$

where A is the surface area, γ is the fluid unit weight, and p_C and h_C are, respectively, the pressure and submerged depth of the centroid of the area. This compressive force acts normal to the plane area. Simple formulas can be derived for the location of this force, but they are easily misused. A more direct approach is to compute the first moment of the pressure distribution about some convenient axis, according to the principles of statics, and equate this to the product of F and the distance to the line of action of F.

An extension of these principles allows the computation of the magnitude and location of a hydrostatic force which acts on a submerged curved surface. The horizontal component of this force is computed from Eq. (6-4), where the area A is now the vertical projection of the curved surface. The vertical force component is basically equal to the weight of the fluid which lies vertically above the curved surface, but some care is needed in applying this principle. When all the fluid lies above the surface, the principle applies exactly. If some fluid also lies beneath a portion of the curved surface, then the *net* upward vertical force is a buoyant force equal to the weight of fluid displaced by the presence of the curved surface. The line of action of each component force and the magnitude and direction of the resultant force are then found by direct application of the principles of statics.

Example 3

A closed circular tank 2 ft in diameter and 6 ft deep, with its axis vertical, contains 4 ft of water. Air at a pressure of 5 psig is pumped into the cylinder. Find the normal force per inch of circumference on the vertical wall of the tank and the distance to the center of pressure from the base of the tank.

Solution.   First we plot a diagram, Fig. 6-3, of the gage pressure exerted on a unit width of the wall. The hydrostatic pressure p_1 at the base of the tank is

$$p_1 = \gamma h = \frac{(62.4)(4)}{144} = 1.73 \text{ psi}$$

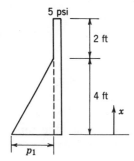

Figure 6-3

The force per inch of circumference equals the area of the pressure diagram, or

$$F = \int p\, dA = 5(6)(12) + \tfrac{1}{2}(1.73)(4)(12)$$
$$= 360 + 41.5 = 401.5 \text{ lb/in.}$$

The distance to the center of pressure x_{cp}, by statics, is

$$x_{cp} = \frac{1}{F} \int px\, dA$$

$$= \frac{1}{401.5} [360(3) + 41.5(\tfrac{4}{3})] = 2.83 \text{ ft} = 34 \text{ in.}$$

CONTINUITY

The law of conservation of mass states that matter is neither created nor destroyed. Applied to a streamtube in incompressible flow, the law requires the flow to vary in a continuous way from cross section to cross section along the streamtube so that at any section

$$Q = \int_A V \, dA = \text{constant} \tag{6-5}$$

where Q is the volume rate of flow and V is the velocity at a point in the cross-sectional streamtube area A. When the velocity is assumed to be constant across the section, we obtain the familiar continuity equation

$$Q = A_1 V_1 = A_2 V_2 \tag{6-6}$$

This equation is widely used in combination with the energy or momentum equations to solve fluid flow problems. In the United States volume rate of flow may commonly be expressed not only in ft^3/sec (cfs) but also in gallons per minute (gpm) and millions of gallons per day (mgd).

Example 4

A fluid flowing steadily through a constant-diameter pipeline has a velocity profile $u(r)$ which varies parabolically across the pipe. Specifically,

$$u = m(1 - s^2)$$

where $s = r/R$. Here m is a constant, r is the distance from the pipe centerline, and R is the pipe radius. What fraction of the total flow in the pipe is flowing between the pipe wall and a distance of 10% of the pipe radius from the wall?

Solution.   For the integration of Eq. (6-5) we choose the differential area shown in Fig. 6-4. Then the total volume rate of flow in the pipe is

$$Q_T = \int_A V \, dA = \int_0^R m(1 - s^2)(2\pi r \, dr)$$

$$= 2\pi m R^2 \int_0^1 (1 - s^2) s \, ds$$

$$= 2\pi m R^2 \left[\frac{s^2}{2} - \frac{s^4}{4} \right]_0^1$$

$$Q_T = \tfrac{1}{2}\pi m R^2$$

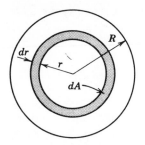

Figure 6-4

The volume rate of flow near the wall is

$$Q = \int_{0.9R}^{R} m(1-s^2)(2\pi r\, dr)$$

$$= 2\pi m R^2 \int_{0.9}^{1} (1-s^2)s\, ds$$

$$= 2\pi m R^2 \left[\frac{s^2}{2} - \frac{s^4}{4}\right]_{0.9}^{1}$$

$$Q = 0.018 \pi m R^2$$

Thus the flow fraction is

$$\frac{Q}{Q_T} = \frac{0.018}{0.5} = 0.036$$

ENERGY EQUATION

For one-dimensional, steady incompressible flow a general energy equation that expresses the changes in energy, *per unit weight* of flowing fluid, between points 1 and 2 is

$$\frac{V_1^2}{2g} + \frac{p_1}{\gamma} + z_1 = \frac{V_2^2}{2g} + \frac{p_2}{\gamma} + z_2 + h_L - E_m \qquad (6\text{-}7)$$

In this equation $V^2/2g$ is the kinetic energy or velocity head, p/γ is the pressure energy or pressure head, and z is the potential energy or elevation head. The head loss h_L represents the loss in energy between points 1 and 2 for any of a variety of causes. The last term E_m is the mechanical energy added to the fluid between the two points; this is accomplished by hydraulic machinery. For pumps, energy is added to the flow and E_m is positive. For turbines, E_m is negative since energy is then extracted from the fluid. When E_m and h_L are both zero, Eq. (6-7) is the classic Bernoulli equation.

The head loss term h_L represents a loss in energy that is due primarily to viscosity and the turbulence in the flow. It may be written as

$$h_L = \sum K \frac{V^2}{2g}$$

where V is a representative velocity for the point where the loss occurs, K is a loss coefficient related to the nature of the loss-producing element, and one sums the effects of all loss elements between the two points. Particularly important is the head loss caused by pipe friction in a pipe of length L and diameter d. In this case

$$h_L = f \frac{L}{d} \frac{V^2}{2g} \qquad (6\text{-}8)$$

The equation is the Darcy-Weisbach equation. The friction factor $f = f(\mathbf{R})$, and the Reynolds number is $\mathbf{R} = Vd\rho/\mu$. For laminar flow in pipes $\mathbf{R} < 2100$ and $f = 64/\mathbf{R}$, but in turbulent flow, when $\mathbf{R} > 4000$, f is also a function of the relative roughness e/d for rough pipes. The height of a representative roughness projection is e. Between the laminar and turbulent zones lies an unpredictable transition zone. Most losses other than that due to pipe friction are termed minor losses, since their magnitude is usually small in comparison to the pipe friction loss in practical problems. A different K value is needed for each type of loss element.

The power produced or expended in a given situation is directly related to the change in energy in the flow. If the weight rate of flow is $Q\gamma$ and the energy change per unit weight of fluid is E, then the power gained or expended is the product $Q\gamma E$. In terms of horsepower (1 hp = 550 ft-lb/sec) the relation is

$$\text{Horsepower} = \frac{Q\gamma E}{550} \qquad (6\text{-}9)$$

For turbines or pumps E is the same as E_m.

Example 5

A Venturi meter with a throat diameter of 6 in. is placed in a 12-in.-diameter pipeline. It meters the flow of oil having a specific gravity $S_0 = 0.80$. A differential manometer containing a fluid of specific gravity $S_M = 3.20$ is connected between the pipe and the throat section and shows a deflection of 2 ft. Above the manometer liquid the tubes are filled with oil (Fig. 6-5). Neglecting energy losses, what is the indicated flow through the meter?

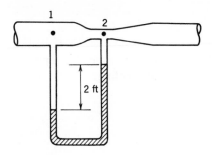

Figure 6-5

Solution. Here we apply the Bernoulli and continuity equations in combination. The Bernoulli equation is

$$\frac{V_1^2}{2g}+\frac{p_1}{\gamma_0}+z_1=\frac{V_2^2}{2g}+\frac{p_2}{\gamma_0}+z_2$$

Since $z_1=z_2$,

$$\frac{p_1-p_2}{\gamma_0}=\frac{V_2^2-V_1^2}{2g}$$

Considering the manometer,

$$p_1+S_0\gamma_w(2)=p_2+S_M\gamma_w(2)$$

and by noting that $S_0\gamma_w=\gamma_0$, we obtain

$$\frac{p_1-p_2}{\gamma_0}=2\left(\frac{S_M}{S_0}-1\right)$$

From Eq. (6-6), the continuity equation is $V_1A_1=V_2A_2$, or

$$V_2=\frac{A_1}{A_2}V_1=\left(\frac{d_1}{d_2}\right)^2 V_1=(\tfrac{12}{6})^2 V_1=4V_1$$

Therefore

$$2\left(\frac{S_M}{S_0}-1\right)=\frac{(4V_1)^2-V_1^2}{2g}=\frac{15V_1^2}{2g}$$

$$V_1^2=\frac{1}{15}\left[2(32.2)(2)\left(\frac{3.20}{0.80}-1\right)\right]=25.8$$

$$V_1=5.08\text{ ft/sec}$$

Finally,

$$Q=A_1V_1=\frac{\pi}{4}(1)^2(5.08)=4.0\text{ cfs}$$

Example 6

A water main with a 24-in. inside diameter carries a flow of 20 cfs. If the friction factor is 0.02 and the pump is 85% efficient, how much horsepower is required to pump the water through 10,000 ft of pipeline?

Solution.   The velocity in the main is

$$V = \frac{Q}{A} = \frac{20}{(\pi/4)(2)^2} = 6.37 \text{ ft/sec}$$

Using Eq. (6-8), the Darcy-Weisbach formula for head loss caused by friction, we obtain

$$h_L = f \frac{L}{d} \frac{V^2}{2g}$$

$$h_L = (0.02) \left(\frac{10,000}{2} \right) \frac{(6.37)^2}{2(32.2)}$$

$$h_L = 63.0 \text{ ft}$$

This head loss is equal to the amount of energy E that must be added per pound of fluid. The net power requirement is

$$\frac{Q\gamma h_L}{550} = \frac{(20)(62.4)(63.0)}{550} = 143 \text{ hp}$$

Since the pump efficiency is only 85%, the total power requirement is $143/0.85 = 168$ hp.

Pipelines of different sizes and lengths commonly occur in combination. When pipes are connected end-to-end in series, the head loss of the combination is the sum of the head losses in the individual pipes, while the flow rate is the same for each element. The situation is reversed for a parallel combination of pipes, that is, a case where two or more pipes are connected between the same two end points. Then the total flow rate is the sum of the individual flow rates, while the head lost is identical for each pipe.

MOMENTUM EQUATION

The steady-flow momentum conservation equation takes the form

$$\mathbf{F}_S + \mathbf{F}_B = \int_S \rho \mathbf{V} V_n \, dA \tag{6-10}$$

This is a vector equation. The first term represents the net external surface force acting *on* a fluid volume which is enclosed by the surface S; this term includes all contributions arising from the surface pressure distribution and from surface viscous stresses acting on S. The second term is the net body force within the control volume; usually this is just the weight of the enclosed fluid acting in the direction of gravity. The right term is the net flux of momentum flowing out of the region enclosed by S. V_n is the fluid velocity component exiting normal (perpendicular) to S. For uniform flow across an entrance section 1 and an exit section 2, we may write

$$\Sigma F_x = \rho Q (V_{2x} - V_{1x})$$
$$\Sigma F_y = \rho Q (V_{2y} - V_{1y})$$

(6-11)

for flow in two dimensions. The volume flow rate is Q. Note that V_x and V_y may be either positive or negative.

Example 7

The 18-in. to 12-in. reducing bend shown in Fig. 6-6 is in a horizontal pipeline conveying oil (of specific gravity $S_0 = 0.90$) at the rate of 10 ft³/sec. The pressure in the line at the 18-in. section is 50 psia when the atmospheric pressure $p_a = 14.7$ psia. Compute the x and y components of thrust caused by the flowing fluid.

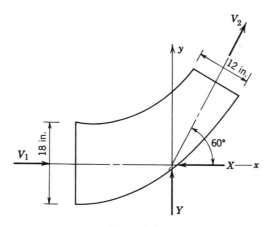

Figure 6-6

Solution. The force components we seek are equal and opposite to the external forces X and Y shown in Fig. 6-6.

The reducing bend and the fluid within it are chosen as a control volume.

Atmospheric pressure acts on the entire control volume surface and thus produces no net force. Employing Eq. (6-11) we find

$$\Sigma F_x = (p_1 - p_a)A_1 - (p_2 - p_a)A_2 \cos 60° - X$$

and

$$\rho Q(V_{2x} - V_{1x}) = \rho Q(V_2 \cos 60° - V_1)$$

Here

$$Q = 10 \text{ cfs}$$

$$V_1 = \frac{Q}{A_1} = \frac{10}{(\pi/4)(\frac{18}{12})^2} = 5.66 \text{ ft/sec}$$

$$V_2 = \frac{Q}{A_2} = \frac{10}{(\pi/4)(\frac{12}{12})^2} = 12.73 \text{ ft/sec}$$

Assuming no energy losses occur in the bend, we apply the Bernoulli equation between the entrance and exit to find p_2:

$$\frac{V_1^2}{2g} + \frac{p_1}{\rho g} + z_1 = \frac{V_2^2}{2g} + \frac{p_2}{\rho g} + z_2$$

Since there is no elevation change $z_1 = z_2$, and

$$p_2 = p_1 + \frac{\rho}{2}(V_1^2 - V_2^2)$$

$$p_2 = 50 + \tfrac{1}{2}(0.90)(1.94)[(5.66)^2 - (12.73)^2]\tfrac{1}{144}$$

$$p_2 = 50 - 0.79 = 49.2 \text{ psia}$$

The momentum equation now becomes

$$(50 - 14.7)(144)\frac{\pi}{4}\left(\frac{18}{12}\right)^2 - (49.2 - 14.7)(144)\frac{\pi}{4}\left(\frac{12}{12}\right)^2 \cos 60° - X$$

$$= (0.90)(1.94)(10)[12.73 \cos 60° - 5.66]$$

$$X = 7020 \text{ lb}$$

The momentum equation in the y direction is

$$\Sigma F_y = \rho Q(V_{2y} - V_{1y})$$

$$Y - (p_2 - p_a)A_2 \sin 60° = \rho Q(V_2 \sin 60° - 0)$$

$$Y - (49.2 - 14.7)(144)\frac{\pi}{4}\left(\frac{12}{12}\right)^2 \sin 60° = (0.90)(1.94)(10)[12.73 \sin 60°]$$

$$Y = 3570 \text{ lb}$$

The thrust components of the fluid are equal and opposite to X and Y, so the x component is 7020 lb to the right and the y component is 3570 lb downward.

OPEN CHANNEL FLOW RATE

The fundamental equation for the flow rate Q in uniform flow as a function of depth of flow and channel characteristics is the Manning equation. In English units it is

$$Q = \frac{1.49}{n} AR^{2/3}S^{1/2} \tag{6-12}$$

To use metric units, replace 1.49 by 1.0. The roughness coefficient n may vary from 0.01 for smooth uniform channels to 0.03 or higher for irregular natural river channels, and S is the channel bottom slope. The cross-sectional area of flow is A, and R is a shape parameter for the channel section. Called the hydraulic radius, $R = A/P$, where P is the wetted perimeter of the cross section. Using this equation to find Q when the other factors are known is straightforward, but solving for the flow depth y, when Q is known, usually requires trial-and-error computations.

Example 8

The trapezoidal channel shown in Fig. 6-7, with $S = 0.0009$ and $n = 0.025$, carries a discharge $Q = 300$ cfs. Compute the flow depth y and the average velocity V.

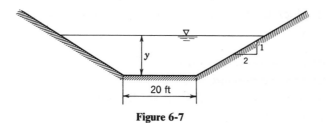

Figure 6-7

Solution. In terms of y, the area A and hydraulic radius R are

$$A = y(20+2y)$$

$$R = \frac{A}{P} = \frac{y(20+2y)}{20+2y\sqrt{5}}$$

The Manning equation then gives

$$300 = \frac{1.49}{0.025}[y(20+2y)] \left|\frac{y(20+2y)}{20+2y\sqrt{5}}\right|^{2/3} (0.0009)^{1/2}$$

or

$$7680 + 1720y = [y(10+y)]^{5/2}$$

This equation must be solved by trial and error.

Trial y	$7680+1720y$	$[y(10+y)]^{5/2}$
3.00	12,840	9,500
3.30	13,356	12,760
3.40	13,528	14,010
3.36	13,460 ~	13,500

The depth of flow is thus $y = 3.36$ ft.

For this depth the area is $A = 3.36[20+2(3.36)] = 89.7$ ft^2 and the average velocity is

$$V = \frac{Q}{A} = \frac{300}{89.7} = 3.34 \text{ ft/sec}$$

PROBLEM 6-1

The anchor block at a bend in a pipeline must be designed primarily to resist forces caused by

(A) friction and acceleration
(B) friction and pressure
(C) pressure and acceleration caused by gravity
(D) static head
(E) pressure and velocity

Solution. The anchor block at a bend in a pipeline must be designed primarily to resist forces caused by pressure and velocity.

Answer is (E)

PROBLEM 6-2

The line showing the pressure head plus the potential head plus the velocity head at any section of a pipe is called the

(A) total head line
(B) energy gradient
(C) energy head line
(D) hydraulic gradient
(E) combined head line

Solution. The line showing the sum of the pressure, potential, and velocity heads is the total head line; it graphically shows the sum of the three terms in the Bernoulli equation.

Answer is (A)

PROBLEM 6-3

Cavitation results from

(A) laminar flow
(B) low soil permeability
(C) turbulent water flow
(D) rough, irregular flow boundaries
(E) excessive pore water pressure in soil

Solution. Cavitation is a high-speed water phenomenon that occurs when the fluid pressure approaches vapor pressure. This can only occur in turbulent flow.

Answer is (C)

PROBLEM 6-4

Capillarity results from

(A) excess pore water pressure
(B) surface tension
(C) low fluid pressures
(D) inadequate compaction
(E) turbulent flow

Solution.   Capillarity results from surface tension.

Answer is (B)

PROBLEM 6-5

The hydraulic jump is utilized for

(A) energy dissipation
(B) pressure regulation
(C) lifting of water
(D) transport of sediment
(E) evaporation rate increase

Solution.   In the hydraulic jump energy is dissipated.

Answer is (A)

PROBLEM 6-6

The primary function of a weir is

(A) wildlife conservation
(B) channel diversion
(C) measurement of discharge
(D) prevention of scour
(E) energy dissipation

Solution.   Weirs are used primarily to measure discharge in channels.

Answer is (C)

PROBLEM 6-7

If the pressure in a plenum is noted as 6 in. of water column, this is most nearly equivalent to

(A) $6 \, lb/ft^2$
(B) $392 \, lb/ft^2$
(C) $9 \, lb/ft^2$
(D) $31 \, lb/ft^2$
(E) $39 \, lb/ft^2$

Solution.

$$p = \gamma z = 62.4 \text{ lb/ft}^3 \times 0.5 \text{ ft} = 31.2 \text{ lb/ft}^2$$

Answer is (D)

PROBLEM 6-8

The value of the coefficient of viscosity of air at 19.2°C is 1.828×10^{-4} poises. The equivalent value, expressed in pounds per foot-second, is closest to

(A) 1.23×10^{-2}
(B) 5.57×10^{-3}
(C) 1.02×10^{-3}
(D) 1.23×10^{-5}
(E) 1.02×10^{-6}

Solution. The poise is equal to 1 g/cm-sec and is named after Poiseuille. Application of conversion factors gives

$$1.828 \times 10^{-4} \frac{\text{g}}{\text{cm-sec}} \times \frac{1 \text{ lb}}{454 \text{ g}} \times \frac{2.54 \times 12 \text{ cm}}{1 \text{ ft}} = 1.227 \times 10^{-5} \frac{\text{lb}}{\text{ft-sec}}$$

Answer is (D)

PROBLEM 6-9

A fluid flows at a constant velocity in a pipe. The fluid completely fills the pipe, and the Reynolds number is such that the flow is just subcritical and laminar. If all other parameters remain unchanged and the viscosity of the fluid is decreased a significant amount, one would generally expect the flow to

(A) not change
(B) become turbulent
(C) become more laminar
(D) increase
(E) temporarily increase

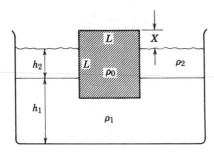

Solution.   Considering the Reynolds number,

$$\mathbf{R} = \frac{VD\rho}{\mu}$$

where V is the average velocity in the pipe, D is the diameter of the pipe, ρ is the density of the fluid flowing, and μ is the viscosity of the fluid flowing, we see that a decrease in μ will increase $\mathbf{R}$ so that it will be greater than the Reynolds number for the transition to turbulent flow, and the flow will become turbulent.

Answer is (B)

PROBLEM 6-10

The transition between laminar and turbulent flow in a pipe usually occurs at a Reynolds number of approximately

(A) 350
(B) 900
(C) 1800
(D) 2100
(E) 3850

Solution.   The lower critical Reynolds number, below which laminar flow always occurs in a pipe, is approximately 2100.

Answer is (D)

PROBLEM 6-11

A cube of wood with sides of length $L = 24$ in. and of density $\rho_0 = 1.55$ slug/ft^3 floats in two fluids, as shown in Fig. 6-8. The heavier fluid is water of density $\rho_1 = 1.94$ slug/ft^3 and depth $h_1 = 36$ in. The lighter fluid is

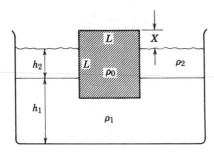

Figure 6-8

oil of density $\rho_2 = 1.65$ slug/ft^3 and forms a layer of thickness $h_2 = 6$ in. Neglecting surface tension, the distance X that the cube projects above the surface of the lighter fluid is nearest to

(A) 3.9 in.
(B) 4.2 in.
(C) 4.5 in.
(D) 4.8 in.
(E) 5.1 in.

Solution. Archimedes' principle tells us that the buoyant force is equal to the weight of the liquid displaced. The weight of the displaced liquid is

$$\text{Volume} \times \text{density} = L^2 h_2 \rho_2 g + L^2 (L - h_2 - X)\rho_1 g$$

for the two liquid layers. The weight of the block ($L^3 \rho_0 g$) is equal to this buoyant force.

$$L^3 \rho_0 g = L^2 g [h_2 \rho_2 + \rho_1 (L - h_2 - X)]$$

$$L \rho_0 = h_2 \rho_2 + \rho_1 L - \rho_1 h_2 - X \rho_1$$

$$X \rho_1 = h_2 \rho_2 + \rho_1 L - \rho_1 h_2 - L \rho_0$$

$$X = h_2 \left(\frac{\rho_2}{\rho_1} - 1 \right) - L \left(\frac{\rho_0}{\rho_1} - 1 \right)$$

$$X = 6 \left(\frac{1.65}{1.94} - 1 \right) - 24 \left(\frac{1.55}{1.94} - 1 \right)$$

$$X = 3.93 \text{ in.}$$

Answer is (A)

PROBLEM 6-12

An "overflow" can brimful of water is suspended from the hook of a spring balance. After a small block of wood is placed in the water, the balance reading is

(A) increased by the weight of the wood block
(B) decreased by the weight of the wood block
(C) decreased by the weight of the water displaced
(D) increased by the weight of the wood above the water surface
(E) unaffected

Solution.   When a floating body is placed in a liquid, it sinks until it displaces its own weight of liquid.

Assuming that the wood floats, the weight of water that will overflow is equal to the weight of the wood block. The balance reading will be unchanged.

Answer is (E)

PROBLEM 6-13

An object weighs 100 lb in air and 25 lb in fresh water.

(1) Its volume in cubic feet is closest to

    (A) 0.75
    (B) 1.1
    (C) 1.2
    (D) 1.3
    (E) 1.5

Solution.

$$\text{Buoyant force} = (\text{weight in air}) - (\text{weight in water})$$

$$= 100 - 25 = 75 \text{ lb}$$

Since the buoyant force is equal to the weight of the water displaced by the object,

$$\text{Volume} = \frac{\text{weight of displaced water}}{\text{unit weight of water}} = \frac{75 \text{ lb}}{62.4 \text{ lb/ft}^2} = 1.2 \text{ ft}^3$$

Answer is (C)

(2) Its specific gravity (relative density) is most nearly

    (A) 0.75
    (B) 1.1
    (C) 1.2
    (D) 1.3
    (E) 1.5

Solution. The buoyant force is also the weight of a volume of water equal to the volume of the object. By definition,

$$\text{Specific gravity } S = \frac{\text{weight in air}}{\text{weight of equal volume of water}} = \frac{100}{75} = 1.3$$

Answer is (D)

PROBLEM 6-14

It is said that Archimedes discovered his principle while seeking to detect a suspected fraud in the construction of a crown. The crown was thought to have been made from an alloy of gold and silver instead of pure gold. The crown had a mass of 1000 g in air and 940 g in pure water. The volume of the alloy equaled the combined volume of the components. If the density of gold = 19.3 g/cm^3 and the density of silver = 10.5 g/cm^3, the mass of gold in the crown is most nearly

(A) 780 g
(B) 810 g
(C) 860 g
(D) 910 g
(E) 940 g

Solution. Archimedes' principle says that an object immersed in a fluid will undergo an apparent loss of weight equal to the weight of the displaced fluid. Since the weight of an object is a nearly constant multiple of its mass on earth, the principle can also be applied directly to masses.

The mass of displaced water is $1000 - 940$ or 60 g. As the density of water is 1 g/cm^3, the volume of displaced water equals the volume of the crown, or 60 cm^3. Let

$$x = \text{mass of gold in the crown}$$
$$1000 - x = \text{mass of silver in the crown}$$

Then the volume of gold plus the volume of silver equals the volume of the crown, or

$$\frac{x}{19.3} + \frac{1000 - x}{10.5} = 60 \qquad 0.0518x + 95.24 - 0.0952x = 60$$

$$0.0434x = 35.24$$

$$\text{Mass of gold} = x = \frac{35.24}{0.0434} = 812 \text{ g}$$

Answer is (B)

PROBLEM 6-15

A uniform solid rod 24 in. long is supported at one end by a string 6 in. above the water. The specific gravity of the rod is $\frac{5}{9}$. Assume the cross section of the rod is small. The length of the immersed section of the rod is most nearly

- (A) 8 in.
- (B) 10 in.
- (C) 12 in.
- (D) 14 in.
- (E) 16 in.

Solution.

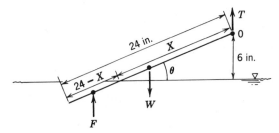

Figure 6-9

There are three forces acting on the rod:

(1) The weight of the rod $W = \gamma V = \frac{5}{9}\gamma_w(24)(A)$, where A = cross-sectional area of the rod.

(2) The buoyant force $F = \gamma_w(24 - X)(A)$.

(3) The tensile force T.

The rod will be in equilibrium if $\sum M_0 = 0$

$$W(12 \cancel{\cos\theta}) - F\left(24 - \frac{24-X}{2}\right)\cancel{\cos\theta} = 0$$

Substituting the values of W and F in this equation, we have

$$12(\tfrac{5}{9}\cancel{\gamma_w})(24\cancel{A}) - \cancel{\gamma_w}(24-X)(\cancel{A})\left(24 - \frac{24-X}{2}\right) = 0$$

$$\frac{12(5)(24)}{9} - (24-X)\left(\frac{24+X}{2}\right) = 0$$

$$160 - \frac{24^2 - X^2}{2} = 0 \qquad X^2 = 256 \qquad X = 16 \text{ in.}$$

Therefore the immersed portion of the rod is 24 in. − 16 in. = 8 in.

Answer is (A)

PROBLEM 6-16

A brass weight of density 8.4 g/cm³ is dropped from the water surface in a tank. The water in the tank is 8.0 m deep. Neglecting water viscosity, the time it takes for the weight to reach the bottom of the tank is closest to

(A) 0.90 sec
(B) 0.96 sec
(C) 1.12 sec
(D) 1.28 sec
(E) 1.36 sec

Solution. Since viscosity directly or indirectly causes all fluid drag, both friction drag (shear) and pressure drag (fore and aft), neglecting viscosity implies neglecting drag forces. Thus the only forces acting on the brass weight are gravity and buoyant forces, both of which are constant. Therefore the acceleration will be constant and the following formula applies:

$$s = \tfrac{1}{2}at^2 \qquad (1)$$

or, solving for the time t to cover the vertical distance s at constant acceleration a,

$$t = \left(\frac{2s}{a}\right)^{1/2} \qquad (2)$$

Using $\sum F = ma$, we have

$$a = \frac{\sum F}{m} = \frac{\gamma_{brass}V - \gamma_{water}V}{\rho_{brass}V} \qquad (3)$$

where γ = specific weight, ρ = mass density, and V = volume of the brass weight. Since $\gamma = \rho g$, Eq. (3) becomes

$$a = \frac{g(\rho_{brass} - \rho_{water})V}{\rho_{brass}V} = g\left(1 - \frac{\rho_{water}}{\rho_{brass}}\right) = 9.81 \text{ m/sec}^2\left(1 - \frac{1}{8.4}\right)$$

$$a = 9.81(0.881) = 8.64 \text{ m/sec}^2$$

Substituting back into Eq. (2)

$$t = \left(\frac{2 \times 8}{8.64}\right)^{1/2} = (1.85)^{1/2} = 1.36 \text{ sec}$$

Answer is (E)

PROBLEM 6-17

A rectangular barge 25 ft wide × 46 ft long × 8 ft deep floats in a canal lock which is 32 ft wide × 60 ft long × 12 ft deep. With no load on the barge other than its own weight, the bottom of the barge is 3 ft beneath the water surface, and the depth of the water in the lock is 7 ft. If a load of steel that weighs 75 tons is added to the barge, the new depth of water in the lock is most nearly

    (A) 7.0 ft
    (B) 7.5 ft
    (C) 8.2 ft
    (D) 9.0 ft
    (E) 10.1 ft

Solution. The barge displacement with no load is $25 \times 46 \times 3 = 3450 \text{ ft}^3$. According to Archimedes, the weight of the steel will equal the weight of the additional water that is displaced. The additional volume of water to be displaced is then $75 \times 2000/62.4 = 2400 \text{ ft}^3$, and the new barge displacement is therefore $3450 + 2400 = 5850 \text{ ft}^3$.

The actual volume of water in the lock is $(32 \times 60 \times 7) - 3450 = 13,440 - 3450 = 13,440 - 3450 = 9990 \text{ ft}^3$. The new *apparent* volume of water in the lock is equal to the actual volume of water plus the displacement of the loaded barge or $9990 + 5850 = 15,840 \text{ ft}^3$, which is equivalent to a new water depth in the lock of $15,840/(32 \times 60) = 8.25 \text{ ft}$.

Answer is (C)

PROBLEMS 6-18

A barge loaded with rocks floats in a canal lock with both the upstream and the downstream gates closed. If the rocks are dumped into the canal lock water with both gates still in the closed position, the water level in the lock will theoretically

(A) rise

(B) rise and then return to original level

(C) fall

(D) fall and then return to original level

(E) remain the same

Solution. With the rocks in the barge, the barge displaces additional water equal to the weight of the rocks. When the rocks are thrown into the canal, they displace their own volume of water. Thus with rocks of specific gravity greater than 1, the water level in the lock will fall.

Answer is (C)

PROBLEM 6-19

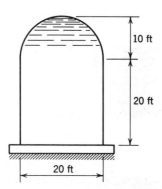

10 ft

20 ft

20 ft

Figure 6-10

A cylindrical water tank with a hemispherical dome has the dimensions shown. The tank is full. The total force exerted by the water on the base of the tank is most nearly

(A) 500,000 lb

(B) 520,000 lb

(C) 550,000 lb

(D) 590,000 lb

(E) 640,000 lb

Solution. The pressure on the bottom of the tank is due to the total head of 30 ft of water. The force is equal to the pressure times the area.

$$p = 30(62.4) = 1872 \text{ lb/ft}^2$$

$$F = pA = 1872\left(\frac{\pi}{4}\right)20^2 = 588,000 \text{ lb}$$

Answer is (D)

PROBLEM 6-20

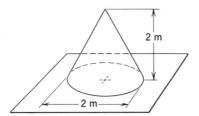

Figure 6-11

A hollow steel cone, with internal dimensions as shown, has a small hole at the apex. The cone is filled with water of weight 9800 N/m³. The minimum weight of the cone W_c that will prevent the water from uplifting the cone and flowing out is closest to

    (A) 61,600 N

    (B) 41,100 N

    (C) 46,200 N

    (D) 20,500 N

    (E) 30,800 N

Solution. The pressure at the base of the cone is $p = \gamma h = (9800 \text{ N/m}^3)(2\text{m}) = 19,600 \text{ N/m}^2$(pascals). The uplift force is $R = pA = (19,600)[(\pi/4)2^2] = 61,600$ N. For vertical equilibrium of the cone and its enclosed water,

$$W_w + W_c = R$$

where W_w = weight of the enclosed water. Here

$$W_w = \gamma_w V_c = \gamma_w \times \tfrac{1}{3}(\text{area of base}) \times h$$
$$= (9800)(\tfrac{1}{3})\pi(2) = 20,500 \text{ N}$$

Hence

$$W_c = 61,600 - 20,500 = 41,100 \text{ N}$$

Answer is (B)

PROBLEM 6-21

In this manometer the unit weight for the first fluid is $\gamma_1 = 56$ lb/ft³, and for the second $\gamma_2 = 99$ lb/ft³. The pressure difference $p_A - p_B$ is most nearly

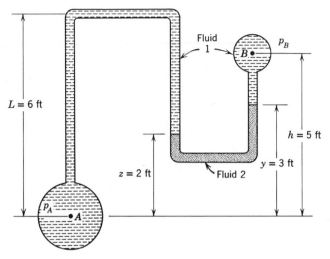

Figure 6-12

(A) $125 \, \text{lb/ft}^2$
(B) $180 \, \text{lb/ft}^2$
(C) $280 \, \text{lb/ft}^2$
(D) $323 \, \text{lb/ft}^2$
(E) $380 \, \text{lb/ft}^2$

Solution.

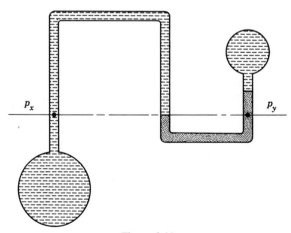

Figure 6-13

$$\Delta p = p_A - p_B$$

$$p_B = p_y - (y - z)\gamma_2 - (h - y)\gamma_1$$

$$p_A = p_x + z\gamma_1 \quad \text{or} \quad p_A - z\gamma_1 = p_x$$

Since $p_x = p_y$,

$$p_A - z\gamma_1 = p_B + (y - z)\gamma_2 + (h - y)\gamma_1$$

$$\Delta p = (h - y + z)\gamma_1 + (y - z)\gamma_2$$

$$\Delta p = (5 - 3 + 2)(56) + (3 - 2)(99)$$

$$\Delta p = 323 \text{ lb/ft}^2$$

Answer is (D)

PROBLEM 6-22

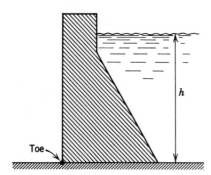

Toe

Figure 6-14

The moment tending to overturn the dam about the toe will increase in proportion to

(A) $h^{1/2}$
(B) h
(C) $h^{3/2}$
(D) h^2
(E) h^3

Solution.

$$L = \text{length of dam}$$

$$\text{Moment} = \frac{\gamma L h^2}{2} \times \frac{h}{3} = \frac{\gamma L h^3}{6}$$

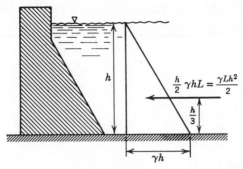

Figure 6-15

The moment will increase in proportion to h^3.

Answer is (E)

PROBLEM 6-23

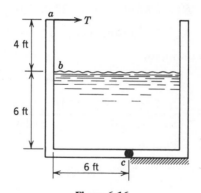

Figure 6-16

The figure shows a cross-sectional view of a 10-ft-long rectangular water tank. The wall of the tank, abc, is hinged at c and supported by a horizontal tie rod at a. The force T in the tie rod is most nearly

- (A) 1000 lb
- (B) 2000 lb
- (C) 7000 lb
- (D) 8000 lb
- (E) 9000 lb

Solution.   A free-body diagram of the hinged wall is shown in Fig. 6-17; it includes the hydrostatic forces R_1 and R_2 and the hinge reactions C_x and C_y.

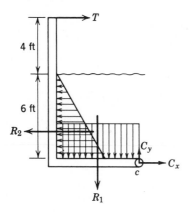

Figure 6-17

The forces R_1 and R_2 are the concentrated forces equivalent to the shaded pressure distributions. By use of Eq. (6-4),

$$R_1 = 10(6)(62.4)(6) = 22,460 \text{ lb}$$

$$R_2 = 10(\tfrac{1}{2})(6)(62.4)(6) = 11,230 \text{ lb}$$

Now apply $\sum M_c = 0$ to obtain

$$2R_2 + 3R_1 - 10T = 0$$

$$T = \frac{2(11,230) + 3(22,460)}{10} = \frac{89,840}{10}$$

$$T = 8984 \text{ lb}$$

Answer is (E)

PROBLEM 6-24

In Fig. 6-18 the gate AB rotates about an axis through B. The gate width is 4 ft. A torque T is applied to the shaft through B. The torque T to keep the gate closed is closest to

(A) 5000 lb-ft
(B) 10,000 lb-ft
(C) 20,000 lb-ft
(D) 30,000 lb-ft
(E) 40,000 lb-ft

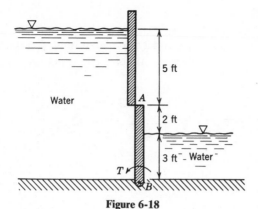

Figure 6-18

Solution. A cross-sectional view of the pressure distribution on gate AB is shown in Fig. 6-19. Also shown are three equivalent concentrated forces

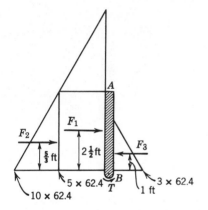

Figure 6-19

acting on the 4-ft-long gate, their locations, and the magnitude of the pressure (lb/ft^2) at three points. The magnitudes of the forces are

$$F_1 = 5(62.4)(5)(4) \quad = 6240 \text{ lb}$$

$$F_2 = \tfrac{1}{2}(5)(62.4)(5)(4) = 3120 \text{ lb}$$

$$F_3 = \tfrac{1}{2}(3)(62.4)(3)(4) = 1123 \text{ lb}$$

Equilibrium around B is ensured by summing moments at B:

$$\sum M_B = 0 = T - 2.5F_1 - \tfrac{5}{3}F_2 + 1F_3$$

$$T = 2.5(6240) + \tfrac{5}{3}(3120) - 1123 = 19{,}677 \text{ lb-ft}$$

Answer is (C)

PROBLEM 6-25

A rectangular gate 1.5 m wide and 2.5 m high is hinged at the top and can be opened at the bottom by pulling on a cable. The water level is 0.6 m over the

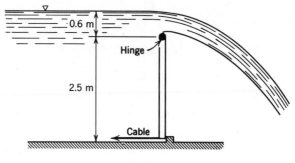

Figure 6-20

top of the gate. Water weighs $9800 \, \text{N/m}^3$. The tension in the cable, in newtons, which is required to open the gate is most nearly

- (A) 8500 N
- (B) 10,500 N
- (C) 35,000 N
- (D) 41,000 N
- (E) 70,000 N

Solution. First construct a free-body diagram of the gate. We use the pressure distribution shown in Fig. 6-21 as a reasonable approximation to

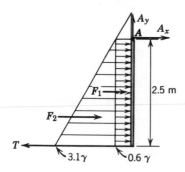

Figure 6-21

the true distribution. Actually the pressure is locally zero at A. Using $\gamma = 9800$ N/m^3 for water, the equivalent concentrated forces are

$$F_1 = (0.6\gamma)(2.5)(1.5) = 2.25\gamma$$
$$F_2 = \tfrac{1}{2}(2.5\gamma)(2.5)(1.5) = 4.69\gamma$$

Now sum moments around point A.

$$\sum M_A = 0 = \tfrac{1}{2}(2.5)F_1 + \tfrac{2}{3}(2.5)F_2 - 2.5T$$
$$0 = \tfrac{1}{2}(2.25\gamma) + \tfrac{2}{3}(4.69\gamma) - T$$
$$T = 41,700 \text{ N}$$

Since the assumed pressure distribution slightly overestimates the true pressure distribution, this force is slightly larger than the true result.

Answer is (D)

PROBLEM 6-26

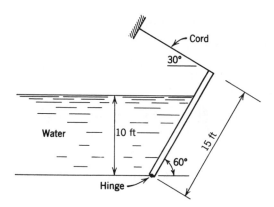

Figure 6-22

The gate is 10 ft wide. The tension in the cord is closest to

    (A) 7000 lb
    (B) 8000 lb
    (C) 9000 lb
    (D) 10,000 lb
    (E) 12,000 lb

Solution. First a free-body diagram of the gate is drawn. The hydrostatic force F acts perpendicularly to the face of the gate at a distance $d/3$ from

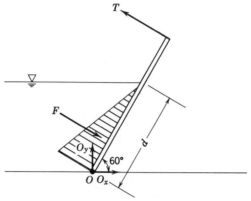

Figure 6-23

point O. From trigonometry $\sin 60° = 10/d$, and $d = 11.55$ ft. The centroid of the submerged portion of the gate is 5 ft below the water surface. Using Eq. (6-4),

$$F = \gamma h_C A = \gamma(5)(10d)$$

Summing moments around O for equilibrium,

$$\Sigma M_O = 0 = 15T - \frac{d}{3}F$$

$$T = \frac{dF}{3(15)} = \frac{50\gamma d^2}{45} = \frac{50}{45}(62.4)(11.55)^2$$

$$T = 9250 \text{ lb}$$

Answer is (C)

PROBLEM 6-27

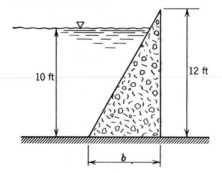

Figure 6-24

Masonry weighs 150 lb/ft³ and the coefficient of friction between the bottom of the dam and the stream bed is 0.4. The required value of b to prevent overturning of the masonry dam is most nearly

(A) 4.6 ft
(B) 4.7 ft
(C) 5.1 ft
(D) 5.5 ft
(E) 5.9 ft

Solution. The problem can be readily solved by first calculating the horizontal and vertical components of the hydrostatic force.

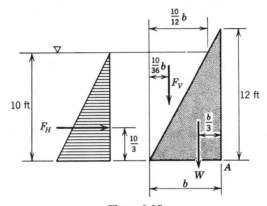

Figure 6-25

$$F_H = \frac{\gamma h}{2} h = \frac{(62.4)(10)}{2}(10) = 3120 \text{ lb}$$

$$F_V = \frac{\gamma h}{2} h = \frac{(62.4)(10)}{2}(\tfrac{10}{12}b) = 260b \text{ lb}$$

$$\sum M_A = 0$$

$$-\tfrac{10}{3}F_H + \tfrac{26}{36}bF_V + \frac{b}{3}W = 0$$

$$-\tfrac{10}{3}(3120) + \tfrac{26}{36}b(260b) + \frac{b}{3}(900b) = 0$$

$$-10{,}400 + 188b^2 + 300b^2 = 0$$

$$b = \left(\frac{10{,}400}{488}\right)^{1/2} = (21.3)^{1/2} = 4.62 \text{ ft}$$

Although a friction factor for sliding is given, the problem asks only for the base width necessary to prevent overturning; hence a check on sliding is not needed.

Answer is (A)

PROBLEM 6-28

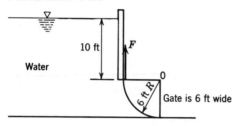

Figure 6-26

(1) The horizontal component of the forces acting on the radial gate, and its line of action above the base, are most nearly

    (A) 6800 lb and 2.00 ft
    (B) 6800 lb and 3.00 ft
    (C) 22,500 lb and 3.00 ft
    (D) 29,230 lb and 2.77 ft
    (E) 29,230 lb and 3.00 ft

Solution. First construct a free-body diagram of the gate and some of the water outside the gate:

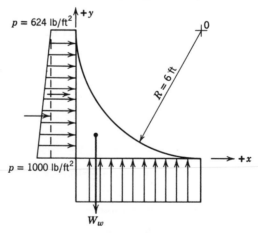

Figure 6-27

At the top of the gate the pressure is $p = 10\gamma = 624 \text{ lb/ft}^2$. At the base of the gate the pressure is $p = 16\gamma = 1000 \text{ lb/ft}^2$.

The horizontal force is

$$F_h = 624(6)(6) + \tfrac{376}{2}(6)(6) = 22{,}460 + 6770 = 29{,}230 \text{ lb}$$

Its line of action is

$$\bar{y} = \frac{\Sigma\, yF}{\Sigma\, F} = \frac{3(22{,}460) + 2(6770)}{29{,}230} = 2.77 \text{ ft above base}$$

Answer is (D)

(2) The vertical component of the forces acting on the gate, and its line of action from point 0, are most nearly

(A) 30,200 lb and 2.55 ft
(B) 33,100 lb and 2.85 ft
(C) 36,000 lb and 3.00 ft
(D) 33,100 lb and 3.15 ft
(E) 30,200 lb and 3.45 ft

Solution.   The net vertical force is

$$F_v = pA - W_w$$

$$F_v = 1000(36) - (62.4)\left(36 - \frac{36\pi}{4}\right)(6) = 36{,}000 - 2900 = 33{,}100 \text{ lb}$$

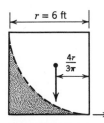

Figure 6-28

The centroid of the water alone, measured from the left edge, is

$$\bar{x} = \frac{\Sigma\, Ax}{A} = \frac{6(6)(3) - \dfrac{\pi}{4}(6)^2\left[6 - \dfrac{4(6)}{3\pi}\right]}{6(6) - \dfrac{\pi}{4}(6)^2}$$

$$\bar{x} = \frac{108 - 97.6}{7.73} = \frac{10.4}{7.73} = 1.35 \text{ ft}$$

The line of action for F_v is therefore

$$\bar{x} = \frac{36{,}000(3) - 2900(1.35)}{33{,}100} = \frac{108{,}000 - 3900}{33{,}100}$$

$$\bar{x} = \frac{104{,}100}{33{,}100} = 3.15 \text{ ft}$$

The distance from point 0 is $6.0 - 3.15 = 2.85$ ft.

Answer is (B)

(3) Neglecting the gate's weight, the force F required to open the gate is most nearly

(A) 18,000 lb
(B) 13,600 lb
(C) 10,000 lb
(D) 4500 lb
(E) 0 lb

Solution. The force to open the gate is found by using $\sum M_0 = 0$

$$6F = -33{,}100(2.85) + 29{,}230(3.23)$$

$$= -94{,}400 + 94{,}400$$

Therefore $F = 0$. This is to be expected since the pressure forces acting directly on the cylinder have lines of action that pass through the center of rotation of the gate.

Answer is (E)

PROBLEM 6-29

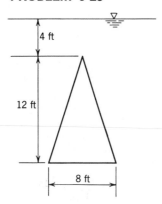

Figure 6-29

Hydrostatic pressure acts on the triangular gate shown in Fig. 6-29. Use the water surface as the reference plane in locating the resultant force.

(1) The resultant force on the gate is closest to

    (A) 24,000 lb
    (B) 30,000 lb
    (C) 32,000 lb
    (D) 36,000 lb
    (E) 48,000 lb

Solution.

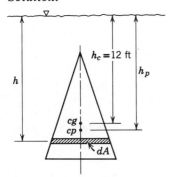

Figure 6-30

Integrate the pressure distribution over the gate to find the force. The force on the differential area dA is

$$dF = \gamma h \, dA$$

$$F = \int dF = \gamma \int h \, dA$$

But $\int h \, dA$ is the moment of the area or $h_c A$; hence

$$F = \gamma h_c A = 62.4(12)(\tfrac{8}{2})(12) = 36,000 \text{ lb}$$

Answer is (D)

(2) The distance from the water surface to the point of application of the resultant is most nearly

    (A) 7.3 ft
    (B) 10.0 ft
    (C) 12.0 ft
    (D) 12.6 ft
    (E) 13.0 ft

Solution. The point of application of the force is located on the vertical centerline of the gate since it is symmetrical. With the water surface as the reference plane, h_p is equal to the moment of the force divided by the force: $h_p = M/F$. $dM = h\,dF$ with $dF = \gamma h\,dA$, so $dM = \gamma h^2\,dA$ and $M = \gamma \int h^2\,dA$. Since $\int h^2\,dA$ is the moment of inertia of the area with respect to the water surface, $M = \gamma I_{ws}$.

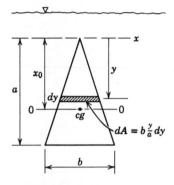

Figure 6-31

We must now determine the moment of inertia of the triangle. Using the transfer equation $I_{ws} = I_0 + Ah_c^2$, we find

$$I_x = \int y^2\,dA = \frac{b}{a}\int_0^a y^3\,dy = \frac{b}{a}\frac{y^4}{4}\Big]_0^a = \frac{ba^4}{4a} = \frac{ba^3}{4}$$

$$I_0 = I_x - Ax_0^2$$

$$I_0 = \frac{8(12)^3}{4} - \frac{8}{2}(12)(8)^2 = 3456 - 3072 = 384\ \text{ft}^4$$

$$I_{ws} = 384 + \tfrac{8}{2}(12)(12)^2$$

$$= 384 + 6912 = 7296\ \text{ft}^4$$

$$h_p = \frac{M}{F} = \frac{\gamma I_{ws}}{F}$$

$$= \frac{(62.4)(7296)}{36,000} = 12.65\ \text{ft}$$

Answer is (D)

PROBLEM 6-30

A vertical water softener is to operate under the following conditions:

    1. Water flow of 307 gpm.
    2. Maximum flow rate of 8.0 gpm/ft^2 of area.
    3. Supply water with a hardness of 12.0 grains/gal.
    4. Softener contains 95.0 ft^3 of exchange resin.
    5. Exchange value of resin is 24,000 grains/ft^3.

(1) The required diameter of the softener is nearest to

    (A) 8 ft
    (B) 7 ft
    (C) 6 ft
    (D) 5 ft
    (E) 4 ft

Solution.

$$Q = AV$$

$$307 \text{ gpm} = \frac{\pi}{4} D^2 \times 8 \text{ gpm/ft}^2 \qquad D = \left[\left(\frac{307}{8} \right) \left(\frac{4}{\pi} \right) \right]^{1/2} = \sqrt{48.9} = 7 \text{ ft}$$

Answer is (B)

(2) The number of gallons of water softened between regenerations is nearest to

    (A) 70,000
    (B) 95,000
    (C) 140,000
    (D) 165,000
    (E) 190,000

Solution.

$$\text{Capacity between regenerations} = \frac{\text{total exchange capacity}}{\text{supply water hardness}}$$

$$= \frac{24,000 \text{ grains/ft}^3 \times 95 \text{ ft}^3}{12 \text{ grains/gal}}$$

$$= 190,000 \text{ gal}$$

Answer is (E)

PROBLEM 6-31

The theoretical head required to push water through a 3-cm-round orifice at 15 m/sec is closest to

(A) 9.8 m
(B) 11.5 m
(C) 12.7 m
(D) 13.4 m
(E) 22.9 m

Solution.

$$h = \frac{v^2}{2g} = \frac{(15)^2}{2(9.81)} = 11.47 \text{ m}$$

Answer is (B)

PROBLEM 6-32

Water is flowing through a pipe. The following data are known:

$$D = 2 \text{ in.} \qquad h_f = 20 \text{ ft}$$
$$p = 70 \text{ lb/in.}^2 \qquad R = 4 \times 10^5$$
$$n = 0.015 \qquad V = 25 \text{ ft/sec}$$

The rate of flow in U.S. gallons per minute is closest to

(A) 200
(B) 245
(C) 310
(D) 770
(E) 980

Solution.

$$Q = VA = V\left(\frac{\pi}{4}\right) D^2 = (25 \text{ ft/sec})\left(\frac{\pi}{4}\right)\left(\frac{2}{12} \text{ ft}\right)^2 = 0.545 \text{ ft}^3/\text{sec}$$

$$= 0.545 \times 60 \text{ sec/min} \times 7.48 \text{ gal/ft}^3 = 245 \text{ gpm}$$

Answer is (B)

PROBLEM 6-33

Water flows through two orifices in the side of a large water tank. The water surface in the tank is constant. The upper orifice is 16 ft above the ground surface, and this stream strikes the ground 8 ft from the base of the tank. The stream from the lower orifice strikes the ground 10 ft from the base of the tank. For convenience, assume $g = 32$ ft/sec².

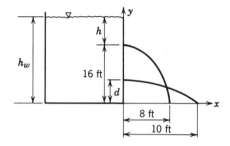

Figure 6-32

(1) The height h_w of the water surface above the ground is

(A) 20 ft
(B) 19 ft
(C) 18 ft
(D) 17 ft
(E) 16 ft

Solution.

For the upper stream, $X = vt$ and $Y = \frac{1}{2}gt^2$ on the trajectory of the stream. Eliminating t, $X^2 = (2v^2/g)Y$, where v is the velocity at the vena contracta of the stream.

$$v^2 = \frac{X^2 g}{2Y} = \frac{(8^2)(32)}{(2)(16)} = 64 \qquad v = 8 \text{ ft/sec}$$

Neglecting friction, the Bernoulli equation may be written as

$$\frac{V_1^2}{2g} + \frac{p_1}{\gamma} + h = \frac{V_2^2}{2g} + \frac{p_2}{\gamma}$$

where subscript 2 refers to the vena contracta and subscript 1 refers to the surface of the water in the tank.

$$V_2 = v = \sqrt{2g}\left(h + \frac{p_1 - p_2}{\gamma} + \frac{V_1^2}{2g}\right)^{1/2}$$

where $p_1 = p_2 =$ atmospheric pressure and $V_1 = 0$. Therefore

$$V_2 = v = \sqrt{2gh}$$

Using the velocity we calculated above for the vena contracta, we find h is

$$8 = \sqrt{2(32)h} \qquad h = 1 \text{ ft}$$

$$h_w = 16 + h = 17 \text{ ft}$$

Answer is (D)

(2) If the two orifices are known to be more than 2 ft apart, the height d of the lower orifice above the ground is closest to

(A) 1.6 ft
(B) 3.8 ft
(C) 8.5 ft
(D) 13.2 ft
(E) 15.4 ft

Solution. The velocity at the vena contracta of the lower orifice will be

$$v = \sqrt{2(32)(17 - d)}$$

Using the equation $X^2 = (2v^2/g)Y$, where now $Y = d$,

$$10^2 = \frac{2[2(32)(17 - d)]}{32}(d) \qquad \text{or} \qquad 100 = 4(17 - d)(d)$$

Solving for d,

$$d^2 - 17d + 25 = 0$$

$$d = \frac{17 \pm \sqrt{17^2 - 4(25)}}{2} = \frac{17 \pm \sqrt{189}}{2} = \frac{17 \pm 13.8}{2}$$

$$d = 15.4 \text{ ft} \qquad \text{or} \qquad d = 1.6 \text{ ft}$$

The first answer must be rejected because it is too near to the upper orifice.

Answer is (A)

PROBLEM 6-34

In a horizontal Venturi meter water flows through a 4 in.-diameter constriction in an 8-in.-diameter pipe. Simple piezometer columns indicate a 10-ft difference in head h between the upstream and constricted sections. Neglecting energy losses, the discharge through the pipe is closest to

    (A) 2.3 ft³/sec
    (B) 3.7 ft³/sec
    (C) 5.1 ft³/sec
    (D) 7.5 ft³/sec
    (E) 9.1 ft³/sec

Solution.

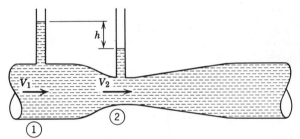

Figure 6-33

The continuity equation is

$$Q = A_1 V_1 = A_2 V_2 \tag{1}$$

The Bernoulli equation is

$$\frac{p_1}{\gamma} + \frac{V_1^2}{2g} + z_1 = \frac{p_2}{\gamma} + \frac{V_2^2}{2g} + z_2 \tag{2}$$

In our case, $z_1 - z_2 = 0$ and $\dfrac{p_1}{\gamma} - \dfrac{p_2}{\gamma} = h$, so the equation reduces to

$$\frac{V_1^2}{2g} + h = \frac{V_2^2}{2g} \tag{3}$$

Squaring Eq. (1),

$$Q^2 = A_1^2 V_1^2 = A_2^2 V_2^2 \qquad V_1^2 = \frac{Q^2}{A_1^2} \qquad V_2^2 = \frac{Q^2}{A_2^2}$$

Substituting these values into Eq. (3),

$$\frac{Q^2}{A_1^2 2g} + h = \frac{Q^2}{A_2^2 2g}$$

Factoring,

$$Q^2\left(\frac{1}{A_1^2} - \frac{1}{A_2^2}\right) = -2gh$$

$$Q^2\left[\left(\frac{A_1}{A_2}\right)^2 - 1\right] = 2ghA_1^2 \qquad Q = A_1\left[\frac{2gh}{(A_1/A_2)^2 - 1}\right]^{1/2}$$

Finally, recalling that the area ratio varies as the square of the diameters, we compute

$$Q = \frac{\pi}{4}\left(\frac{8}{12}\right)^2\left[\frac{2(32.2)(10)}{(8/4)^4 - 1}\right]^{1/2} = 2.29 \text{ ft}^3/\text{sec}$$

Answer is (A)

PROBLEM 6-35

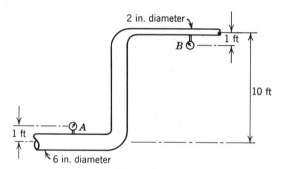

Figure 6-34

Gage A indicates a pressure of 25 psi, and Gage B indicates a pressure of 15 psi. Assume that losses are negligible in the transition from the 6-in. pipe to the 2-in. pipe. The rate of flow in gallons per minute for this system is closest to

(A) 1000
(B) 650
(C) 400
(D) 310
(E) 290

Solution. The Bernoulli equation is

$$\frac{p_1}{\gamma}+\frac{V_1^2}{2g}+z_1=\frac{p_2}{\gamma}+\frac{V_2^2}{2g}+z_2 \tag{1}$$

Also, 1 ft of water $= 0.433$ psi. Each of the terms can be evaluated with reference to a datum plane passing through the 6-in.-diameter pipe.

$$p_6 = 25 \text{ psig} + 1 \text{ ft water} = 25 + 0.433 = 25.4 \text{ psig} \qquad z_6 = 0$$

$$p_2 = 15 \text{ psig} - 1 \text{ ft water} = 15 - 0.433 = 14.6 \text{ psig} \qquad z_2 = 10 \text{ ft}$$

$$\frac{25.4}{0.433}+\frac{V_6^2}{(2)(32.2)}+0=\frac{14.6}{0.433}+\frac{V_2^2}{(2)(32.2)}+10 \tag{2}$$

$$58.7 + 0.0155\,V_6^2 = 33.7 + 0.0155\,V_2^2 + 10$$

$$V_2^2 - V_6^2 = \frac{58.7 - 43.7}{0.0155} = 968 \text{ ft}^2/\text{sec}^2 \tag{3}$$

The continuity equation shows

$$Q = A_6 V_6 = A_2 V_2 \tag{4}$$

Using

$$A_2 = \frac{\pi}{4}\left(\frac{2}{12}\right)^2 = \frac{\pi}{144} = 0.0218 \text{ ft}^2$$

and

$$A_6 = \frac{\pi}{4}\left(\frac{6}{12}\right)^2 = \frac{9\pi}{144} = 0.1963 \text{ ft}^2$$

$$V_6 = \frac{A_2}{A_6}V_2 = \frac{0.0218}{0.1963}V_2 = 0.111 V_2 \qquad V_6^2 = 0.0123 V_2^2$$

Substituting this relation into Eq. (3),

$$V_2^2 - 0.0123\,V_2^2 = 968 \text{ ft}^2/\text{sec}^2$$

$$V_2 = \left(\frac{968}{0.9877}\right)^{1/2} = (980)^{1/2} = 31.31 \text{ ft/sec}$$

Now using Eq. (4),

$$Q = A_2 V_2 = (0.0218)(31.31) = 0.68 \text{ ft}^3/\text{sec}$$

But the problem asks for the rate of flow in gallons per minute. Here

$$Q = 0.68 \text{ cfs} \times 7.48 \text{ gal/ft}^3 \times 60 \text{ sec/min} = 305 \text{ gpm}$$

Answer is (D)

PROBLEM 6-36

In Fig. 6-35 the pipe is of uniform diameter. The gage pressure at A is 20 psi and at B is 30 psi. If the liquid has a specific weight (weight density) of 30 lb/ft^3, the head loss and direction of flow are

(A) 18 ft, from A to B
(B) 29 ft, from A to B
(C) 18 ft, from B to A
(D) 29 ft, from B to A
(E) 7 ft, from A to B

Figure 6-35

Solution. Assume the direction of flow is from B to A. The Bernoulli equation is

$$\frac{p_B}{\gamma}+\frac{V_B^2}{2g}+z_B=\frac{p_A}{\gamma}+\frac{V_A^2}{2g}+z_A+h_L$$

Since the diameter is uniform, the continuity equation $Q = AV$ tells us that $V_B = V_A$. Therefore the Bernoulli equation reduces to

$$\frac{p_B}{\gamma}+z_B=\frac{p_A}{\gamma}+z_A+h_L$$

$$\frac{30(144)}{30}+0=\frac{20(144)}{30}+30\text{ ft}+h_L$$

$$144=96+30+h_L\qquad \text{head loss } h_L=18\text{ ft}$$

Since h_L is positive, the assumed direction of flow (from B to A) is correct.

Answer is (C)

PROBLEM 6-37

A 15-cm I.D. pipe discharges water at an elevation 10 m below the surface of a reservoir. If the total head loss to the point of discharge is 5 m, the discharge in cubic meters per second is most nearly

   (A) 0.18
   (B) 0.22
   (C) 0.25
   (D) 0.32
   (E) 0.70

Solution.

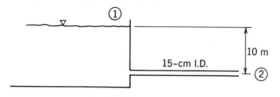

Figure 6-36

The Bernoulli equation is

$$\frac{p_1}{\gamma}+\frac{V_1^2}{2g}+z_1=\frac{p_2}{\gamma}+\frac{V_2^2}{2g}+z_2+h_L$$

p_1 and V_1 are zero at the free surface of the water. p_2 is zero since the water discharges at atmospheric pressure. Using point 2 as the reference elevation, $z_2 = 0$. Thus the Bernoulli equation reduces to

$$z_1=\frac{V_2^2}{2g}+h_L$$

$$10=\frac{V_2^2}{2g}+5 \qquad V^2=10g$$

$$V=[10(9.81)]^{1/2}=9.90 \text{ m/sec}$$

$$Q=AV=\frac{\pi}{4}(0.15)^2(9.90)=0.175 \text{ m}^3/\text{sec}$$

Answer is (A)

PROBLEM 6-38

Friction head in a pipe carrying water varies

   (A) inversely with gravity squared
   (B) directly with diameter

(C) inversely with diameter

(D) directly with velocity

(E) inversely with the coefficient of friction f

Solution. The formula for friction losses in a pipe is

$$h_L = f \frac{L}{D} \frac{V^2}{2g}$$

From the formula we see that h_L varies inversely with diameter.

Answer is (C)

PROBLEM 6-39

Figure 6-37 shows the plan view of a horizontal branching and converging pipeline. The main and its branches are circular pipes in which the flow is laminar. The main delivers fluid at a rate Q_0 to the branch lines. These lines have diameters and effective lengths of $d_1, d_2, L_1,$ and L_2, respectively. The only head loss is due to pipe friction. The friction factor for laminar flow in a circular pipe is inversely proportional to the Reynolds number. For the conditions $Q_0 = 9$, $d_2 = 2d_1$, and $L_2 = 2L_1$, the flow rate Q_1 is nearest to

(A) 1

(B) 2

(C) 3

(D) 4

(E) 5

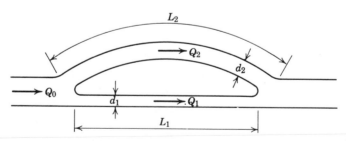

Figure 6-37

Solution. Head loss varies directly with velocity head and pipe length and inversely with pipe diameter. With a coefficient of proportionality f (the friction factor) the equation becomes

$$h_L = f \frac{L}{D} \frac{V^2}{2g}$$

The problem indicates that $f \propto 1/\text{Reynolds number}$. Since the Reynolds number for a given fluid is directly proportional to Vd,

$$h_L \propto \frac{1}{Vd} \frac{L}{d} \frac{V^2}{2g} = \frac{LV}{2d^2 g} \propto \frac{LV}{d^2}$$

The head loss in each branch must be identical, so

$$\frac{L_1 V_1}{d_1^2} = \frac{L_2 V_2}{d_2^2}$$

Substituting $d_2 = 2d_1$ and $L_2 = 2L_1$

$$\frac{L_1 V_1}{d_1^2} = \frac{2L_1 V_2}{(2d_1)^2} \qquad \frac{V_1}{V_2} = \frac{2L_1 d_1^2}{4L_1 d_1^2} = \frac{1}{2}$$

$$Q = AV \qquad Q_0 = Q_1 + Q_2$$

Therefore

$$Q_0 = Q_1 + Q_2 = \frac{\pi}{4} d_1^2 V_1 + \frac{\pi}{4}(2d_1)^2(2V_1) = 9 \qquad Q_0 = \frac{\pi}{4} d_1^2 V_1 + \frac{8\pi}{4} d_1^2 V_1 = 9$$

$$Q_1 + 8Q_1 = 9 \qquad Q_1 = 1$$

Answer is (A)

PROBLEM 6-40

A 10-in. pipeline which carries a flow of 5 ft^3/sec branches into a 6-in. line 500 ft long and an 8-in. line 1000 ft long. The 6-in. and 8-in. pipes rejoin and continue as a 10-in. line. The Darcy friction factor is 0.022 for both pipes. The flow rate in cubic feet per second in the 8-in. line is closest to

(A) 1.5
(B) 2.0
(C) 2.5
(D) 3.0
(E) 3.5

Solution. Where a pipeline branches and then rejoins, the head loss in each branch must be equal.

$$h_6 = \left[f \frac{L}{d} \frac{V^2}{2g} \right]_6 \qquad h_8 = \left[f \frac{L}{d} \frac{V^2}{2g} \right]_8$$

Since f and $2g$ are constants and $h_6 = h_8$,

$$\left[\frac{L}{d} V^2 \right]_6 = \left[\frac{L}{d} V^2 \right]_8 \qquad \text{or} \qquad \frac{500}{\frac{1}{2}} V_6^2 = \frac{1000}{\frac{2}{3}} V_8^2$$

$$V_6^2 = \frac{(1000)(\frac{1}{2})}{(500)(\frac{2}{3})} V_8^2 = 1.5 V_8^2 \qquad V_6 = (1.5 V_8^2)^{1/2} = 1.22 V_8$$

Continuity states that $Q_{10} = Q_6 + Q_8 = 5 \text{ ft}^3/\text{sec}$ and also $Q = AV$.

$$5 = \frac{\pi}{4} (\tfrac{1}{2})^2 V_6 + \frac{\pi}{4} (\tfrac{2}{3})^2 V_8 \qquad 5 = 0.196 V_6 + 0.349 V_8$$

Using $V_6 = 1.22 V_8$,

$$5 = 0.196(1.22 V_8) + 0.349 V_8 = 0.588 V_8$$

$$V_8 = \frac{5}{0.588} = 8.50 \text{ ft/sec}$$

$$Q_8 = A_8 V_8 = (0.349)(8.50) = 2.97 \text{ ft}^3/\text{sec}$$

Answer is (D)

PROBLEM 6-41

A pump requires 100 hp to pump water (with a specific gravity of 1.0) at a certain capacity to a given elevation. What horsepower is required if the capacity and elevation conditions are the same but the fluid pumped has a specific gravity of 0.8?

    (A) 60 hp
    (B) 80 hp
    (C) 100 hp
    (D) 125 hp
    (E) 130 hp

Solution.

$$\text{Power} = \text{work per unit time} = \frac{\text{weight} \times \text{height}}{\text{time}}$$

If the specific gravity of the fluid is reduced from 1.0 to 0.8, then the weight is reduced by this ratio. Consequently, the horsepower will be reduced similarly to 80 hp.

$$\text{Answer is (B)}$$

PROBLEM 6-42

The head loss through a 24-in. butterfly valve is approximately 0.3 of a velocity head. If water is being pumped through the valve at a rate of 30 ft^3/sec for 6 months out of the year, what is the annual energy charge for pumping through the valve if the cost of energy is $0.07 per kilowatt hour?

(A) $260
(B) $280
(C) $330
(D) $380
(E) $420

Solution.

$$Q = AV \qquad V = \frac{Q}{A} = \frac{30}{(\pi/4)2^2} = 9.55 \text{ ft/sec}$$

$$h_L = 0.3 \frac{V^2}{2g} = 0.3 \frac{9.55^2}{(2)(32.2)}$$

$$= 0.425 \text{ ft}$$

$$\text{Lost hp} = \frac{Q\gamma h_L}{550} = \frac{(30)(62.4)(0.425)}{550} = 1.45 \text{ hp}$$

Since 1 hp = 0.746 kw, the cost per year is

$$(1.45)(0.746)(\tfrac{365}{2})(24)(0.07) = \$330$$

$$\text{Answer is (C)}$$

PROBLEM 6-43

Water is supplied by pumping for a certain industrial application. Owing to fluctuations in the water demand, two identical pumps are connected in parallel. A plot of their performance is given in Fig. 6-38.

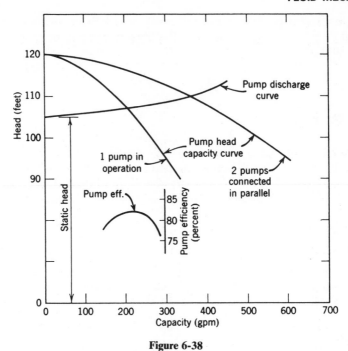

Figure 6-38

(1) The maximum combined pump capacity in gpm when both pumps are working is closest to

(A) 180
(B) 210
(C) 285
(D) 360
(E) 600

Solution.   The maximum pump capacity is given by the intersection of the pump head and pump discharge curves:

One pump: Range of capacity 0–210 gpm
Two pumps: Range of capacity 0–360 gpm

Answer is (D)

(2) The maximum horsepower required when both pumps are working is

closest to

(A) 6.2
(B) 6.9
(C) 12.3
(D) 15.4
(E) 18.2

Solution.

$$\text{Work rate} = \frac{110 \text{ ft} \times 360 \text{ gpm} \times 8.33 \text{ lb/gal}}{0.81 \text{ eff} \times 33{,}000 \text{ ft-lb/min}}$$

$$= 12.34 \text{ hp total for both pumps}$$

Answer is (C)

PROBLEM 6-44

Water falling from a height of 120 ft at the rate of 1000 ft³/min drives a turbine that is directly connected to an electric generator. The generator rotates at 120 rpm. If the total resisting torque due to friction and other losses is 250 lb-ft and the water leaves the turbine blades with a velocity of 15 ft/sec, the horsepower developed by the generator is nearest to

(A) 207
(B) 214
(C) 221
(D) 228
(E) 235

Solution. The exit velocity head is $V^2/2g = 15^2/(2)(32.2) = 3.5$ ft, so the net head is $120 - 3.5 = 116.5$ ft.

$$\text{Power} = Q\gamma h = (1000)(62.4)(116.5) = (7.27)10^6 \text{ ft-lb/min}$$

From this we must deduct the friction and other losses to obtain the generator output.

$$\text{Lost power} = 2\pi NT = (2\pi)(120)(250) = 188{,}500 \text{ ft-lb/min}$$

$$\text{Generator horsepower} = \frac{7{,}270{,}000 - 188{,}500}{33{,}000} = 214.6 \text{ hp}$$

Answer is (B)

PROBLEM 6-45

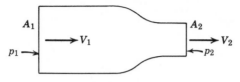

Figure 6-39

The nozzle shown has a large cross-sectional area $A_1 = 4\,\text{ft}^2$ acted on by pressure $p_1 = 23\,\text{lb/in.}^2$ and a small area $A_2 = 1\,\text{ft}^2$ acted on by $p_2 = 10\,\text{lb/in.}^2$ With water flowing into the nozzle at velocity $V_1 = 8\,\text{ft/sec}$ and out at $V_2 = 32\,\text{ft/sec}$, the external force required to hold the nozzle stationary is nearest to

    (A) 10,300 lb
    (B) 11,000 lb
    (C) 11,800 lb
    (D) 12,500 lb
    (E) 13,200 lb

Solution. The force can be found directly by applying the momentum principle, Eq. (6-11), in the direction of flow. Upon recognizing that F_1 and

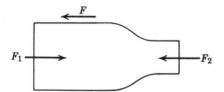

Figure 6-40

F_2 are simply the product of the appropriate pressures and areas, and that the discharge Q can be written as $A_1 V_1$, the result from Eq. (6-11) is

$$p_1 A_1 - p_2 A_2 - F = \rho A_1 V_1 (V_2 - V_1)$$

or

$$F = p_1 A_1 - p_2 A_2 - \rho A_1 V_1 (V_2 - V_1)$$

$$F = 23(144)(4) - 10(144)(1) - 1.94(4)(8)(32 - 8)$$

$$F = 10{,}300\,\text{lb}$$

Answer is (A)

PROBLEM 6-46

An open channel moves a given discharge most efficiently when the water is flowing

    (A) at critical depth
    (B) over a sharp crested weir
    (C) so that the Reynolds number is 4200
    (D) so that the depth equals one-half the width
    (E) at optimum energy gradient

Solution. An open channel moves a given discharge most efficiently when the water is flowing at critical depth.

$$\text{Answer is (A)}$$

PROBLEM 6-47

A 6-ft-diameter pipe has a depth of flow of 5.6 ft. The hydraulic radius of the pipe for this depth of flow is nearest to

    (A) 1.40 ft
    (B) 1.50 ft
    (C) 1.75 ft
    (D) 2.80 ft
    (E) 3.00 ft

Solution. By definition,

$$\text{Hydraulic radius} = \frac{\text{flow area}}{\text{wetted perimeter}} = \frac{A}{P}$$

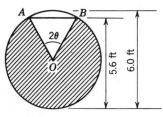

Figure 6-41

From Fig. 6-41 we can see that determination of A and P both depend on θ. From the figure

$$\theta = \cos^{-1}\left(\frac{2.6}{3.0}\right) = 30°$$

and therefore $BC = 3.0 \cos \theta = 1.5$ ft

$$\text{Area of } \triangle OAB = 2(2.6)(1.5)\tfrac{1}{2} = 3.9 \text{ ft}^2$$

$$\text{Shaded area} = \pi R^2\left(\tfrac{300}{360}\right) = \pi(3)^2\left(\tfrac{5}{6}\right)$$

$$= 23.6 \text{ ft}^2$$

$$A = \text{Area of wetted section} = 3.9 + 23.6 = 27.5 \text{ ft}^2$$

$$P = \text{Wetted perimeter} = \pi D\left(\tfrac{300}{360}\right) = \pi(6.0)\tfrac{5}{6}$$

$$= 15.71 \text{ ft}$$

$$\text{Hydraulic radius } R = \frac{A}{P} = \frac{27.5}{15.71} = 1.75 \text{ ft}$$

Answer is (C)

PROBLEM 6-48

A rectangular concrete-lined channel ($n = 0.013$) 30 ft wide flows 4 ft deep on a slope of 0.0009. The discharge is nearest to

(A) 870 ft³/sec
(B) 890 ft³/sec
(C) 1040 ft³/sec
(D) 1300 ft³/sec
(E) 2800 ft³/sec

Solution. The Manning equation for open channel flow is

$$Q = \frac{1.49}{n} AR^{2/3}S^{1/2}$$

Here $n = 0.013$, $S = 0.0009$, and $A = (30)(4) = 120$ ft². The hydraulic radius $R = A/P$, and since the wetted perimeter $P = 30 + 2(4) = 38$ ft, $R =$

$120/38 = 3.16$ ft. Hence

$$Q = \frac{1.49}{0.013}(120)(3.16)^{2/3}(0.0009)^{1/2}$$

$$Q = \frac{1.49}{0.013}(120)(2.15)(0.03)$$

$$Q = 887 \text{ ft}^3/\text{sec}$$

Answer is (B)

PROBLEM 6-49

The tendency of a free liquid surface to contract is called

(A) elasticity
(B) adhesion
(C) cohesion
(D) capillarity
(E) surface tension

Solution. This phenomenon is called surface tension.

Answer is (E)

7 Thermodynamics

Thermodynamics studies energy; it normally concentrates primarily on thermal and mechanical energy transfer. Engineering thermodynamics applies this knowledge to the analysis and design of a myriad of engineering devices, including engines of all kinds. Being both broadly based and practically useful, thermodynamics is founded on a small number of fundamental laws. We shall first review some of these basic principles and then look into the application of thermodynamics to some engineering problems of general interest.

We begin by reviewing the concepts of absolute temperature and absolute pressure which are used in gas law calculations both in thermodynamics and in chemistry. The first and second laws of thermodynamics are then examined. Some properties of gases and that important working fluid, steam, are reviewed. This is followed by an examination of some elements of compression processes and cycles. The chapter concludes with a brief review of some heat transfer principles.

In this field quantities are routinely expressed in a diversity of unit systems, and more than normal care must be exercised to ensure that the units used within an equation are consistent.

ABSOLUTE TEMPERATURE AND PRESSURE

The temperature scales in common use around the world are relative temperature scales. Two such scales, called the Celsius or Centigrade and the Fahrenheit scales, have in the past been constructed by assuming a linear temperature variation between arbitrarily chosen numerical values assigned to the freezing and boiling points for water. The Celsius scale was divided into 100 parts between 0°C (freezing) and 100°C (boiling); the Fahrenheit scale had 180 divisions between 32°F and 212°F.

278

An absolute temperature scale can be associated with each of the two foregoing relative temperature scales. The lower end of each absolute scale is called absolute zero and is the lowest temperature that can exist. One of these absolute scales, the Kelvin scale, assigns the value 273°K to the freezing point of water; the other, the Rankine scale, assigns the value of 492°R to this point. Also $1K° = 1C°$ and $1R° = 1F°$. The conversion from one temperature scale to another is easily accomplished by deriving the relation from Fig. 7-1.

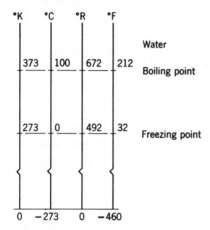

Figure 7-1

Many pressure gages register differential pressures, usually the difference in pressure between the point of interest and the surrounding atmosphere; this is a gage pressure. In many problems in thermodynamics, however, it is important to use absolute pressures that are measured relative to a perfect vacuum. Assuming that atmospheric pressure is known, we can easily convert values from one system to the other since the absolute pressure is equal to the sum of the gage pressure and the local atmospheric pressure. At sea level, atmospheric pressure is normally 14.7 psia.

THERMAL EQUILIBRIUM

A body is in thermal equilibrium when it is not exchanging thermal energy in the form of heat with another body. Directly related to this is the zeroth law of thermodynamics: two bodies each in thermal equilibrium with a third body will also be in thermal equilibrium with each other. On the other hand, if two bodies are placed together when they are not in thermal equilibrium, heat transfer will take place from one body to the other so that equilibrium will eventually be established. The amount of heat transfer Q is

$$Q = mc(T_2 - T_1) \qquad (7\text{-}1)$$

where m is the mass of the body, T_2 and T_1 are, respectively, the final and initial temperatures of the body, and c is the specific heat (or specific heat capacity) of the body (the amount of heat transfer per unit mass per degree). For nongases c is relatively independent of the nature of the heat transfer process. For gases this is not true; the specific heats c_p and c_v are normally given for a constant-pressure and a constant-volume process for a gas. In some cases a change of phase occurs during the heat transfer process. Then the constant that is characteristic of the phase change is the latent heat L or specific latent heat l, and the heat transfer is $Q = L = ml$.

Example 1

In a thermos flask are 90 g of water and 10 g of ice in equilibrium at a temperature of 0°C. A 100-g piece of metal with a specific heat of 0.40 cal/g-°C and a temperature of 100°C is dropped into the flask. What is the final equilibrium temperature, assuming no heat loss or gain to or from the surroundings? (The latent heat of fusion of water is 80 cal/g.)

Solution.   Since the system within the thermos is thermally insulated from its surroundings, the heat lost by the metal must equal the heat gained by the ice and water. Also, the ice and water are initially in equilibrium so that all the ice will melt before the water temperature is raised. Stated mathematically with T_e as the unknown equilibrium temperature, Eq. (7-1) gives

$$(100 \text{ g})(0.40 \text{ cal/g-}°\text{C})(100°\text{C} - T_e)$$

$$= (10 \text{ g})(80 \text{ cal/g}) + (90 \text{ g} + 10 \text{ g})(1.0 \text{ cal/g-}°\text{C})T_e$$

or

$$4000 - 40T_e = 800 + 100T_e$$

$$140T_e = 3200$$

$$T_e = 22.9°\text{C}$$

FIRST LAW OF THERMODYNAMICS

This law is a statement that energy is conserved. It is a very general law. For different classes of problems the law may be written in varying mathematical forms for ease of application. In all cases, however, the equations state that during a given process the net amount of heat Q transferred into a system is equal to the net work output W of the system plus the change in the system internal energy ΔE, that is,

$$Q = W + \Delta E \qquad (7\text{-}2)$$

Other definitions or sign conventions may also be selected here; what is important is that the net energy change of the system is the difference between the amount of energy entering and leaving the system. In this equation the internal energy is a property of the system, while heat and work generally are not properties. One exception, however, is the adiabatic process ($Q = 0$) where the net work is identified with the change in internal energy.

Most constant-mass or nonflow processes can be analyzed directly by Eq. (7-2), but for flow processes the terms are usually modified to make more explicit the different contributing energy terms. The foregoing work term for a flow process represents both shaft work and the flow work pv, where p is the fluid pressure and v is the fluid specific volume (reciprocal of the density ρ); let W now represent shaft work only. We also split the internal energy per unit mass into kinetic energy per unit mass $V^2/2$, potential energy per unit mass gz and specific internal energy u and at this time introduce the enthalpy $h = u + pv$. Since these terms are often not all in the same units, care in converting to common units is necessary, as the problems will show. The steady flow energy equation, written on a unit time basis, is then

$$\left(\frac{V^2}{2} + gz + h\right)_1 \dot{m} + Q = \left(\frac{V^2}{2} + gz + h\right)_2 \dot{m} + W \qquad (7\text{-}3)$$

The mass rate of flow $\dot{m}$ is the same in steady flow for the entering and exiting stream. The subscript 1 denotes terms related to the entering fluid stream and the subscript 2 refers to the exiting stream. For additional fluid streams, additional groups of terms must be added. For only one stream entering and exiting, Eq. (7-3) may be divided by $\dot{m}$ to express the first law on a unit mass basis.

SECOND LAW OF THERMODYNAMICS

The second law of thermodynamics, unlike the first law, is not a conservation law. One statement of the second law, following Kelvin and Planck, says that a system cannot operate cyclically and produce a net work output while exchanging heat only at one fixed temperature. Systems whose operations violate the second law are impossible. Related to the second law are the concepts of reversibility, entropy, and thermal efficiency.

A reversible process is one where *both* the system *and* its surroundings can be restored exactly to a prior state; all other processes are irreversible. Strictly speaking, no real process is reversible, but it is a useful concept and many actual processes closely approximate this condition. One system property, the entropy S, is defined in terms of a reversible heat transfer process as

$$S_B - S_A = \Delta S_{AB} = \int_A^B dS = \int_A^B \frac{dQ_{rev}}{T} \tag{7-4}$$

between any system states A and B. This definition of entropy in terms of the reversible heat transfer dQ_{rev} applies regardless of whether the actual process is reversible or irreversible. From Eq. (7-4) we note that a reversible, adiabatic ($dQ = 0$) process is isentropic ($S =$ constant). In general, any process satisfying two conditions, those of being reversible and adiabatic or isentropic, also satisfies the third condition. Finally, in all nonisentropic processes the overall entropy of the process must increase, according to the second law.

The efficiency η of a cyclical process is the ratio of the net work output to the total heat input. If Q_1 is the heat input and Q_2 is the heat output of a cycle, then the first law shows the thermal efficiency of the cycle is

$$\eta = 1 - \frac{Q_2}{Q_1} \tag{7-5}$$

The maximum thermal efficiency occurs for a reversible engine; the second law then shows the thermal efficiency to be $\eta = 1 - T_2/T_1$, where T_1 and T_2 are, respectively, the (absolute) temperature of the heat received and rejected.

Example 2

A mass flow of 200 lbm of air per minute is passing through a steady flow machine. Entrance conditions are: pressure $= 400$ psia, specific volume $= 1.387 \text{ ft}^3/\text{lbm}$, temperature $= 1040°F$, velocity $= 100 \text{ ft/sec}$. Exit conditions are: pressure $= 20$ psia, specific volume $= 9.25 \text{ ft}^3/\text{lbm}$, temperature $= 40°F$, velocity $= 110 \text{ ft/sec}$. The transferred heat given up by the air is 40 Btu/lbm. The entrance and exit connections are at the same elevation. For air, $c_p = 0.24 \text{ Btu/lbm-°R}$.

Find the shaft horsepower. Is the work done on or by the air?

Solution. We apply the steady flow energy equations, Eq. (7-3), to this process. Since the equation is written on a unit time basis, Q and W represent heat transfer and work per unit time. The entrance and exit elevations are identical; their net effect is therefore zero. Hence Eq. (7-3) is

$$W = Q + \dot{m}\left(h_1 - h_2 + \frac{V_1^2 - V_2^2}{2}\right)$$

The entrance and exit temperatures are $T_1 = 1040 + 460 = 1500°R$, $T_2 = 40 + 460 = 500°R$. It can be shown for perfect gases that $h = c_p T$, so that

$$h_1 - h_2 = c_p(T_1 - T_2)$$

$$h_1 - h_2 = (0.24 \text{ Btu/lbm-°R})(1500 - 500)°R = 240 \text{ Btu/lbm}$$

In computing the velocity terms, we must convert from mechanical to thermal units:

$$\tfrac{1}{2}(V_1^2 - V_2^2) = \frac{\tfrac{1}{2}(100^2 - 110^2) \text{ ft}^2/\text{sec}^2}{(32.2 \text{ lbm-ft/lbf-sec}^2)(778 \text{ ft-lbf/Btu})} = -0.042 \frac{\text{Btu}}{\text{lbm}}$$

Hence

$$W = \left(-40 \frac{\text{Btu}}{\text{lbm}}\right)\left(200 \frac{\text{lbm}}{\text{min}}\right) - \left(200 \frac{\text{lbm}}{\text{min}}\right)\left(240 - 0.042\right)\frac{\text{Btu}}{\text{lbm}}$$

$$W = 40,000 \frac{\text{Btu}}{\text{min}}$$

$$W = \frac{\left(40,000 \frac{\text{Btu}}{\text{min}}\right)\left(\frac{1}{60} \frac{\text{min}}{\text{sec}}\right)\left(778 \frac{\text{ft-lb}}{\text{Btu}}\right)}{\left(550 \frac{\text{ft-lb}}{\text{sec-hp}}\right)}$$

$$W = 943 \text{ hp done by the air}$$

STEAM TABLES

In thermodynamic analyses it is regularly necessary to know the properties of fluids which do not obey a simple equation of state. This is the case for water in the liquid and vapor (steam) states, since it is so commonly used as the working fluid in machines. Steam tables provide these data in tabular form;* much of the same data is alternately displayed in thermodynamic

* J. H. Keenan et al., *Steam Tables* (New York: Wiley, 1969). Abridged versions of such tables appear in many thermodynamics textbooks.

charts such as the Mollier diagram, which is an enthalpy-entropy chart for steam.

For liquid-vapor mixtures of water, called unsaturated steam, the value of any extensive property of the mixture, such as the specific volume v, enthalpy h, entropy s, or internal energy u, is the sum of individual property values for the liquid and vapor phases. As an example, consider specific volume. Let the subscript f indicate the saturated liquid state and the subscript g denote the saturated vapor state. The difference in specific volume between the two saturated states is v_{fg}, that is,

$$v_{fg} = v_g - v_f \qquad (7\text{-}6)$$

For a mass fraction x (called quality) of vapor in a mixture, the mixture specific volume v_x is found by summing the liquid and vapor fractional components; thus

$$v_x = (1-x)v_f + xv_g$$
or $\qquad (7\text{-}7)$
$$v_x = v_f + xv_{fg}$$

In many problems the steam quality x is not given directly but instead is determinable from two stated properties. Once x is found, the other fluid properties can also be computed in the same manner as v_x.

Example 3

One pound of a mixture of steam and water at 160 psia is contained in a rigid vessel. Heat is added to the vessel until the contents are at 560 psia and 600°F. Determine the quantity of heat, in Btu's, added to the tank contents.

Solution.   Using the tables for steam at 560 psia and 600°F, we find the final enthalpy to be $h_2 = 1293.4$ Btu/lbm, and the vapor is in a superheated state. The associated specific volume is $v_2 = 1.0224$ ft^3/lbm. Since the containing vessel is rigid, this is also the specific volume at 160 psia when the process began.

At 160 psia, $v_f = 0.01815$ ft^3/lbm and $v_g = 2.834$ ft^3/lbm. Hence

$$v_1 = (1-x)v_f + xv_g$$

$$1.0224 = (1-x)(0.0182) + x(2.834)$$

The steam quality is

$$x = 0.357$$

Then the original enthalpy was

$$h_1 = (1-x)h_f + xh_g$$
$$h_1 = (1-0.357)(335.93) + (0.357)(1195.1)$$
$$h_1 = 642.7 \text{ Btu/lbm}$$

Since no work can be done on a fluid in a rigid container, the heat added, according to the first law, is

$$Q = E_2 - E_1 = u_2 - u_1$$

Internal energy can be found from enthalpy $h = u + pv$, and the result is

$$Q = h_2 - h_1 - (p_2 - p_1)v = 1293.4 - 642.7 - (560 - 160)(144)(1.0224)/778$$
$$Q = 575 \text{ Btu}$$

GAS LAW RELATIONS

The behavior of many common gases is reasonably well described by a set of simple equations. These equations are called gas laws, and the gases that are assumed to obey these laws exactly are called perfect (or ideal) gases.

The equation of state for a mass m of a gas is

$$pV = mRT \qquad (7\text{-}8)$$

Here V is the volume occupied by the gas, and p and T are the absolute pressure and temperature. R is a gas constant related to the universal gas constant $\bar{R}$ by the relation $R = \bar{R}/M$, M being the molecular weight of the gas. Numerically, $\bar{R} = 1545$ ft-lbf/lbm-mole-°R, and for air $R = 53.35$ ft-lbf/lbm-°R. From the state equation we obtain Charles' law $p/T = $ constant for a constant-volume process and Boyle's law $pV = $ constant for a constant-temperature process. For a given mass of gas they may be combined to give $pV/T = $ constant, which is sometimes called the universal gas law. (Additional problems involving gas laws are presented in Chapter 8.)

Some other simple relations are useful in describing further the behavior of perfect gases. For a perfect gas the specific heats c_p and c_v are each constant, and it can be shown that $c_p - c_v = R$. Also, the internal energy and enthalpy are each directly proportional to the absolute temperature; specifically $u = c_v T$ and $h = c_p T$. By evaluating Eq. (7-4) for different paths, several different expressions for the entropy change of a perfect gas between two points can be derived. One relation is

$$s_2 - s_1 = c_p \ln\left(\frac{v_2}{v_1}\right) + c_v \ln\left(\frac{p_2}{p_1}\right) \tag{7-9}$$

Equation (7-9) is useful in learning about compression processes. Using the specific heat ratio $k = c_p/c_v$, Eq. (7-9) can be rearranged to show that an isentropic or reversible adiabatic compression process satisfies the relation

$$pV^k = \text{constant} \tag{7-10}$$

By using the equation of state, the alternative relations $TV^{k-1} = \text{constant}$ and $Tp^{(1-k)/k} = \text{constant}$ can be derived. The isentropic compression process is actually a special case of the more general polytropic process described by $pV^n = \text{constant}$, where n is not equal to k. The polytropic process is generally not isentropic.

Example 4

The air pressure in an automobile tire was checked at a service station and found to be 30 psig when the temperature was 65°F. Later the same tire was checked again, and the pressure gage read 35 psi. Assuming that the atmospheric pressure of 14.7 psi did not change, what was the new temperature of the air in the tire?

Solution.   If we assume that the volume of the tire remained constant, a special case of the universal gas law, called Charles' law, may be applied. The law shows $p/T = \text{constant}$ for a constant-volume process. Here

$$p_1 = 30 + 14.7 = 44.7 \text{ psia}$$

$$p_2 = 35 + 14.7 = 49.7 \text{ psia}$$

$$T_1 = 65 + 460 = 525°R$$

and T_2 is to be found. Thus

$$\frac{p_1}{T_1} = \frac{p_2}{T_2}$$

$$T_2 = T_1\left(\frac{p_2}{p_1}\right) = 525\left(\frac{49.7}{44.7}\right) = 584°R$$

$$T_2 = 584 - 460 = 124°F$$

Example 5

Five pounds of a gas initially at 0.0 psig and 60°F are compressed. Later it is found that the increase in pressure was 1900% and the decrease in volume was 90%. The barometric pressure was 24.44 in. of mercury during the compression process. Find

(a) the value of n in $pV^n = \text{constant}$
(b) the final gage pressure and final volume if the initial volume was 10 ft^3.

Solution. (a) From $pV^n = \text{constant}$,

$$\frac{p_2}{p_1} = \left(\frac{V_1}{V_2}\right)^n$$

Here $p_2 = 19p_1$ and $V_2 = 0.10V_1$, so that

$$19 = \left(\frac{1}{0.10}\right)^n = (10)^n$$

Taking logarithms,

$$\log_{10}(19) = n \log_{10}(10) = n$$

Hence

$$n = 1.279$$

(b) $$V_2 = 0.10V_1 = 0.10(10) = 1.0 \text{ ft}^3$$

The initial pressure $p_1 = 0.0 \text{ psig} = 24.44$ in. of mercury on the absolute scale.

$$24.44 \text{ in. of mercury} = \frac{24.44}{12} \frac{(62.4)(13.55)}{144} = 12.0 \text{ psia}$$

The final pressure is therefore $p_2 = 19p_1 = 19(12.0) = 228 \text{ psia}$, or

$$p_2 = 228 - 12 = 216 \text{ psig}$$

CYCLES

Cyclical processes play a significant role in engineering thermodynamics. The characteristics of several basic cycles are briefly reviewed here. Each is an idealized cycle that is composed entirely of reversible processes.

The Carnot cycle is a reversible four-process cycle consisting of alternating isothermal heat transfer processes and adiabatic compression or expansion processes. Since the second law shows that a reversible cycle is the most efficient of all cycles operating between two given temperature levels, the simple Carnot cycle is often used as a standard against which the performance characteristics of all other cycles may be measured.

The Otto, Diesel, and Brayton cycles are four-process gas cycles. If air is the working fluid, they are called air-standard cycles. In all three cycles every second process is an isentropic expansion or compression process. The Otto cycle is used for reciprocating engines; the other two processes in this cycle are constant-volume heat transfer processes. The Diesel cycle is thermodynamically like the Otto cycle except that the idealized heat supply process is presumed to occur at constant pressure. In the Brayton cycle both heat transfer processes occur at constant pressure; it is the standard cycle for gas turbines.

The idealized cycle which describes the operation of steam power plants is the Rankine cycle, which, of course, is a vapor cycle. The cycle consists, in turn, of an isentropic expansion process for the steam, a constant-pressure heat transfer process, an isentropic compression of the liquid, and a constant-pressure heat transfer process to the liquid. The first process could occur at any of several states (see Fig. 7-2).

State diagrams for these basic cycles are given in Fig. 7-2. A variable listed beside a process line indicates that the variable is constant during the process. The extra line on the Rankine cycle plots divides the vapor and liquid phases.

HEAT TRANSFER

Engineers often want to know not only the amount but also the rate of heat transfer during a process. For this reason the various heat transmission modes are outlined here. All heat transfer occurs by some combination of the three mechanisms of conduction, convection, and radiation.

Conductive heat transfer occurs in the absence of mass transfer. The rate of heat transfer q, in Btu/hour, is given in this case by Fourier's law

$$q = -kA\frac{dT}{dx} \qquad (7\text{-}11)$$

where k is the thermal conductivity in Btu/hr-ft-°F, A is the cross-sectional area in square feet, and dT/dx is the temperature gradient in the direction of heat flow.

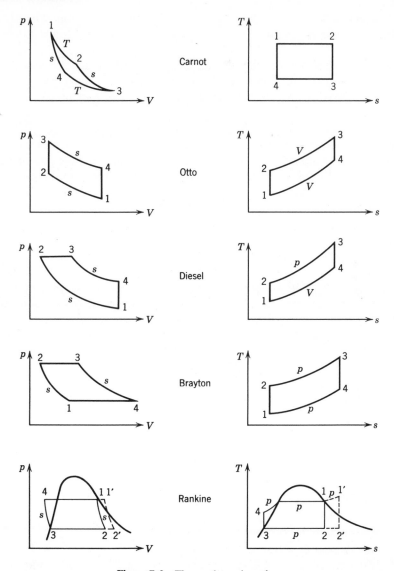

Figure 7-2 Thermodynamic cycles

Convective heat transfer involves the mass transfer of fluid from regions of high temperature to regions of lower temperature. The process is described by Newton's cooling law as

$$q = hA \, \Delta T \tag{7-12}$$

Here h is called the convective heat transfer coefficient or film coefficient and is measured in Btu/hr-ft^2-°F. ΔT is the temperature difference causing the process.

Radiation depends on the transmission of electromagnetic waves to achieve heat transfer. The rate of emission of heat is governed by the Stefan-Boltzmann law for black-body radiation, which may be written

$$q = A\sigma(T_1^4 - T_2^4) \qquad (7\text{-}13)$$

This expression applies for a small body having an area A which emits radiation at a temperature T_1 (absolute) to some point which is at temperature T_2. The Stefan-Boltzmann constant is σ. For nonblack bodies this constant is replaced by $e\sigma$, e being the dimensionless emissivity of the particular body.

Example 6

A masonry wall has a 4-in.-thick facing wall bonded to an 8-in.-thick concrete backing. On a day when the room temperature is 68°F and the outside temperature is 12°F, the inner surface temperature of the concrete is 57°F and the outer surface temperature of the brick is 19°F. See Fig. 7-3. The thermal conductivities for brick and concrete are 0.36 and

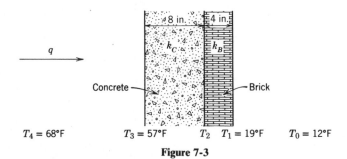

$T_4 = 68°F$ $T_3 = 57°F$ T_2 $T_1 = 19°F$ $T_0 = 12°F$

Figure 7-3

0.68 Btu/hr-ft-°F, respectively. Determine
   (a) the overall heat transfer coefficient.
   (b) the convective heat transfer coefficients for the concrete and brick walls.

Solution. (a) Here we assume a steady-state heat flow. A combination of conduction and convection is occurring. By analogy to Eqs. (7-11) and (7-12), an overall heat transfer coefficient U may be defined by the equation

$$q = UA\,\Delta T = UA(T_4 - T_0)$$

From Eqs. (7-11) and (7-12) we may write several expressions for the ratio q/A:

$$\frac{q}{A} = h_C(T_4 - T_3) = \frac{k_C}{L_C}(T_3 - T_2) = \frac{k_B}{L_B}(T_2 - T_1) = h_B(T_1 - T_0)$$

Using the two conduction expressions, we can solve for the unknown temperature T_2 and then determine q/A. Then U can be computed. Thus

$$\frac{q}{A} = \frac{0.68}{\frac{8}{12}}(57 - T_2) = \frac{0.36}{\frac{4}{12}}(T_2 - 19)$$

$$1.02(57 - T_2) = 1.08(T_2 - 19)$$

$$T_2 = 37.4°F$$

and

$$\frac{q}{A} = 1.02(57 - T_2) = 1.02(57 - 37.4)$$

$$\frac{q}{A} = 20 \text{ Btu/hr-ft}^2$$

Therefore

$$U = \frac{q}{A}\frac{1}{\Delta T} = 20\frac{1}{(68 - 12)}$$

$$U = 0.357 \text{ Btu/hr-ft}^2\text{-°F}$$

(b) The convective heat transfer coefficients can now be found.

$$\frac{q}{A} = h_C(T_4 - T_3) = h_B(T_1 - T_0)$$

$$20 = h_C(68 - 57) = h_B(19 - 12)$$

For the concrete,

$$h_C = \tfrac{20}{11} = 1.82 \text{ Btu/hr-ft}^2\text{-°F}$$

For the brick,

$$h_B = \tfrac{20}{7} = 2.86 \text{ Btu/hr-ft}^2\text{-°F}$$

PROBLEM 7-1

The temperature 45°C is equal to

(A) 45°F
(B) 57°F
(C) 113°F
(D) 81°F
(E) 25°F

Solution. The melting point of ice and the boiling point of water are, respectively, 0°C = 32°F and 100°C = 212°F. By direct proportion,

$$45°C = 32°F + 45\left(\frac{212 - 32}{100}\right)$$

$$= 113°F$$

This solution is equivalent to using the formula

$$°C = \tfrac{5}{9}(°F - 32)$$

Answer is (C)

PROBLEM 7-2

How much heat is required to raise 10 g of water from 0°C to 1°C?

(A) 10 Btu
(B) 1 Btu
(C) 1 cal
(D) 5 J
(E) 10 cal

Solution. One cal will raise the temperature of 1 g of water 1°C. Hence 10 g of water will require 10 cal.

Answer is (E)

PROBLEM 7-3

The amount of heat required to raise the temperature of 1 ft^3 of water by 1°F is nearest to

 (A) 65,800 J
 (B) 48,500 J
 (C) 28,300 J
 (D) 15,700 J
 (E) 62 J

Solution.　One Btu will raise the temperature of 1 lb of water by 1°F, and water weighs 62.4 lb/ft³; hence 62.4 Btu are required. The SI equivalent is

$$(62.4 \text{ Btu})\left(778 \frac{\text{ft-lb}}{\text{Btu}}\right)\left(0.3048 \frac{\text{m}}{\text{ft}}\right)\left(4.45 \frac{\text{N}}{\text{lb}}\right) = 65,800 \text{ N-m}$$

or 65,800 joules (J).

Answer is (A)

PROBLEM 7-4

An adiabatic process is one in which

 (A) the pressure is constant
 (B) internal energy is constant
 (C) no work is done
 (D) no heat is transferred
 (E) friction is not considered

Solution.　By definition, an adiabatic process is one in which no heat is transferred into or out of the system.

Answer is (D)

PROBLEM 7-5

How much boiling water is required to melt 1000 g of ice at 0°C and produce a mixture at 20°C?

 (A) 200 g
 (B) 250 g
 (C) 800 g
 (D) 1000 g
 (E) 1250 g

Solution.   The amount of heat lost by the boiling water must equal the heat gained by the ice and cold water.

Let l = heat of fusion of ice = 80 cal/g

c = heat capacity of water = 1 cal/g-°C

M_w = mass of boiling water at 100°C

M_i = mass of ice = 1000 g

ΔT_w = decrease in hot water temperature

ΔT_i = increase in temperature for ice

Then

$$M_w c \, \Delta T_w = l M_i + M_i c \, \Delta T_i$$

$$M_w(1)(100-20) = 80(1000) + (1000)(1)(20-0)$$

$$80 M_w = 100{,}000$$

$$M_w = 1250 \text{ g of boiling water}$$

Answer is (E)

PROBLEM 7-6

A mass of 0.2 kg of a metal having a temperature of 100°C is plunged into 0.04 kg of water at 20°C. The temperature of the water and metal becomes 48°C. The latent heat of ice at 0°C is 335 kJ/kg, and the specific heat capacity of water is 4.19 kJ/kg-°C.

(1) The specific heat capacity of the metal is closest to

   (A) 0.25 kJ/kg-°C
   (B) 0.30 kJ/kg-°C
   (C) 0.40 kJ/kg-°C
   (D) 0.45 kJ/kg-°C
   (E) 0.75 kJ/kg-°C

Solution.   Assuming no heat loss to the surroundings, we equate the amount of heat lost by the metal and the amount of heat gained by the water. We employ the formula $Q = Mc \, \Delta T$ and we use the subscripts m for metal and w for water. Then

$$M_m c_m \, \Delta T_m = M_w c_w \, \Delta T_w$$

$$0.2 c_m (100 - 48) = 0.04(4.19)(48 - 20)$$

The specific heat capacity of the metal is therefore

$$c_m = 0.451 \text{ kJ/kg-°C}$$

Answer is (D)

(2) The number of kg of ice at 0°C that can be melted by 0.2 kg of this metal at 100°C is nearest to

(A) 0.015
(B) 0.024
(C) 0.027
(D) 0.036
(E) 0.045

Solution. Again we equate the heat lost by the metal and the heat gained by the ice. Let l = latent heat of ice. We obtain

$$M_m c_m \, \Delta T_m = l M_i$$

$$0.2(0.451)(100 - 0) = 335 M_i$$

Hence

$$M_i = 0.027 \text{ kg}$$

Answer is (C)

PROBLEM 7-7

The melting rate of snow will normally be accelerated when warm rain falls. The following data are provided:

Snow	Rain
Depth 50 in.	Amount 2 in.
Water content 40%	Temperature 68°F
Temperature 32°F (melting point)	
Heat of fusion 144.0 Btu/lb of water	

Assume the weight of melted snow and rain water to be $62.4 \, lb/ft^3$. The percentage of the snow pack that will be melted by the rain is nearest to

(A) 1.0
(B) 2.5
(C) 4.0
(D) 5.0
(E) 10.0

Solution. We first calculate the amount of heat available in the 68°F rain. Then we calculate the amount of snow at 32°F that will be converted to water at 32°F in absorbing this amount of heat.

The quantity of heat available above 32°F, per square foot of surface, is

$$\tfrac{2}{12}(62.4)(68 - 32) = 374.4 \, Btu/ft^2$$

This amount of heat will melt an amount of snow equivalent to $374.4/144.0 = 2.6$ lb of water, or

$$\left(\frac{2.6}{62.4}\right)\left(\frac{1}{0.4}\right) = 0.104 \, ft^3 \text{ of snow/ft}^2 \text{ of surface}$$

This equals $0.104 \times 12 = 1.25$ in. of snow. The percentage of the 50-in. snow pack that is melted is

$$\frac{1.25}{50}(100) = 2.5\%$$

Answer is (B)

PROBLEM 7-8

The first law of thermodynamics may be referred to as the principle of the conservation of

(A) mass
(B) momentum
(C) heat
(D) energy
(E) enthalpy

Solution. The first law of thermodynamics is a statement of the principle of energy conservation which relates the net heat Q and net work W to the internal energy change ΔE of a system.

Answer is (D)

PROBLEM 7-9

A perfect gas is contained in a piston-and-cylinder machine. Within the machine the pressure of the gas is always directly proportional to its volume. Initially the gas is at a pressure of 15 psia and a volume of 1 ft³. Heat is transferred reversibly to the gas until its pressure is 150 psia. If the movement of the piston is frictionless, the work done by the gas, expressed in Btu's, is most nearly

(A) 25
(B) 80
(C) 135
(D) 200
(E) 275

Solution. The gas pressure p is directly proportional to its volume V, or $p = KV$. The initial pressure p_0 is $p_0 = 15(144)$ lb/ft² absolute when the initial volume $V_0 = 1$ ft³. Hence

$$15(144) = K(1) \qquad K = 2160 \text{ lb/ft}^5$$

When the final pressure $p_1 = 150(144)$ lb/ft² absolute, the volume is

$$V_1 = \frac{p_1}{K} = \frac{150(144)}{2160} = 10 \text{ ft}^3$$

The work done is

$$W = \int_{V_0}^{V_1} p\, dV = \int_1^{10} KV\, dV = 2160 \left[\frac{V^2}{2} \right]_1^{10}$$

$$W = 2160(\tfrac{1}{2})(10^2 - 1) = 106{,}900 \text{ ft-lb}$$

The work done by the gas, in Btu's, is

$$W = \frac{106{,}900}{778} = 137.4 \text{ Btu}$$

Answer is (C)

PROBLEM 7-10

Steam enters a turbine at a velocity of 100 ft/sec and an enthalpy of 1410 Btu/lbm; it leaves the turbine at 390 ft/sec and an enthalpy of 990 Btu/lbm. Heat is lost to the surroundings at a rate of 22 Btu/lbm of steam. If the steam flow rate is 75,000 lbm/hr, the power output in kilowatts is closest to

(A) 6400
(B) 6500
(C) 6800
(D) 8700
(E) 11,800

Solution. The turbine work output can be found by using the steady flow energy equation written on a unit mass basis and neglecting changes in potential energy. Using the subscript notation $i =$ inlet, $e =$ exit, the equation is

$$W = (h_i - h_e) + \frac{V_i^2 - V_e^2}{2} + Q$$

$$W = (1410 - 990)\frac{Btu}{lbm} + \frac{\left(100^2 - 390^2\right)\frac{ft^2}{sec^2}}{2\left(32.2\frac{lbm\text{-}ft}{lbf\text{-}sec^2}\right)\left(778\frac{ft\text{-}lbf}{Btu}\right)} - 22\frac{Btu}{lbm}$$

$$W = 420 - 2.84 - 22 = 395.2 \text{ Btu/lbm}$$

The heat term is negative because it is a loss. The power output P is equal to work times mass rate of flow. Since the equivalence between kilowatts and Btu/hr is

$$1 \text{ kW} = \frac{\left(1000\frac{N\text{-}m}{sec}\right)(60^2)\frac{sec}{hr}}{\left(4.448\frac{N}{lbf}\right)\left(0.3048\frac{m}{ft}\right)\left(778\frac{ft\text{-}lbf}{Btu}\right)} = 3413\frac{Btu}{hr}$$

$$P = \frac{(395.2)(75,000)}{3413} = 8680 \text{ kW}$$

Answer is (D)

PROBLEM 7-11

Fluid enters a turbine with a velocity of 2 m/sec and an enthalpy of 900 Btu/lbm and leaves with an enthalpy of 850 Btu/lbm and a velocity of 100 m/sec. Heat losses are 1200 W, and the mass rate of flow is 1 kg/sec.

The inlet to the turbine is 3 m higher than the outlet. The maximum theoretical power, in kilowatts, that can be developed by the turbine is nearest to

(A) 48
(B) 53
(C) 80
(D) 105
(E) 116

Solution. Only the enthalpies are not already in metric units; let us first convert from Btu/lbm to J/kg:

$$1\frac{Btu}{lbm} = \left(1\frac{Btu}{lbm}\right)\left(778\frac{ft\text{-}lbf}{Btu}\right)\left(\frac{0.3048\ m}{ft}\right)\left(\frac{4.45\ N}{lbf}\right)\left(\frac{2.20\ lbm}{kg}\right)$$

$$= 2320\ J/kg$$

Now we may employ the steady flow energy equation, written on a unit mass basis, which is

$$\frac{V_1^2}{2} + gz_1 + h_1 + \frac{dQ}{dm} = \frac{V_2^2}{2} + gz_2 + h_2 + \frac{dW}{dm}$$

Here z = elevation above a chosen datum, and the subscripts 1 and 2 represent the turbine entrance and exit, respectively.

$$\frac{dQ}{dm} = \frac{dQ/dt}{dm/dt} = \frac{500\ W}{1\ kg/sec} = 500\ J/kg$$

$$\frac{(2^2 - 100^2)\left(\frac{m}{sec}\right)^2}{(1\ kg\text{-}m/N\text{-}sec^2)} + \frac{(9.81\ m/sec^2)(3\ m)}{(1\ kg\text{-}m/N\text{-}sec^2)}$$

$$+ (900 - 850)\frac{Btu}{lbm}\left(2320\frac{J/kg}{Btu/lbm}\right) - 1200\frac{J}{kg} = \frac{dW}{dm}$$

$$\frac{dW}{dm} = 104.8\ kJ/kg$$

The work per unit time, or power, is therefore

$$\frac{dW}{dt} = \left(\frac{dW}{dm}\right)\left(\frac{dm}{dt}\right) = \left(104.8\frac{kJ}{kg}\right)\left(1\frac{kg}{sec}\right) = 104.8\ kW$$

Answer is (D)

PROBLEM 7-12

Entropy

(A) remains constant during an irreversible process
(B) is independent of temperature
(C) is a maximum at absolute zero
(D) is a measure of unavailable energy
(E) is the reciprocal of enthalpy

Solution.   Entropy is a measure of unavailable energy.

Answer is (D)

PROBLEM 7-13

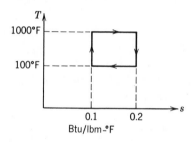

Figure 7-4

In the temperature-entropy diagram in Fig. 7-4, the thermal efficiency of the process is nearest to

(A) 40%
(B) 60%
(C) 80%
(D) 90%
(E) 93%

Solution.

Temperatures in this problem must be expressed in an absolute system, in this case degrees Rankine.

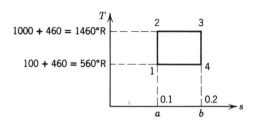

Figure 7-5

Thermal efficiency is defined as

$$\eta = \frac{W}{Q} = \frac{\text{net work}}{\text{heat added}} = \frac{\text{Area 1-2-3-4}}{\text{Area } a\text{-2-3-}b} = \frac{(0.2-0.1)(1460-560)}{(0.2-0.1)(1460)}$$

$$\eta = \tfrac{900}{1460} = 0.616 = 61.6\%$$

Answer is (B)

PROBLEM 7-14

A Carnot engine uses steam (a vapor) as the thermodynamic medium. 1000 Btu/min is supplied by a source at 500°F. The temperature of the refrigerator is 120°F. The output of the engine, in horsepower, is closest to

(A) 6.6
(B) 9.3
(C) 12.6
(D) 14.2
(E) 17.9

Solution.   Converting to degrees Rankine,

$$T_1 = 500 + 460 = 960°R \qquad T_2 = 120 + 460 = 580°R$$

Figure 7-6 is a temperature-entropy diagram for this Carnot cycle. The

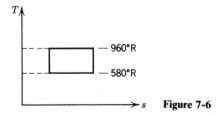

Figure 7-6

thermal efficiency η is

$$\eta = \frac{T_1 - T_2}{T_1} = \frac{960 - 580}{960}$$

$$\eta = 0.396$$

The net work output is then

$$\eta Q_1 = (0.396)\left(1000 \frac{\text{Btu}}{\text{min}}\right) = 396 \frac{\text{Btu}}{\text{min}}$$

The equivalent horsepower is

$$hp = \frac{\left(396 \frac{\text{Btu}}{\text{min}}\right)\left(778 \frac{\text{ft-lb}}{\text{sec}}\right)}{\left(60 \frac{\text{sec}}{\text{min}}\right)\left(550 \frac{\text{ft-lb}}{\text{sec-hp}}\right)} = 9.34 \text{ hp}$$

Answer is (B)

PROBLEM 7-15

Which one of the following is an extensive property?

   (A) temperature
   (B) pressure
   (C) mass
   (D) specific heat capacity
   (E) specific volume

Solution.   Intensive properties of a thermodynamic system are independent of the mass of the system; extensive properties depend on the amount
of mass present. Clearly, mass itself is an extensive property.

Answer is (C)

PROBLEM 7-16

For a single component system, the number of properties required to define uniquely a phase is

(A) 1
(B) 2
(C) 3
(D) 4
(E) 5

Solution. The required number is 2. As an example, the state of an ideal gas is completely defined if two of the three quantities (p, V, T) appearing in the gas law are known, for then the third variable and other quantities can be found.

Answer is (B)

PROBLEM 7-17

A room experiences a heat gain of 100,000 Btu/hr and must be maintained at 80°F. If $c_p = 0.24$ Btu/lbm-°R and $R = 53.3$ ft-lbf/lbm-°R for air, the number of cubic feet per minute of 64°F air required to maintain the desired temperature is nearest to

(A) 26,000
(B) 8200
(C) 5800
(D) 4100
(E) 450

Solution. Entering the room ($p = 14.7$ psi, $T = 64°F = 524°R$) the mass of 1 ft^3 of air is, according to the gas law, approximately

$$m = \frac{pV}{RT} = \frac{\left(14.7\frac{\text{lbf}}{\text{in}^2}\right)\left(144\frac{\text{in}^2}{\text{ft}^2}\right)(1 \text{ ft}^3)}{\left(53.3\frac{\text{ft-lbf}}{\text{lbm-}°R}\right)(524°R)} = 0.0758 \text{ lbm}$$

The heat capacity C of 1 ft^3 of air is therefore

$$C = (0.24 \text{ Btu/lbm-}°R)(0.0758 \text{ lbm}) = 0.018 \text{ Btu/}°R$$

Here $\Delta T = 80 - 64 = 16°F$, and hence the amount of heat to be removed is $100,000/60 = 1667$ Btu/min. Hence the required volume of air is

$$V = \frac{1667}{16(0.018)} = 5800 \text{ ft}^3/\text{min}$$

Answer is (C)

PROBLEM 7-18

At the place where Piccard started his ascent in the stratosphere balloon, the temperature was 17°C and the pressure 640 mm of Hg. At the highest altitude reached, the temperature was −48°C and the pressure 310 mm of Hg. None of the gas was vented. The fractional part of its total capacity to which the balloon was filled before ascending so that it would be fully expanded at the highest altitude reached is most nearly

(A) 0.56
(B) 0.62
(C) 0.68
(D) 0.74
(E) 0.80

Solution. Assuming air to be a perfect gas, here we can apply the universal gas law

$$\frac{p_1 V_1}{T_1} = \frac{p_2 V_2}{T_2}$$

In this equation absolute temperatures and pressures must be used. Initially, $T_1 = 17°C = 17 + 273 = 290°K$, $p_1 = 640$ mm of Hg, and V_1 is unknown. At the highest altitude, $T_2 = -48°C = -48 + 273 = 225°K$, $p_2 = 310$ mm of Hg, and $V_2 =$ total capacity of balloon. Hence

$$\frac{V_1}{V_2} = \frac{p_2}{p_1} \frac{T_1}{T_2} = \left(\frac{310}{640}\right)\left(\frac{290}{225}\right) = 0.624$$

and the balloon therefore was filled to 0.624 or five-eighths of its capacity before it was launched.

Answer is (B)

PROBLEM 7-19

An automobile tire registered a gage pressure of 28 psi when the temperature was 70°F. After driving for a while, the gage pressure was found to be
31 psi. Assume that the barometric pressure was constant at 14.3 psi and
that the tire volume was constant. The temperature of the tire, in °F, at the
time of the second reading was most nearly

(A) 75
(B) 78
(C) 90
(D) 108
(E) 127

Solution. Assuming air to behave as a perfect gas while undergoing a
constant-volume process, we may apply Charles' law $p_1/T_1 = p_2/T_2$. In this
equation we must use absolute pressures and temperatures.

$$p_1 = 28 + 14.3 = 42.3 \text{ psia}$$
$$p_2 = 31 + 14.3 = 45.3 \text{ psia}$$
$$T_1 = 70 + 460 = 530°R$$

Thus

$$T_2 = T_1 \frac{p_2}{p_1} = 530\left(\frac{45.3}{42.3}\right) = 568°R$$

$$T_2 = 568 - 460 = 108°F$$

Answer is (D)

PROBLEM 7-20

Nitrogen is pumped into a 10-ft³ tank until the pressure gage reads 185.3 psi
and the temperature of the gas is 200°F. The tank is cooled until the
temperature of the nitrogen is 80°F. Assume atmospheric pressure of
14.7 psia. The following data are also available:

Gas	Chemical Formula	Molecular Weight	R ft-lbf lbm-°R	c_p Btu lbm-°R	c_v Btu lbm-°R	k $\frac{c_p}{c_v}$
Nitrogen	N_2	28.0	55.1	0.248	0.177	1.40

(1) The amount of heat removed from the nitrogen gas is nearest to

   (A) 156 Btu
   (B) 168 Btu
   (C) 219 Btu
   (D) 236 Btu
   (E) 555 Btu

Solution. We first determine the mass of the nitrogen in the tank by using the equation of state for a perfect gas $pV = mRT$, where p and T must be in absolute units.

$$m = \frac{pV}{RT} = \frac{(185.3 + 14.7)(12^2)(10)}{(55.1)(200 + 460)} = 7.92 \text{ lbm}$$

We can find the heat removed from the equation $Q = mc(T_2 - T_1)$. For this constant-volume process, the specific heat $c = c_v$, the specific heat at constant volume. Hence

$$Q = mc_v(T_2 - T_1) = (7.92)(0.177)(80 - 200) = -168.2 \text{ Btu}$$

The minus sign indicates heat removed.

Answer is (B)

(2) The final gage pressure, in psi, is nearest to

   (A) 65
   (B) 74
   (C) 149
   (D) 152
   (E) 164

Solution. Using the gas law at constant volume (Charles' law) $V_1 = V_2$,

$$\frac{p_2}{p_1} = \frac{T_2}{T_1}$$

or

$$p_2 = (185.3 + 14.7)\left(\frac{80 + 460}{200 + 460}\right) = 200\left(\frac{540}{660}\right) = 163.6 \text{ psia}$$

$$p_2 = 163.6 - 14.7 = 148.9 \text{ psig}$$

Answer is (C)

PROBLEM 7-21

Two moles of oxygen at 50 psia and 40°F are in a container that is connected by a valve to a second container filled with 5 moles of nitrogen at 30 psia and 140°F. The valve is opened, and adiabatic mixing occurs. The following data are also supplied:

Gas	Chemical Formula	Molecular Weight	R $\dfrac{\text{ft-lbf}}{\text{lbm-°R}}$	c_p $\dfrac{\text{Btu}}{\text{lbm-°R}}$	c_v $\dfrac{\text{Btu}}{\text{lbm-°R}}$	k $\dfrac{c_p}{c_v}$
Nitrogen	N_2	28.0	55.1	0.248	0.177	1.40
Oxygen	O_2	32.0	48.3	0.219	0.157	1.39

(1) The equilibrium temperature, in °F, of the mixture is closest to

(A) 90
(B) 95
(C) 100
(D) 105
(E) 110

Solution.   First we determine the initial volume of each gas using the state equation $pV = mRT$.

$$2 \text{ lb-moles of oxygen} = 2(32) = 64 \text{ lbm oxygen}$$

$$5 \text{ lb-moles of nitrogen} = 5(28) = 140 \text{ lbm nitrogen}$$

$$N_2: \quad V = \frac{mRT}{p} = \frac{(140)(55.1)(140+460)}{(30)(12^2)} = 1071 \text{ ft}^3$$

$$O_2: \quad V = \frac{mRT}{p} = \frac{(64)(48.3)(40+460)}{(50)(12^2)} = 215 \text{ ft}^3$$

An adiabatic process is one in which no net heat transfer to the surroundings occurs, so the heat loss by the nitrogen in mixing is gained by the oxygen. The mixing is a constant-volume process so that c_v, the specific heat at constant volume, is the proper specific heat value to use. The heat transfer Q in this process is $Q = mc_v(T_2 - T_1)$. Let T be the final, or equilibrium, temperature of the mixture. Then

Heat lost by the nitrogen $= 140(0.177)(140 - T) = 3470 - 24.8T$

Heat gained by the oxygen $= 64(0.157)(T - 40) = 10.0T - 402$

$$3470 - 24.8T = 10.0T - 402$$

$$34.8T = 3870$$

The equilibrium temperature is

$$T = 111°F$$

Answer is (E)

(2) The equilibrium pressure, in psia, of the mixture is nearest to

(A) 30.0
(B) 33.3
(C) 36.7
(D) 40.0
(E) 43.3

Solution.   For a mixture of gases the pressure P of the mixture is equal to the sum of the partial pressures of the constituent gases (Dalton's law). Since T and V are now common to both gases and $p = mRT/V$,

$$P = p_{N_2} + p_{O_2} = [140(55.1) + 64(48.3)]\left(\frac{111 + 460}{1071 + 215}\right)$$

$$= 4800 \text{ lb/ft}^2 \text{ abs} = 4800/144 = 33.3 \text{ psia}$$

Answer is (B)

PROBLEM 7-22

The temperature at which the vapor in a mixture starts to condense when the mixture is cooled at constant pressure is called the

(A) dry-bulb temperature
(B) wet-bulb temperature
(C) dew point
(D) relative humidity
(E) specific humidity

Solution. The dry-bulb temperature is measured with an ordinary thermometer. The wet-bulb temperature is measured by a thermometer covered
with a liquid film. Relative humidity is the ratio of the amount of vapor
present to the amount of vapor required to produce saturation at the same
temperature, whereas specific humidity is normally the mass ratio of water
vapor to air in a mixture.

Answer is (C)

PROBLEM 7-23

A closed cylinder contains 3.00 ft^3 of dry air at a temperature of 60°F and a
gage pressure of 10 psi. A piston very suddenly reduces the enclosed volume
to 2.00 ft^3.

(1) The new temperature of the air, in °F, is most nearly

(A) 71
(B) 104
(C) 128
(D) 152
(E) 177

Solution. We assume the air behaves as an ideal gas. From the description
of the piston movement we may also reasonably assume the volume change
to be adiabatic.

$$p_1 = 10 \text{ psig} = 10 + 14.7 = 24.7 \text{ psia} \qquad p_2 = ?$$
$$T_1 = 60°F = 60 + 460 = 520°R \qquad T_2 = ?$$
$$V_1 = 3.00 \text{ ft}^3 \qquad V_2 = 2.00 \text{ ft}^3$$

The specific heat ratio for air is $k = 1.4$. From the above assumptions it can
be shown that both TV^{k-1} and pV^k are constant during the compression
process.

$$\frac{T_1}{T_2} = \left(\frac{V_2}{V_1}\right)^{k-1}$$

$$T_2 = T_1\left(\frac{V_1}{V_2}\right)^{k-1} = 520\left(\frac{3.00}{2.00}\right)^{1.4-1}$$

$$T_2 = 520(1.5)^{0.4} = 520(1.176)$$

$$T_2 = 612°R = 612 - 460 = 152°F$$

Answer is (D)

(2) The new gage pressure of the air, in psi, is closest to

(A) 17.6
(B) 18.4
(C) 22.6
(D) 28.9
(E) 30.7

Solution.

$$\frac{p_1}{p_2} = \left(\frac{V_2}{V_1}\right)^k$$

$$p_2 = p_1\left(\frac{V_1}{V_2}\right)^k = 24.7\left(\frac{3.00}{2.00}\right)^{1.4} = 24.7(1.764)$$

$$p_2 = 43.6 \text{ psia} = 43.6 - 14.7 = 28.9 \text{ psig}$$

Answer is (D)

PROBLEM 7-24

Air is compressed polytropically so that the quantity $pV^{1.4}$ is constant. If 0.02 m^3 of air at atmospheric pressure (101.3 kN/m^2) and 4°C are compressed to a gage pressure of 405 kN/m^2, the final temperature of the air, in °C, is closest to

(A) 155
(B) 165
(C) 175
(D) 185
(E) 195

Solution.   One approach is to use the compression relation to determine the final volume and then employ the gas law to obtain the temperature.
     From the compression relation we have

$$(101.3)(0.02)^{1.4} = (405 + 101.3)V^{1.4}$$

$$(101.3)(0.00418) = (506)V^{1.4}$$

$$V^{1.4} = 8.37(10^{-4}) \qquad V = 6.34(10^{-3}) \text{ m}^3$$

Using the gas law,

$$\frac{p_1 V_1}{T_1} = \frac{p_2 V_2}{T_2}$$

$$\frac{101.3(0.02)}{4+273} = \frac{.(506)(0.00634)}{T_2}$$

$$T_2 = 439°K = 439 - 273 = 166°C$$

Answer is (B)

PROBLEM 7-25

The volume in the single cylinder of an air compressor is $0.57 \, \text{ft}^3$ at the beginning of the compression stroke with air at atmospheric pressure. The piston compresses the air polytropically to 69.8 psig according to the law $pV^{1.35} = \text{constant}$. At the end of the compression stroke the volume is nearest to

(A) $0.10 \, \text{ft}^3$
(B) $0.12 \, \text{ft}^3$
(C) $0.14 \, \text{ft}^3$
(D) $0.16 \, \text{ft}^3$
(E) $0.18 \, \text{ft}^3$

Solution. Assuming atmospheric pressure to be 14.7 psia, we have

$$p_1 V_1^{1.35} = p_2 V_2^{1.35}$$

or

$$V_2 = V_1 \left(\frac{p_1}{p_2}\right)^{1/1.35}$$

$$V_2 = (0.57)\left(\frac{14.7}{14.7+69.8}\right)^{1/1.35}$$

$$V_2 = (0.57)(0.174)^{1/1.35} = (0.57)(0.274)$$

$$V_2 = 0.156 \, \text{ft}^3$$

Answer is (D)

PROBLEM 7-26

If an ideal gas is compressed from a lower pressure to a higher pressure at a constant temperature, which of the following is true?

(A) the work required is zero

(B) the volume remains constant

(C) the volume varies inversely with the absolute pressure

(D) heat is being absorbed

(E) none of these

Solution. For an isothermal compression the pressure-volume relation is governed by the universal gas law, which gives $pV = $ constant.

Answer is (C)

PROBLEM 7-27

Air is taken from the atmosphere at 14.7 psia and 70°F and delivered to a tank in which the pressure is 100 psia and 300°F. The amount of heat removed per unit mass (Btu/lbm) from the air during the compression and delivery process is closest to

(A) 13

(B) 45

(C) 60

(D) 95

(E) 125

Solution. For a polytropic process it can be shown that

$$\frac{T_2}{T_1} = \left(\frac{p_2}{p_1}\right)^{(n-1)/n}$$

Here

$$p_1 = 14.7 \text{ psia} \qquad\qquad p_2 = 100 \text{ psia}$$

$$T_1 = 70 + 460 = 530°R \qquad T_2 = 300 + 460 = 760°R$$

$$\frac{760}{530} = \left(\frac{100}{14.7}\right)^{1-1/n}$$

$$1.434 = (6.80)^{1-1/n}$$

Taking the logarithm of each term in the equation,

$$\ln{(1.434)} = \left(1 - \frac{1}{n}\right)\ln{(6.80)}$$

$$0.360 = \left(1 - \frac{1}{n}\right)(1.92) \qquad n = 1.23$$

The work done is

$$W = \int p\,dV = \frac{p_2 V_2 - p_1 V_1}{1 - n} = \frac{R(T_2 - T_1)}{1 - n} = \frac{(53.3)(760 - 530)}{(1 - 1.23)(778)} = -68.5\,\frac{\text{Btu}}{\text{lbm}}$$

The heat transfer Q can now be found by using the first law of thermodynamics. For air, $c_p = 0.24$ Btu/lbm-°F.

$$Q = W + (h_2 - h_1)$$
$$Q = W + c_p(T_2 - T_1)$$
$$Q = -68.5 + 0.24(760 - 530)$$
$$Q = -13.3 \text{ Btu/lbm}$$

The minus sign indicates that the heat is transferred from, not to, the air.

Answer is (A)

PROBLEM 7-28

Air enters an engine in which the expansion ratio is 5 at a temperature of 70°F. The expansion is polytropic in accordance with $pV^{1.36} = C$ and the specific heat capacity for this process is given by $c_n = c_v(k - n)/(1 - n)$. The amount of heat added to the air, in Btu/lbm, during the expansion process is closest to

(A) 0.6
(B) 1.2
(C) 4.5
(D) 10.0
(E) 40.0

Solution. For polytropic compression of a perfect gas

$$\frac{T_2}{T_1} = \left(\frac{V_1}{V_2}\right)^{n-1}$$

or

$$T_2 = (70+460)(\tfrac{1}{5})^{1.36-1} = 297°R$$

For air, $c_v = 0.1715$ Btu/lbm-°R and $k = 1.4$. Thus

$$c_n = c_v \frac{k-n}{1-n} = (0.1715)\frac{1.4-1.36}{1-1.36} = -0.0191 \frac{\text{Btu}}{\text{lbm-°R}}$$

The heat added is therefore

$$Q = c_n(T_2 - T_1) = -0.0191(297-530) = 4.45 \text{ Btu/lbm}$$

Answer is (C)

PROBLEM 7-29

Air having an initial pressure $p_i = 14$ psia, temperature $T_i = 80°F$, and volume $V_i = 28.6$ ft^3 is compressed isentropically in a nonflow process to a final pressure $p_f = 120$ psia.

(1) The final volume, in cubic feet, is nearest to

(A) 15.5
(B) 12.1
(C) 6.2
(D) 3.7
(E) 1.4

Solution. For an isentropic process $pV^k = $ constant. For air, $k = 1.4$. Hence

$$\frac{p_f}{p_i} = \left(\frac{V_i}{V_f}\right)^k$$

$$V_f = V_i\left(\frac{p_i}{p_f}\right)^{1/k}$$

$$V_f = 28.6(\tfrac{14}{120})^{1/1.4} = 6.16 \text{ ft}^3$$

Answer is (C)

(2) The final temperature, in °F, is nearest to

(A) 815
(B) 540
(C) 370
(D) 200
(E) 190

Solution. Also for an isentropic process (with $T_i = 80 + 460 = 540°R$)

$$\frac{T_f}{T_i} = \left(\frac{p_f}{p_i}\right)^{(k-1)/k}$$

$$T_f = 540(\tfrac{120}{14})^{(1.4-1)/1.4}$$

$$T_f = 540(8.57)^{0.286}$$

$$T_f = 998°R = 998 - 460 = 538°F$$

Answer is (B)

(3) The increase in internal energy, in Btu, is nearest to

(A) 80
(B) 85
(C) 110
(D) 155
(E) 220

Solution. To determine the change in internal energy ΔU, we must find the mass of the air in this process. Using the equation of state $pV = mRT$, we have

$$m = \frac{p_i V_i}{RT_i} = \frac{\left(14\,\frac{\text{lbf}}{\text{in.}^2}\right)\left(144\,\frac{\text{in.}^2}{\text{ft}^2}\right)(28.6\,\text{ft}^3)}{\left(53.35\,\frac{\text{ft-lbf}}{\text{lbm-}°R}\right)(540°R)}$$

$$m = 2.00\,\text{lbm}$$

Now $\Delta U = mc_v(T_f - T_i)$. Using $c_v = 0.1715$ Btu/lbm-°R for air,

$$\Delta U = (2.00)(0.1715)(998 - 540) = 157\,\text{Btu}$$

Answer is (D)

PROBLEM 7-30

One hundred gal/min of kerosene is to be heated from 85°F to 195°F (zero vaporization) in an exchanger using 30 psia steam of 96% quality. The heat losses to the surrounding air have been estimated to be 3% of the heat transferred from the condensing steam to the kerosene. For kerosene, specific gravity = 0.82, specific heat = 0.52 Btu/lbm-°F. For saturated water at 30 psia, $h_f = 218.8$ Btu/lbm, $v_f = 0.0170$ ft^3/lbm. For saturated steam, $h_g = 1164.0$ Btu/lbm, $v_g = 13.76$ ft^3/lbm. If the steam condensate leaves at its saturation point, the amount of steam, in lbm/hr, that is required by the exchanger is nearest to

  (A) 2590
  (B) 2670
  (C) 2700
  (D) 3260
  (E) 3310

Solution. One gallon of kerosene has a mass of

$$\left(62.4 \frac{\text{lbm}}{\text{ft}^3}\right)\left(\frac{1 \text{ ft}^3}{7.48 \text{ gal}}\right)(0.82) = 6.84 \text{ lbm/gal}$$

The heat from the condensing steam Q_{st} will be equal to the heat gained by the kerosene Q_k plus 3% of Q_k due to the heat losses. The amount of heat gained by the kerosene is

$$Q_k = \dot{m}c_p \,\Delta T = \left(100 \frac{\text{gal}}{\text{min}} \times 60 \frac{\text{min}}{\text{hr}} \times 6.84 \frac{\text{lbm}}{\text{gal}}\right)\left(0.52 \frac{\text{Btu}}{\text{lbm-°F}}\right)(195-85)°F$$

$$= 23.5 \times 10^5 \text{ Btu/hr}$$

Therefore

$$Q_{st} = 1.03Q_k = 24.2 \times 10^5 \text{ Btu/hr}$$

We must now calculate the heat transfer Q'_{st} per lbm of steam, which is equal to the difference in enthalpies before and after the process. At 30 psia and a steam quality $x = 96\% = 0.96$, the initial enthalpy is

$$h_x = xh_g + (1-x)h_f$$

$$= 0.96(1164.0) + (1 - 0.96)(218.8) = 1117 + 8.75$$

$$= 1125.8 \text{ Btu/lbm}$$

Thus

$$Q'_{st} = (h_x - h_f) = (1125.8 - 218.8) = 907.0 \text{ Btu/lbm}$$

The required amount of steam in pounds per hour is then

$$\frac{Q_{st}}{Q'_{st}} = \frac{24.2 \times 10^5 \text{ Btu/hr}}{907.0 \text{ Btu/lbm}} = 2670 \frac{\text{lbm}}{\text{hr}}$$

Answer is (B)

PROBLEM 7-31

Steam enters the blades of a turbine at 300 psia and 1000°F. The discharge is at 3 psia.

(1) The work done, in Btu/lbm of steam, for an isentropic expansion through the turbine is nearest to

(A) 430
(B) 445
(C) 460
(D) 475
(E) 490

Solution. The solution of this problem requires the use of either a Mollier diagram or steam tables. Using a Mollier diagram (Fig. 7-7),

$$h_1 = 1524 \text{ Btu/lbm} \quad \text{and} \quad h_2 = 1067 \text{ Btu/lbm}$$

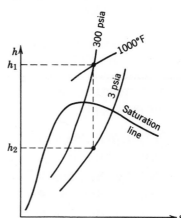

Figure 7-7

for isentropic expansion. The work W is

$$W = h_1 - h_2 = 457 \text{ Btu/lbm}$$

Answer is (C)

(2) If the efficiency of the turbine is 90%, the quality of the exhaust is nearest to

(A) 99%
(B) 98%
(C) 97%
(D) 96%
(E) 95%

Solution. If the turbine efficiency is 90%, the actual work done W' is

$$W' = 0.9W = 0.9(457) = 411 \text{ Btu/lbm}$$

Also, $W' = h_1 - h_2' = 1524 - h_2'$, so that

$$h_2' = 1524 - 411 = 1113 \text{ Btu/lbm}$$

On the diagram at the intersection of the lines $h_2' = 1113$ Btu/lbm and $p = 3$ psia, we read the steam quality x as

$$x = 99\%$$

Answer is (A)

PROBLEM 7-32

A small steam turbine is supplied with steam at 1000 psia and 100% quality, and exhausts at 14.7 psia. The turbine uses 40 lbm of steam per hour for each horsepower delivered at the turbine shaft. Heat losses from the turbine to its surroundings are negligible. The entropy per pound of the exhaust steam is closest to

(A) 1.76 Btu/lbm-°R
(B) 1.72 Btu/lbm-°R
(C) 1.68 Btu/lbm-°R
(D) 1.64 Btu/lbm-°R
(E) 1.60 Btu/lbm-°R

Solution. From steam tables it is found that the initial enthalpy of the steam is $h_1 = 1191.8 \, \text{Btu/lbm}$. Since heat losses Q are negligible and the work done per unit mass w is

$$\frac{W}{\dot{m}} = w = \frac{1 \, \text{hp}}{40 \, \text{lbm steam/hr}} = (1 \, \text{hp}) \frac{2545 \, \text{Btu/hr-hp}}{40 \, \text{lbm/hr}} = 63.6 \frac{\text{Btu}}{\text{lbm}}$$

the first law yields

$$0 = Q = W + \dot{m}(h_2 - h_1)$$

$$h_2 = h_1 - w = 1191.8 - 63.6 = 1128.2 \, \text{Btu/lbm}$$

The enthalpy h_2 and the exhaust pressure $p_2 = 14.7 \, \text{psia}$ determine the final thermodynamic state of the steam. The steam tables show that the steam quality at this state is less than 100%. Consequently we must interpolate in the tables to find the final entropy.

$$\begin{array}{ll} h_g = 1150.4 & s_g = 1.7566 \\ \underline{h_f = 180.1} & \underline{s_f = 0.3120} \\ h_{fg} = 970.3 & s_{fg} = 1.4446 \end{array}$$

Interpolating,

$$\frac{s_2 - s_g}{s_{fg}} = \frac{h_2 - h_g}{h_{fg}}$$

$$s_2 = 1.7566 + \frac{1.4446}{970.3}(1128.2 - 1150.4)$$

$$= 1.7566 - 0.0331 = 1.7235 \, \text{Btu/lbm-°R}$$

Answer is (B)

PROBLEM 7-33

Three lbm of steam expand isentropically (nonflow) from $p_1 = 300 \, \text{psia}$ and $T_1 = 700°\text{F}$ to $T_2 = 200°\text{F}$. Additional data are as follows:

Initial conditions:

$p_1 = 300 \, \text{psia}$
$T_1 = 700°\text{F}$
$v_1 = $ specific volume of steam $= 2.227 \, \text{ft}^3/\text{lbm}$
$h_1 = $ specific enthalpy of steam $= 1368.3 \, \text{Btu/lbm}$
$s_1 = $ specific entropy of steam $= 1.6751 \, \text{Btu/lbm-°R}$

Final conditions:

$T_2 = 200°F$
$h_f = $ enthalpy of water $= 167.99$ Btu/lbm
$h_{fg} = $ change of enthalpy during evaporation of water $= 977.9$ Btu/lbm
$s_f = $ entropy of saturated water $= 0.2938$ Btu/lbm-°R
$s_{fg} = $ change of entropy during evaporation of water
  $= 1.4824$ Btu/lbm-°R
$v_g = $ specific volume of steam $= 33.64$ ft^3/lbm
$p_2 = $ final pressure $= 11.53$ psia

(1) The final quality of the steam in this process is most nearly

(A) 99%
(B) 97%
(C) 95%
(D) 93%
(E) 91%

Solution. In an isentropic (reversible, adiabetic) process, no heat is transferred ($Q = 0$), and there is no change in entropy ($\Delta s = 0$). Since the steam is initially superheated,

$$s_1 = s_2 = s_f + x s_{fg} = 0.2938 + x(1.4824) = 1.6751$$

The final quality of the steam is then

$$x = \frac{1.6751 - 0.2938}{1.4824} = 0.932 = 93.2\%$$

Answer is (D)

(2) The work done in this process is most nearly

(A) 230 Btu
(B) 290 Btu
(C) 460 Btu
(D) 700 Btu
(E) 870 Btu

Solution. The first law states, per pound, $q = \Delta u + w$. Since $q = 0$,

$$w = -\Delta u = u_1 - u_2$$

If h_2 and v_2 are found, then both specific energies can be calculated by the relation $u = h - pv$.

$$h_2 = h_f + xh_{fg} = 167.99 + 0.932(977.9) = 1079 \text{ Btu/lbm}$$

$$v_2 = xv_g + (1 - x)v_f$$

In this case v_f is unknown, but since $1 - x = 0.068$ and v_f is normally much smaller than v_g, it is a good approximation to neglect the term $(1 - x)v_f$ and compute v_2 as

$$v_2 = xv_g = 0.932(33.64) = 31.4 \text{ ft}^3$$

Now we can solve for u_1 and u_2:

$$u_1 = h_1 - p_1 v_1 = 1368.3 - \frac{300(144)(2.227)}{778} = 1244.6 \text{ Btu/lbm}$$

$$u_2 = h_2 - p_2 v_2 = 1079 - \frac{11.53(144)(31.4)}{778} = 1012 \text{ Btu/lbm}$$

Hence

$$w = u_1 - u_2 = 1244.6 - 1012 = 232.6 \text{ Btu/lbm}$$

or, for 3 lbm,

$$W = 3(232.6) = 697.8 \text{ Btu}$$

Answer is (D)

PROBLEM 7-34

One lbm of air completes a reversible cycle consisting of the following two processes: (1) From a volume of 2 ft^3 and a temperature of 40°F, the air is compressed adiabatically to half the original volume. (2) Heat is then added at constant pressure until the original volume is reached.

(1) The work done in the first process is closest to

(A) 12,600 ft-lbf
(B) 21,300 ft-lbf
(C) 27,600 ft-lbf
(D) 29,800 ft-lbf
(E) 38,600 ft-lbf

Solution. Plots of the p-V and T-s planes for this process are shown in Fig. 7-8. For a reversible adiabatic process the heat transfer $Q = 0$. The temperature change is given by

$$\frac{T_B}{T_A} = \left(\frac{V_A}{V_B}\right)^{k-1} = (2.0)^{0.4} = 1.32$$

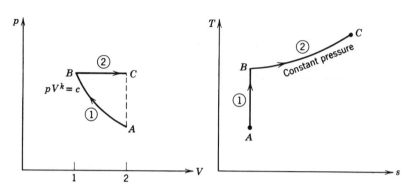

Figure 7-8

since $k = 1.4$ for air. $T_A = 40 + 460 = 500°R$ so that $T_B = 1.32(500) = 660°R$. The work done is (also see problem 7-27)

$$W = m\frac{R}{k-1}(T_B - T_A)$$

$$W = (1 \text{ lbm})\left(\frac{53.35\,\dfrac{\text{ft-lbf}}{\text{lbm-°R}}}{1.4-1}\right)(660 - 500)°R$$

$$W = 21,300 \text{ ft-lbf}$$

Answer is (B)

(2) The heat added in the second process is closest to

(A) 80 Btu
(B) 113 Btu
(C) 143 Btu
(D) 160 btu
(E) 170 Btu

Solution. At constant pressure we may use the gas law to obtain

$$\frac{V_B}{V_C} = \frac{T_B}{T_C}$$

$$T_C = T_B \frac{V_C}{V_B} = 660(\tfrac{2}{1}) = 1320°R$$

The heat transfer Q, using $c_p = 0.24$ Btu/lbm-°R for air, is

$$Q = mc_p(T_C - T_B) = (1)(0.24)(1320 - 660)$$

$$Q = 158.4 \text{ Btu}$$

Answer is (D)

PROBLEM 7-35

A mass of 0.5 kg of air, to be considered a perfect gas, is contained in a cylinder and piston machine at an initial volume of 0.03 m³ and at a pressure of 700 kN/m² (state 1). The gas is expanded reversibly at constant temperature to a volume of 0.06 m³ (state 2). The gas is then compressed reversibly and adiabatically to a volume of 0.03 m³ (state 3). It is then returned to its initial state (state 1) by a reversible constant-volume process.

(1) The net change in entropy for this cycle is most nearly

(A) −3.1 J/kg-°K
(B) 0
(C) 1.6 J/kg-°K
(D) 3.1 J/kg-°K
(E) 4.8 J/kg-°K

Solution. A T-s diagram of the cycle is shown in Fig. 7-9. For this

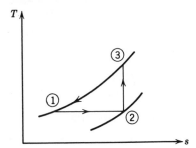

Figure 7-9

reversible cyclical process, the change in entropy Δs is zero, as is shown in the diagram.

Answer is (C)

(2) The net heat transferred from the gas is most nearly

   (A) 2.2 kJ
   (B) 4.4 kJ
   (C) 8.9 kJ
   (D) 16.8 kJ
   (E) 21.0 kJ

Solution.   The air mass $m = 0.5$ kg. At state 1, $p_1 = 700$ kN/m^2 and $V_1 = 0.03$ m^3. Also, $V_2 = 0.06$ m^3, $V_3 = 0.03$ m^3. Using the equation of state,

$$T_1 = \frac{p_1 V_1}{mR} = \frac{(7 \times 10^5 \text{ N/m}^2)(0.03 \text{ m}^3)}{(0.5 \text{ kg})(286.8 \text{ N-m/kg-°K})} = 146.4°\text{K}$$

For a constant-temperature process the gas law gives

$$p_2 = p_1 \left(\frac{V_1}{V_2}\right) = (700 \text{ kN/m}^2)(\tfrac{1}{2}) = 350 \text{ kN/m}^2$$

Then T_3 can be found from the relation

$$\frac{T_3}{T_2} = \left(\frac{V_2}{V_3}\right)^{k-1} = \left(\frac{2}{1}\right)^{1.4-1} = 1.32$$

and

$$T_3 = 1.32 T_2 = 1.32(146.4) = 193.2°\text{K}$$

The net heat transfer per unit mass is

$$\oint_{\text{cycle}} dQ = {}_1Q_2 + {}_2Q_3 + {}_3Q_1$$

$$= T_1 R \ln\left(\frac{V_2}{V_1}\right) + 0 + c_v(T_1 - T_3)$$

$$= (146.4°\text{K})(286.8 \text{ N-m/kg-°K}) \ln\left(\frac{0.06}{0.03}\right)$$

$$+ (716 \text{ N-m/kg-°K})(146.4 - 193.2)°\text{K}$$

$$= 29.1 - 33.5 = -4.4 \text{ kJ/kg}$$

For the 0.5 kg mass, the result is $(-4.4 \text{ kJ/kg})(0.5 \text{ kg}) = -2.2 \text{ kJ}$. The negative sign indicates the heat is transferred away from the gas.

Answer is (A)

PROBLEM 7-36

A gas having a molecular weight of 36 and initially at 50 psia and 1200°F is expanded adiabatically in a turbine to a pressure of 15 psia. The specific heat ratio k of the gas is 1.25, and the isentropic efficiency of the turbine is 0.70. The gas is expanded at a steady rate of 1 lbm/sec, and the gas velocity is low at both the inlet to and the exhaust from the turbine. The horsepower developed by the turbine is most nearly

(A) 71
(B) 84
(C) 97
(D) 110
(E) 137

Solution. The isentropic efficiency $\eta = W_A/W_s = 0.7$, where W_A is the actual work done and W_s is the work that would be done if the process were isentropic. We shall determine W_s and thus find W_A. For an adiabatic expansion we have

$$\frac{T_2}{T_1} = \left(\frac{p_2}{p_1}\right)^{(k-1)/k}$$

$$T_2 = (1200+460)(\tfrac{15}{50})^{(1.25-1)/1.25}$$

$$= (1660)(0.3)^{0.2} = 1660(0.786) = 1305°\text{R}$$

Neglecting changes in kinetic and potential energy for an ideal gas that undergoes a steady flow, adiabatic process, the isentropic work W_s is

$$W_s = \dot{m}c_p(T_1-T_2) = \dot{m}\frac{kR}{k-1}(T_1-T_2)$$

where $R = \bar{R}/M$ with $\bar{R}$ = universal gas constant and M = molecular weight.

$$R = \frac{1545}{36} = 42.9\,\frac{\text{ft-lbf}}{\text{lbm-°R}}$$

$$W_s = (1)\frac{(1.25)(42.9)}{1.25-1}(1660-1305)$$

$$W_s = 76{,}100 \text{ ft-lbf}$$

The actual work done is

$$W_A = \eta W_s = \frac{0.7(76,100)}{550 \dfrac{\text{ft-lbf}}{\text{sec-hp}}}$$

$$W_A = 96.9 \text{ hp}$$

Answer is (C)

PROBLEM 7-37

A turbine, which is part of a Rankine cycle, receives steam at a pressure of 180 psia and 400°F from the boiler. The turbine exhausts at a pressure of 5 psia. Assume that the turbine operates adiabatically and reversibly. Neglect pump work.

(1) The heat efficiency of the cycle is closest to

   (A) 0.17
   (B) 0.20
   (C) 0.23
   (D) 0.26
   (E) 0.28

Solution.   Since the turbine process is both adiabatic and reversible, it is also isentropic, so that $s_1 = s_2$.

From the tables for superheated steam at 180 psia and 400°F we find $h_1 = 1214.0$ Btu/lbm and $s_1 = 1.5745$ Btu/lbm-°R. At 5 psia, the tables give

$$h_f = 130.13 \text{ Btu/lbm} \qquad h_{fg} = 1001.0 \text{ Btu/lbm}$$

$$s_f = 0.2347 \text{ Btu/lbm-°R} \qquad s_{fg} = 1.6094 \text{ Btu/lbm-°R}$$

The steam quality x is determinable from the equation

$$s_1 = s_2 = s_f + x s_{fg}$$

$$1.5745 = 0.2347 + x(1.6094) \qquad \text{or} \qquad x = 0.832 = 83.2\%$$

Now we can find the final enthalpy h_2:

$$h_2 = h_f + x h_{fg}$$

$$h_2 = 130.13 + (0.832)(1001.0)$$

$$h_2 = 962.96 \text{ Btu/lbm}$$

The heat efficiency $\eta = \dfrac{\text{net work output}}{\text{total heat input}}$

$$\eta = \frac{h_1 - h_2}{h_1 - h_f} = \frac{1214.0 - 962.96}{1214.0 - 130.13} = 0.23 = 23\%$$

Answer is (C)

(2) The number of lbm steam per horsepower-hour is closest to

(A) 8
(B) 10
(C) 12
(D) 14
(E) 16

Solution. The steam rate $w = \dfrac{\text{work/hp-hr}}{\text{work/lbm steam}}$

$$w = \frac{2545}{h_1 - h_2} = \frac{2545}{1214.0 - 962.96} = 10.1 \text{ lbm steam/hp-hr}$$

Answer is (B)

PROBLEM 7-38

A nozzle is designed to expand air from 100 psia and 90°F to 20 psia. Assume an isentropic expansion and a zero initial velocity. The air flow rate is 3 lbm/sec, and the molecular weight of air is 29.

(1) The exit velocity, in feet per second, is nearest to

(A) 280
(B) 880
(C) 1320
(D) 1560
(E) 2040

Solution. Applying the first law to the nozzle flow,

$$\frac{V_1^2}{2} + h_1 = \frac{V_2^2}{2} + h_2$$

Noting that the initial velocity $V_1 = 0$,

$$V_2 = [2(h_1 - h_2)]^{1/2}$$

which for an ideal gas can be expressed as

$$V_2 = \left[2c_p T_1 \left(1 - \frac{T_2}{T_1} \right) \right]^{1/2}$$

Since the expansion is isentropic and the initial and final pressures are known, we shall use the isentropic relation $Tp^{(1-k)/k} = \text{constant}$ and $k = 1.4$ for air to obtain

$$V_2 = \left\{ 2c_p T_1 \left[1 - \left(\frac{p_2}{p_1} \right)^{(k-1)/k} \right] \right\}^{1/2}$$

With the aid of some conversion factors we find

$$V_2 = \left\{ 2 \left(32.2 \frac{\text{lbm-ft}}{\text{lbf-sec}^2} \right) \left(778 \frac{\text{ft-lbf}}{\text{Btu}} \right) \left(0.24 \frac{\text{Btu}}{\text{lbm-}^\circ\text{R}} \right) (90 + 460)^\circ\text{R} \right.$$
$$\left. \times \left[1 - \left(\frac{20}{100} \right)^{(1.4-1)/1.4} \right] \right\}^{1/2}$$

$$V_2 = \{6.61(10^6)[1 - 0.631] \, \text{ft}^2/\text{sec}^2\}^{1/2} = 1562 \, \text{ft/sec}$$

$$\text{Answer is (D)}$$

(2) The correct exit cross-sectional area, in in.^{2}, is closest to

    (A) 2.00
    (B) 1.75
    (C) 1.50
    (D) 1.25
    (E) 1.00

Solution. Since $\dfrac{T_2}{T_1} = \left(\dfrac{p_2}{p_1} \right)^{(k-1)/k} = 0.631$,

$$T_2 = 0.631(90 + 460) = 347^\circ\text{R}$$

By the equation of state, the specific volume v_2 is

$$v_2 = \frac{RT_2}{p_2} = \frac{\bar{R}T_2}{Mp_2} = \frac{(1545)(347)}{(29)(20)(144)}$$

$$v_2 = 6.42 \, \text{ft}^3/\text{lbm}$$

The mass rate of flow is 3 lbm/sec; thus conservation of mass requires

$$w = \frac{A_2 V_2}{v_2}$$

or

$$A_2 = \frac{w v_2}{V_2} = \frac{3(6.42)}{1562} = 0.0123 \text{ ft}^2$$

$$A_2 = (0.0123)(144) = 1.77 \text{ in.}^2$$

Answer is (B)

PROBLEM 7-39

Figure 7-10 is a schematic diagram of a typical refrigeration system showing the major components, as well as the state of the refrigerant between the components (that is, liquid or vapor). Assume the refrigeration system has a 10-ton capacity while operating with a 40°F suction temperature and a condensing temperature of 102°F. Also assume the heat added to the vapor

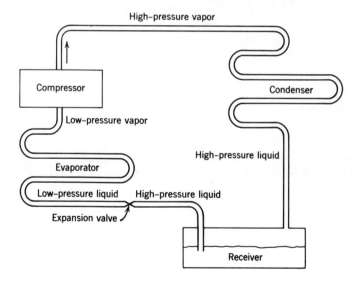

Figure 7-10

during compression is 29.6 Btu/min per ton of refrigeration, and a ton of refrigeration equals 12,000 Btu/hr.

The amount of water, in gallons per minute, required for condensing if a 15°F temperature rise is allowed is nearest to

    (A) 12
    (B) 14
    (C) 16
    (D) 18
    (E) 20

Solution.   The heat rejected by the system must equal the heat entering the system plus the work done on the system. The heat in is

$$\left(\frac{12,000 \text{ Btu/hr}}{1 \text{ ton}}\right)\left(\frac{1}{60}\frac{\text{hr}}{\text{min}}\right)(10 \text{ ton}) = 2000 \frac{\text{Btu}}{\text{min}}$$

The heat rejected is then $2000 + 29.6(10) = 2296$ Btu/min. This heat must be accepted by the cooling water with a coolant temperature rise of 15°F. Since 1 Btu will raise 1 lb of water 1°F, the flow rate of cooling water must be $2296/15 = 153$ lb/min of water. One gal of water weighs

$$\left(62.4 \frac{\text{lb}}{\text{ft}^3}\right)\left(231 \frac{\text{in.}^3}{\text{gal}}\right)\left(\frac{1}{1728}\frac{\text{ft}^3}{\text{in.}^3}\right) = 8.34 \frac{\text{lb}}{\text{gal}}$$

Hence $153/8.34 = 18.4$ gal/min of cooling water are required.

Answer is (D)

PROBLEM 7-40

A counterflow heat exchanger is operating with the hot liquid entering at 400°F and leaving at 327°F. The cool liquid enters at 100°F and leaves at 283°F. The logarithmic mean temperature difference between the hot and cold liquids is nearest to

    (A) 172
    (B) 166
    (C) 152
    (D) 136
    (E) 120

Solution.   A diagram of the process is shown in Fig. 7-11. The logarithmic

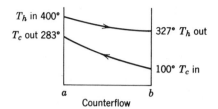

T_h in 400°

T_c out 283°

327° T_h out

100° T_c in

a   *b*

Counterflow

Figure 7-11

mean temperature difference ΔT_m, as used in heat exchanger calculations, is
defined as

$$\Delta T_m = \frac{\Delta T_a - \Delta T_b}{\ln (\Delta T_a / \Delta T_b)}$$

Here

$$\Delta T_a = 400 - 283 = 117°$$

$$\Delta T_b = 327 - 100 = 227°$$

Hence

$$\Delta T_m = \frac{117 - 227}{\ln (117/227)} = \frac{-110}{-0.663} = 166°F$$

Answer is (B)

PROBLEM 7-41

A windowless wall of a house 8 ft high by 15 ft long is of wood frame
construction with a $\frac{3}{4}$-in. stucco exterior, 3 in. of insulation, and a $\frac{1}{2}$-in.
plasterboard interior. The outside temperature is 100°F and the inside
temperature is 75°F. The conductivities of the stucco and the insulation are
12.00 and 0.27, respectively. The conductance of the plasterboard is 2.82.
(Ignore the effects of the studs and both the inside and outside surface
conductances.)

The heat flow through the wall, in Btu/hr, is most nearly

(A) 270
(B) 268
(C) 266
(D) 264
(E) 262

Solution.   A cross section of the wall is shown in Fig. 7-12. The basic heat

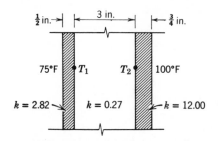

Figure 7-12

conduction equation is

$$Q = k\frac{A}{L}(T_a - T_b)\,\frac{\text{Btu}}{\text{hr}}$$

where k = conductivity in Btu-in./ft²-°F-hr, A = wall area in square feet, L = wall thickness in inches, and $T_a - T_b$ = temperature difference in degrees Fahrenheit. For steady heat flow through the wall, the heat flow Q is identi·al in each layer of material. Thus

$$(100 - T_2) = \frac{0.75}{12}\frac{Q}{A}$$

$$(T_2 - T_1) = \frac{3}{0.27}\frac{Q}{A}$$

$$(T_1 - 75) = \frac{0.5}{2.82}\frac{Q}{A}$$

Adding these three equations gives

$$(100 - 75) = \frac{Q}{A}\left[\frac{0.75}{12} + \frac{3}{0.27} + \frac{0.5}{2.82}\right]$$

Using $A = 8(15) = 120\ \text{ft}^2$, we have

$$(25)(120) = Q(0.06 + 11.11 + 0.18)$$

$$Q = \frac{(25)(120)}{11.35} = 264\,\frac{\text{Btu}}{\text{hr}}$$

Answer is (D)

PROBLEM 7-42

One end of a copper bar of cross-sectional area 4 cm² and length 80 cm is kept in steam at 1 atm pressure; the other end is in contact with melting ice. The thermal conductivity of copper is $k = 420$ J/m-sec-°C, and the specific latent heat of ice is 335 kJ/kg.

The amount of ice, in grams, that will be melted in 10 min if the sides of the copper rod are insulated is closest to

(A) 35.0

(B) 37.5

(C) 40.0

(D) 42.5

(E) 45.0

Solution.

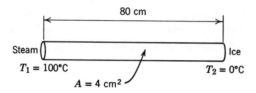

Figure 7-13

The rate at which heat is transferred from the hot end to the cold end of the rod is

$$Q = k\frac{A}{L}(T_1 - T_2)$$

$$Q = \left(\frac{420 \text{ J}}{\text{m-sec-°C}}\right)\left(\frac{4}{80} \text{ cm}\right)\left(\frac{1 \text{ m}}{100 \text{ cm}}\right)(100 - 0)\text{°C} = 21\frac{\text{J}}{\text{sec}}$$

In 10 min the total heat transferred is

$$\left(21\frac{\text{J}}{\text{sec}}\right)\left(60\frac{\text{sec}}{\text{min}}\right)(10 \text{ min}) = 12{,}600 \text{ J}$$

The amount of ice melted is therefore

$$\frac{12{,}600 \text{ J}}{335 \text{ kJ/kg}} = 37.6 \text{ g}$$

Answer is (B)

8 Chemistry

Chemistry, as a branch of science, is primarily concerned with kinds of matter and the changes that occur when they are brought together. Chemistry problems cover a multitude of topics, some of which require specific recall. This collection of chemistry problems will emphasize the review of basic principles rather than the acquisition of specific fragments of information.

GAS LAW RELATIONS

For ease of comparison, gas volumes often are given at a temperature of 0°C and 760 mm of mercury (1 atm or 14.7 psia). These values are referred to as *standard temperature and pressure* (STP) or *standard conditions.*

Boyle's law and Charles' law are often combined to form the universal gas law, which is

$$\frac{p_1 V_1}{T_1} = \frac{p_2 V_2}{T_2}$$

Here the subscripts represent two different states. Values of absolute temperature and pressure are required in this relation.

At the same temperature and pressure, equal volumes of different gases contain equal numbers of molecules. One g-mole (gram molecular weight) of any gas contains Avogadro's number 6.02×10^{23} molecules, and at standard conditions it occupies a volume of 22.4 liters.

If different gases are mixed together, each component of the mixture acts as if it alone were present in the container. The pressure exerted by each component in the mixture is proportional to its mole concentration in the mixture. The total pressure of the mixture of gases is equal to the sum of the pressures of the components. Called Dalton's law, this can be written

$$p_T = p_1 + p_2 + p_3 + \cdots + p_n.$$

The partial pressure ratio (p_1/p_T) is equal to the mole fraction (n_1/n_T) of that component, or $p_1/p_T = n_1/n_T$, where $n_1 = $ moles of gas component 1 and $n_T = $ total moles of gas. For additional discussion on gas laws, refer to Chapter 7.

Example 1

A candle is burned under an inverted beaker until the flame dies out. A sample of the mixture of perfect gases in the beaker after the flame has burnt out contains 8.30×10^{20} molecules of nitrogen, 0.70×10^{20} molecules of oxygen, and 0.50×10^{20} molecules of CO_2. If the total pressure of the mixture is 760 mm of mercury, how many moles of gas are present, and what is the partial pressure of each gas?

Solution. *Law of partial pressures*: The pressure exerted by each component in a gaseous mixture is proportional to its concentration in the mixture, and the total pressure of the gas is equal to the sum of those of its components. Therefore

$$p = p_{N_2} + p_{O_2} + p_{CO_2} = 760 \text{ mm of mercury}$$

The number of moles of each gas is equal to the number of molecules divided by Avogadro's number (6.02×10^{23}):

Gas	No. molecules	Divided by	Moles of gas
N_2	8.30×10^{20}	6.02×10^{23}	13.79×10^{-4}
O_2	0.70×10^{20}	6.02×10^{23}	1.16×10^{-4}
CO_2	0.50×10^{20}	6.02×10^{23}	0.83×10^{-4}

Total moles gas $= 15.78 \times 10^{-4}$

$$p_{N_2} = p \frac{\text{moles of } N_2}{\text{total moles of gas}} = (760) \frac{13.79 \times 10^{-4}}{15.78 \times 10^{-4}} = 664 \text{ mm of mercury}$$

$$p_{O_2} = (760) \frac{1.16 \times 10^{-4}}{15.78 \times 10^{-4}} = 56 \text{ mm of mercury}$$

$$p_{CO_2} = (760) \frac{0.83 \times 10^{-4}}{15.78 \times 10^{-4}} = 40 \text{ mm of mercury}$$

CHEMICAL BALANCE CALCULATIONS

Frequently a problem describes a chemical reaction and then the quantities of the various components are to be calculated. The starting point is a balanced chemical equation. Using the atomic weights of the elements, the molecular weights of the components of the chemical equation are determined. The coefficients in front of the formulas provide the relative numbers of each kind of molecule. The relative weights of the various components are thus known. The values may be set in a ratio to determine the quantities of each component in a specified situation.

Example 2

One of the principal scale-forming constituents of water is soluble calcium bicarbonate, $Ca(HCO_3)_2$. This substance may be removed by treating the water with lime, $Ca(OH)_2$, in accordance with the following reaction:

$$Ca(HCO_3)_2 + Ca(OH)_2 \rightarrow 2CaCO_3 + 2H_2O$$

Atomic weights: $Ca = 40 \qquad H = 1 \qquad C = 12 \qquad O = 16$

Determine the kilograms of lime required to remove 1 kg of calcium bicarbonate.

Solution. Molecular weights:

$$Ca(HCO_3)_2 = 40 + (1 + 12 + 3 \times 16)(2) = 162$$

$$Ca(OH)_2 = 40 + (16 + 1)(2) = 74$$

$$CaCO_3 = 40 + 12 + (16)(3) = 100$$

$$H_2O = 2(1) + 16 = 18$$

From the balanced chemical equation we read that 1 mole of $Ca(HCO_3)_2$ plus 1 mole of $Ca(OH)_2$ combines to produce 2 moles of $CaCO_3$ and 2 moles of H_2O. Since a mole is a molecular weight of a substance in any desired weight units, here we will use 1 mole = 1 kg-molecular weight (in other problems 1 mole might be a gram-molecular weight or some other unit).

$$Ca(HCO_3)_2 + Ca(OH)_2 \rightarrow 2CaCO_3 + 2H_2O$$

$$162 \text{ kg} + 74 \text{ kg} \rightarrow 2(100) \text{ kg} + 2(18) \text{ kg}$$

From the balanced equation we see that 1 mole of lime combines with 1 mole of calcium bicarbonate. Thus 74 kg of lime combine with 162 kg of calcium bicarbonate.

For 1 kg of calcium bicarbonate,

$$(74/162), \text{ or } 0.457, \text{ kg of lime is required.}$$

COMBUSTION

Combustion is a specialized and somewhat more complex example of a chemical balance calculation. In simple complete combustion a hydrocarbon fuel (one containing carbon and hydrogen) is burned in the presence of the theoretically correct amount of oxygen to produce carbon dioxide (CO_2) and water (H_2O) as the combustion products. The source of the oxygen is air. (Air = 21% oxygen and 79% nitrogen by volume.) Often this type of problem is readily solved by a balanced calculation if two additional items are recalled:

1. One g-mole of any gas at standard conditions occupies a volume of 22.4 liters.
2. For 1 liter of oxygen, $100/21 = 4.76$ liters of air are required.

Example 3

Propane (C_3H_8) is completely burned in air with carbon dioxide (CO_2) and water (H_2O) being formed. If 15 lb of propane are burned per hour, how many cubic feet per hour of dry CO_2 are formed at 70°F and atmospheric pressure?

$$\text{Atomic weights: } C = 12 \qquad H = 1 \qquad O = 16$$

Solution. The equation for the reaction is

$$C_3H_8 + 5O_2 \rightarrow 3CO_2 + 4H_2O \qquad \text{(ignoring the } N_2 \text{ in the air)}$$

Molecular weights:

$$C_3H_8 = 36 + 8 = 44 \qquad CO_2 = 12 + 32 = 44$$

1 lb-mole = 1 lb-molecular weight.
Since 44 lb of propane produce 3(44) lb of carbon dioxide, 15 lb of propane produce 45 lb of carbon dioxide. Using the equation of state

$$pV = mRT$$

where $p = 14.7 \times 144 = 2117 \ \mathrm{lb/ft^2}$ abs

$m = 45 \ \mathrm{lbm} \ CO_2$

$$R = \frac{1545 \ \text{ft-lbf}}{\text{lbm-mole} \ °R} \times \frac{1}{\text{mol wt of} \ CO_2 = 44} = 35.1 \ \text{ft-lbf/lbm-}°R$$

$T = 70°F + 460 = 530°R$

$$V = \frac{mRT}{p} = \frac{45 \times 35.1 \times 530}{2117} = 395 \ \mathrm{ft^3/hr} \ CO_2$$

Example 4

The heating value of natural gas is $1000 \ \mathrm{Btu/ft^3}$. If a furnace has an output efficiency of 90%, what should be the supply of air in cubic feet per minute for an output of 100,000 Btu/hr? Assume the oxygen content of air is 21% by volume and that natural gas is essentially methane, which combines with oxygen as follows:

$$CH_4 + 2O_2 \rightarrow CO_2 + 2H_2O$$

Solution.   The input of natural gas is $100{,}000/(0.90 \times 1000) = 111 \ \mathrm{ft^3/hr}$. From the balanced equation we see that 1 mole of CH_4 combines with 2 moles of O_2. For $111 \ \mathrm{ft^3/hr}$ of natural gas, twice this amount, or $222 \ \mathrm{ft^3/hr}$, of pure oxygen is required. Since air is 21% by volume oxygen, the amount required is

$$\frac{222}{0.21} = 1058 \ \mathrm{ft^3/hr} \qquad \frac{1058}{60} = 17.6 \ \mathrm{ft^3} \ \text{of air per minute}$$

MOLAR SOLUTIONS

A molar solution designation is an important method of describing the relative amount of a component of a solution in quantitative form. A molar solution contains 1 g-mole of the solute in 1 liter of solution.

Example 5

How much NaCl is there in a 1 M solution? How much in a 0.5 M solution?

Atomic weights: $Na = 23$ $Cl = 35.5$

Solution. A $1 M$ solution of NaCl would contain $23 + 35.5 = 58.5 \text{g}$ of NaCl in 1 liter of solution. Similarly, a $0.5 M$ solution of NaCl would contain 29.25 g/liter.

pH VALUE

The pH value of a solution is the negative logarithm of the hydrogen ion concentration:

$$pH = -\log_{10}[H^+]$$

where $[H^+]$ means hydrogen ion concentration in g-moles per liter. In neutral solutions the pH is 7. If the pH is less than 7, the solution is acidic; if greater than 7, the solution is basic.

PROBLEM 8-1

The atomic number of an atom is derived from the number of

(A) protons
(B) electrons
(C) neutrons
(D) mesons
(E) neutrinos

Solution. The atomic number of an atom is the number of protons in the nucleus.

Answer is (A)

PROBLEM 8-2

A mu meson particle has a mass approximately

(A) that of an electron
(B) that of a proton
(C) that of a neutron
(D) between an electron and a proton
(E) between a proton and an alpha particle

Solution. Pi and mu mesons are short-lived elementary nuclear particles produced by bombardment. They decay spontaneously and have mass between those of an electron and a proton.

Answer is (D)

PROBLEM 8-3

The mass number of an atom is the number of

(A) electrons in the atom

(B) protons in the nucleus

(C) neutrons in the nucleus

(D) protons and neutrons in the nucleus

(E) electrons and neutrons in the nucleus

Solution. The mass number of an atom is the number of protons and neutrons in the nucleus.

Answer is (D)

PROBLEM 8-4

An alpha particle consists of

(A) an electron

(B) a proton

(C) a neutron

(D) a proton and a neutron

(E) two protons and two neutrons

Solution. Alpha particles are helium nuclei and therefore have two protons and two neutrons.

Answer is (E)

PROBLEM 8-5

If 5 g of mass were entirely converted to energy and the velocity of light equals 3×10^{10} cm/sec, the amount of energy liberated would be

(A) 15×10^3 joules
(B) 75×10^3 joules
(C) 22×10^{13} joules
(D) 45×10^{13} joules
(E) 45×10^{20} joules

Solution. The energy liberated in ergs is found from the famous energy equation

$$E = mc^2$$

where $m =$ converted mass in grams

$c =$ speed of light in centimeters per second

$$E = 5(3 \times 10^{10})^2 = 45 \times 10^{20} \text{ ergs}$$

$$1 \text{ joule} = 10^7 \text{ erg}$$

$$E = 45 \times 10^{13} \text{ joules}$$

Answer is (D)

PROBLEM 8-6

The halogen group includes the elements

(A) Na, Ca, K
(B) F, Cl, Br, I
(C) Au, Ag, Pt
(D) Mg, Mn, Mo
(E) C, O, H

Solution. The halogen group of elements ("the salt producers") contains fluorine (F), chlorine (Cl), bromine (Br), and iodine (I).

Answer is (B)

PROBLEM 8-7

The group of metals which is comprised of lithium, sodium, potassium, rubidium, and cesium forms a closely related family known as

(A) the rare earth group
(B) the metals of the fourth outer group
(C) the alkali metals
(D) the elements of the inner group
(E) the metals of Group VIII

Solution. These metals are all members of Group I_A—the alkali metals.

Answer is (C)

PROBLEM 8-8

The substance that does not belong in the following group is

(A) lye
(B) digestive fluid
(C) sour milk
(D) sulfuric acid
(E) HCl

Solution. Lye is a base, NaOH. The other four are acids.

Answer is (A)

PROBLEM 8-9

The number that expresses the oxidation state of an atom of an element or of a group of atoms is called the

(A) indicator
(B) displacement factor
(C) electrolyte
(D) valence
(E) conductance

Solution. Valence is the number of electron pair bonds which an atom shares with other atoms. Valence often is used synonymously with oxidation state of an atom or ion in inorganic chemistry.

Answer is (D)

PROBLEM 8-10

Which of the following statements is false?

(A) The atomic weight of oxygen is 16.
(B) The *pascal* is the pressure or stress of one newton per square meter.
(C) Zero on the absolute temperature scale is approximately $-273°C$.
(D) The *coulomb* is the quantity of electricity transported in one second by a current of one ampere.
(E) Alcohol has a freezing point of $-32°C$.

Solution. The freezing point of methyl alcohol is $-94°C$; the freezing point of ethyl alcohol is even lower.

Answer is (E)

PROBLEM 8-11

The chemical formula for ethyl alcohol (ethanol) is

(A) C_2H_6
(B) CH_4
(C) CH_3OH
(D) C_2H_5OH
(E) C_3H_7OH

Solution. The formulas given are

(A) ethane
(B) methane
(C) methyl alcohol
(D) ethyl alcohol
(E) propyl alcohol

Answer is (D)

PROBLEM 8-12

The ion responsible for the chemistry of bases in water is

(A) H^+
(B) Na^+

(C) H_3O^+

(D) OH^-

(E) HOH^-

Solution. Hydroxyl ion (OH^-) is the characteristic base of aqueous solutions.

Answer is (D)

PROBLEM 8-13

In the earth's crust (which includes a 10-mile shell and the atmosphere above it), aside from oxygen and silicon, what is the most abundant element found?

(A) aluminum

(B) iron

(C) calcium

(D) sodium

(E) hydrogen

Solution.

Oxygen	49.2%	Calcium	3.4%
Silicon	25.7	Sodium	2.4
Aluminum	7.4	Hydrogen	1.0
Iron	4.7		

(Although different sources show slightly different percentages, they agree with this order of ranking.)

Answer is (A)

PROBLEM 8-14

The principal metal applied to iron or steel in the galvanizing process is

(A) nickel

(B) chromium

(C) cadmium

(D) mercury

(E) zinc

Solution. Zinc is the principal metal applied to iron or steel in the galvanizing process.

Answer is (E)

PROBLEM 8-15

Aside from nitrogen and oxygen, which of the following gases is the most plentiful in dry air?

(A) argon

(B) carbon dioxide

(C) neon

(D) helium

(E) krypton

Solution.

Nitrogen	78.03%	Neon	0.0015%
Oxygen	20.99	Helium	0.0005
Argon	0.94	Krypton	0.00011

Carbon dioxide 0.023 to 0.050%

Answer is (A)

PROBLEM 8-16

Oxygen is converted into ozone by

(A) great heat

(B) high pressure

(C) electric discharge

(D) a catalyst

(E) none of these processes

Solution. Ozone is formed by passing an electric discharge through oxygen.

Answer is (C)

PROBLEM 8-17

The phenomenon known as corrosion in metals is by nature

    (A) chemical
    (B) organic
    (C) electrochemical
    (D) inorganic
    (E) electrical

Solution.   The phenomenon is electrochemical. It is often called electrolytic corrosion.

Answer is (C)

PROBLEM 8-18

Radioactivity is a property of all elements with atomic numbers greater than

    (A) 48
    (B) 62
    (C) 83
    (D) 88
    (E) 90

Solution.   Radioactivity is the property of naturally occurring elements with atomic numbers greater than 83.

Answer is (C)

PROBLEM 8-19

A certain process has three factors which always occur: (1) a current flow; (2) substances that are formed at each electrode; and (3) matter that is transported through a solution or molten salt. This process is called

    (A) evaporation
    (B) decomposition
    (C) hydrolysis
    (D) condensation
    (E) electrolysis

Solution. The process is called electrolysis.

Answer is (E)

PROBLEM 8-20

The process in which a solid changes directly to the gaseous state is called

(A) sublimation
(B) homogenization
(C) crystallization
(D) vaporization
(E) distillation

Solution. The process is sublimation. Solid CO_2, iodine, and naphthalene are examples of materials that sublime.

Answer is (A)

PROBLEM 8-21

Tetraethyl lead is used as an antiknock compound in gasoline. The formula for tetraethyl lead is

(A) CH_4Pb
(B) C_2H_5Pb
(C) $(C_2H_5)_4Pb_4$
(D) $(C_2H_5)_4Pb$
(E) $(C_2H_4)_4Pb$

Solution. Ethyl is C_2H_5; hence tetraethyl lead would be $(C_2H_5)_4Pb$, or, as it is often written, $Pb(C_2H_5)_4$.

Answer is (D)

PROBLEM 8-22

Carbon dioxide "snow" from a fire extinguisher puts out a fire because it

(A) lowers the kindling temperature of the material
(B) displaces the supply of oxygen

(C) raises the kindling temperature of the material

(D) cools the material below its kindling temperature

(E) acts as a catalyst to induce a chemical reaction

Solution. Carbon dioxide "snow" smothers a fire by keeping the oxygen away from the fire.

Answer is (B)

PROBLEM 8-23

Hardness in water supplies is primarily due to the solution in it of

(A) carbonates and sulfates of calcium and magnesium

(B) alum

(C) soda ash

(D) sodium sulfate

(E) sodium chloride

Solution. The compounds of calcium and magnesium cause hardness. These may include carbonates, bicarbonates, sulfates, and chlorides.

Answer is (A)

PROBLEM 8-24

Hydrogen-free carbon in the form of coke is burned with complete combustion, using 26.4% excess air. Assume the air contains 79 mole % nitrogen and 21 mole % oxygen.

Atomic weights: $C = 12$ $O = 16$ $N = 14$

If 100 moles of carbon are burned, the number of moles of CO_2 formed is nearest to

(A) 100

(B) 133

(C) 267

(D) 300

(E) 400

Solution.

$$C + O_2 \rightarrow CO_2$$

This tells us that 100 moles of carbon, when burned with 100 moles of O_2, forms 100 moles of CO_2.

Answer is (A)

PROBLEM 8-25

Helium gas and hydrogen gas are used in lighter-than-air craft. The lifting power of helium as compared to the same volume of hydrogen is nearest to

(A) 25%
(B) 50%
(C) 75%
(D) 90%
(E) 100%

Solution. In equal volumes of different gases, at the same temperature and pressure, the numbers of molecules present are equal. Thus density is proportional to molecular weight.

$$H_2 = 2 \times 1.008 = 2.016$$

$$He = 4.003$$

But the lifting power of gas is proportional to the difference between its density and the density of air. Since air has an average molecular weight of about 29, the lifting power of hydrogen is $29 - 2 = 27$ and of helium $29 - 4 = 25$.

Thus the ratio of lifting power

$$\frac{He}{H_2} = \frac{25}{27} = 93\%$$

Answer is (D)

PROBLEM 8-26

Pure water has a hydrogen-ion concentration in moles per liter of approximately

(A) 1.0
(B) 0.1

(C) 0.001

(D) 0.00001

(E) 0.0000001

Solution. Pure water is neutral (pH = 7), though slightly lowered by dissolved CO_2.

$$pH = -\log[H^+] = 7$$

Since $-(-7)\log 10 = 7$, the hydrogen-ion concentration $[H^+] = 10^{-7}$

Answer is (E)

PROBLEM 8-27

The common anesthetic chloroform $CHCl_3$ is a colorless liquid boiling at 61.2°C.

Atomic weights: $C = 12$ $H = 1$ $Cl = 35.5$

The proportion by weight of chlorine it contains is nearest to

(A) 20%

(B) 35%

(C) 60%

(D) 75%

(E) 90%

Solution. Proportion of each element:

$$
\begin{array}{ll}
1 \text{ atom of carbon} & = \quad 12 \\
1 \text{ atom of hydrogen} & = \quad 1 \\
3 \text{ atoms of chlorine} = 3(35.5) & = 106.5 \\
\hline
\text{Molecular weight} & = 119.5
\end{array}
$$

$$\text{Percent chlorine by weight} = \frac{3(35.5)}{119.5} \times 100 = 89\%.$$

Answer is (E)

PROBLEM 8-28

The hydroxyl ion concentration of 1×10^{-11} moles/liter in water can be expressed by what pH number?

(A) 3
(B) 5
(C) 7
(D) 11
(E) 14

Solution.

$$pOH = -\log[OH] = -\log 10^{-11} = 11$$

For water, $pH + pOH = 14$; therefore, $pH = 14 - 11 = 3$.

Answer is (A)

PROBLEM 8-29

The volume of 1 g-mole of oxygen at 0°C and 1 atmosphere of pressure is closest to

(A) 1.0 liter
(B) 16.0 liters
(C) 22.4 liters
(D) 0.1 meter^3
(E) 1.0 meter^3

Solution. The volume of 1 mole of any gas at standard conditions (0°C and 1 atm of pressure) is 22.4 liters.

Answer is (C)

PROBLEM 8-30

Which one of the following chemical equations is balanced? The valences of the various elements or radicals are as follows:

$$H^+ \quad Na^+ \quad Ag^+ \quad Ca^{++} \quad Zn^{++} \quad C^{++++}$$

$$OH^- \quad Cl^- \quad NO_3^- \quad O^{--} \quad CO_3^{--} \quad SO_4^{--}$$

(A) $H_2 + O_2 \rightarrow H_2O$

(B) $NaCl + 2AgNO_3 \rightarrow NaNO_3 + AgCl$

(C) $CaO + H_2O \rightarrow Ca(OH)_3$

(D) $CaSO_4 + Na_2CO_3 \rightarrow CaCO_3 + Na_2SO_4$

(E) $ZnCO_3 \rightarrow ZnO + 2CO_2$

Solution. The correct equations are

(A) $2H_2 + O_2 \rightarrow 2H_2O$

(B) $NaCl + AgNO_3 \rightarrow NaNO_3 + AgCl$

(C) $CaO + H_2O \rightarrow Ca(OH)_2$

(D) correct as given

(E) $ZnCO_3 \rightarrow ZnO + CO_2$

Answer is (D)

PROBLEM 8-31

Which one of the following chemical equations is not balanced? The valences of the various elements or radicals are as follows:

$$Ag^+ \quad H^+ \quad Ca^{++} \quad Ba^{++} \quad Pb^{++} \quad C^{++++}$$

$$C^- \quad Cl^- \quad NO_3^- \quad HCO_3^- \quad O^{--} \quad SO_4^{--} \quad CO_3^{--}$$

(A) $2AgNO_3 + CaCl_2 \rightarrow \underline{2AgCl} + Ca(NO_3)_2$

(B) $BaO + H_2SO_4 \rightarrow \underline{BaSO_4} + H_2O$

(C) $CaCO_3 + HCl \rightarrow CaCl + H_2O + CO_2$

(D) $PbCl_2 + H_2SO_4 \rightarrow \underline{PbSO_4} + 2HCl$

(E) $Ca(HCO_3)_2 \rightarrow CaCO_3 + H_2O + CO_2$

Solution. Equation (C) should be

$$CaCO_3 + 2HCl \rightarrow CaCl_2 + H_2O + CO_2$$

Answer is (C)

PROBLEM 8-32

The action of nitric acid on silver gives silver nitrate, water, and nitric oxide gas. The balanced equation for this reaction is

(A) $Ag + HNO_3 \rightarrow AgNO_3 + NO + H_2O$
(B) $2Ag + H_2NO_3 \rightarrow AgNO_3 + NO + H_2O$
(C) $3Ag + 4HNO_3 \rightarrow 3AgNO_3 + NO + 2H_2O$
(D) $Ag + 4HNO_3 \rightarrow AgNO_3 + 4NO + 8H_2O$
(E) $Ag + 2H_2NO_3 \rightarrow AgNO_3 + NO + 2H_2O$

Solution.

$$4H^+ + 3e^- + NO_3^- \rightarrow NO + 2H_2O$$
$$3Ag^0 - 3e^- \qquad \rightarrow 3Ag^+$$

$$3Ag + 4HNO_3 \quad \rightarrow 3AgNO_3 + NO + 2H_2O$$

Answer is (C)

PROBLEM 8-33

The specific gravity of a substance is the weight of a given volume of the substance divided by the weight of an equal volume of water. If 2 kg of a liquid with a specific gravity of 0.800 are mixed with 1 kg of water, the specific gravity of the mixture is nearest to

(A) 0.833
(B) 0.850
(C) 0.857
(D) 0.867
(E) 1.000

Solution. The equivalent volume of liquid $= \dfrac{2}{0.8} + 1 = 3.5$. The weight of liquid $= 2 + 1 = 3$ kg.

$$\text{Specific gravity} = \frac{\text{Wt of liquid}}{\text{Wt of equiv. volume of water}}$$

$$= \frac{3}{3.5(1)} = 0.857$$

Answer is (C)

PROBLEM 8-34

How much water is required to set 100 kg of plaster of Paris? The formulas are

plaster of Paris, $CaSO_4 \cdot \frac{1}{2}H_2O$

set form of plaster of Paris (gypsum), $CaSO_4 \cdot 2H_2O$

Atomic weights: $Ca = 40.1 \qquad S = 32.1 \qquad O = 16 \qquad H = 1$

Molecular weights:

$$CaSO_4 \cdot \tfrac{1}{2}H_2O = 40.1 + 32.1 + 4(16) + \tfrac{1}{2}(2 + 16) = 145.2$$

$$H_2O = 2(1) + 16 = 18$$

The amount of water is closest to

(A) 20 kg
(B) 25 kg
(C) 35 kg
(D) 90 kg
(E) 100 kg

Solution.

$$2(CaSO_4 \cdot \tfrac{1}{2}H_2O) + 3H_2O \rightarrow 2(CaSO_4 \cdot 2H_2O)$$

For 1 molecule of $CaSO_4 \cdot \frac{1}{2}H_2O$, $1\frac{1}{2}$ molecules of H_2O are needed to form $CaSO_4 \cdot 2H_2O$. Thus for 145.2 kg of plaster of Paris, $\frac{3}{2}(18) = 27$ kg of water are needed to form gypsum.

$$\frac{100}{145.2} = \frac{X}{27}$$

The amount of water X to set 100 kg of plaster of Paris is 18.6 kg.

Answer is (A)

PROBLEM 8-35

The number of moles of sulfuric acid required to dissolve 10 moles of aluminum according to the reaction

$$2Al + 3H_2SO_4 = Al_2(SO_4)_3 + 3H_2$$

is

(A) 3
(B) 6
(C) 10
(D) 15
(E) 30

Solution.

$$\frac{10 \text{ moles Al}}{2\text{Al}} = \frac{X \text{ moles } H_2SO_4}{3H_2SO_4} \qquad X = 15 \text{ moles } H_2SO_4$$

Answer is (D)

PROBLEM 8-36

C_7H_{16} burns completely to form water and carbon dioxide. If 10 grams of it are burned, how many liters of carbon dioxide are formed at standard conditions?

Atomic weights: $C = 12 \qquad O = 16 \qquad H = 1$

Molecular weight of $C_7H_{16} = 7(12) + 16(1) = 100$

The amount of CO_2 is nearest to

(A) 15 liters
(B) 20 liters
(C) 22.4 liters
(D) 45 liters
(E) 90 liters

Solution.   Basis: 1 g-mole C_7H_{16} burned. By inspection: 1 g-mole of C_7H_{16} will yield, on complete combustion, 7 g-moles of CO_2, occupying $7 \times 22.4 = 156.8$ liters at standard conditions.

Since 100 g (1 g-mole) of C_7H_{16} produce 156.8 liters of CO_2, 10 g will produce one-tenth that amount, or 15.7 liters of CO_2 at standard conditions.

Answer is (A)

PROBLEM 8-37

How many kilograms of limestone, 80% $CaCO_3$, would be required to make 1000 kg of CaO?

$$\text{Atomic weights: } Ca = 40 \qquad C = 12 \qquad O = 16$$

The quantity of limestone required is nearest to

(A) 1000 kg
(B) 1400 kg
(C) 1800 kg
(D) 2200 kg
(E) 2600 kg

Solution.

$$CaCO_3 \rightarrow CaO + CO_2$$

Thus 1 mole of $CaCO_3$ yields 1 mole of CaO.
  Molecular weights:

$$CaCO_3: \ 40 + 12 + 3(16) = 100$$

$$CaO: \ 40 + 16 = 56$$

So 100 kg of pure $CaCO_3$ yields 56 kg of CaO. For 1000 kg of CaO, the quantity of limestone required is

$$\frac{1000}{56} \times \frac{100}{0.8} = 2232 \text{ kg}$$

Answer is (D)

PROBLEM 8-38

If silver sells for $120 per kilogram, what is the maximum amount you could pay for a tank containing 1000 kg of silver nitrate solution that is 60% water by weight? The cost of recovering the silver from the silver nitrate solution is $1000.

$$\text{Atomic weights: } Ag = 108 \qquad N = 14 \qquad O = 16$$

The maximum amount that could be paid for the silver nitrate solution is closest to

(A) $20,000
(B) $30,000
(C) $40,000
(D) $47,000
(E) $48,000

Solution.

$$\text{Pure AgNO}_3 = 1000(0.40) = 400 \text{ kg}$$

$$\text{Molecular weight: AgNO}_3 = 108 + 14 + 3(16) = 170$$

$$\text{Zn} + 2\text{AgNO}_3 \rightarrow 2\text{Ag} + \text{Zn(NO}_3)_2$$

$$170 \text{ kg} \rightarrow 108 \text{ kg}$$

$$\text{Weight of Ag in 400 kg of AgNO}_3 = \frac{400 \times 108}{170} = 254.12 \text{ kg}$$

$$\text{Value of the silver nitrate solution} = 254.12 \text{ kg} \times \$120 - \$1000$$

$$= \$29,494.40$$

Answer is (B)

PROBLEM 8-39

The number of liters of oxygen necessary to burn 10 liters of H_2S gas according to the reaction $2H_2S + 3O_2 \rightarrow 2H_2O + 2SO_2$ is

(A) 3
(B) 6
(C) 10
(D) 15
(E) 30

Solution. Using a properly balanced formula we could easily calculate the molecular weights of the gases and of the resulting products. This is not necessary in this problem because we know that equal numbers of molecules of gases occupy equal volumes.

Here we have 2 molecules of hydrogen sulfide and 3 molecules of oxygen reacting. We see, therefore, that 2 volumes of H_2S will react with 3 volumes of O_2.

For 10 liters of H_2S, $\frac{3}{2} \times 10 = 15$ liters of oxygen will be required.

Answer is (D)

PROBLEM 8-40

A compound was found to contain 42.12 wt % carbon, 51.45 wt % oxygen, and 6.43 wt % hydrogen. Its molecular weight was determined to be approximately 340. What is the chemical formula of the compound?

Atomic weights: $C = 12$ $H = 1$ $O = 16$

(A) $C_4H_6O_5$
(B) C_6H_8
(C) $C_8H_{10}O_7$
(D) $C_{10}H_{20}O_8$
(E) $C_{12}H_{22}O_{11}$

Solution.   One molecular weight of the compound contains:

C	$42.12\% \times 340 = 143.21 \div 12 = 11.93$	or	12 gram-atoms of C
O	$51.45\% \times 340 = 174.93 \div 16 = 10.93$	or	11 gram-atoms of O
H	$6.43\% \times 340 = 21.86 \div 1 = 21.86$	or	22 gram-atoms of H

100.00   340.00

The formula of the compound contains:

12C   11O   22H

We would recognize this as the formula for plain sugar, which would be correctly written $C_{12}H_{22}O_{11}$.

Answer is (E)

PROBLEM 8-41

Calculate the molecular weight of a gas, 1.13 liters of which, collected over water at 24°C and 754 mm pressure, weighed 1.251 g when deprived of the aqueous vapor. The vapor pressure of water at 24°C is 22.2 mm. The gas constant $\bar{R} = 0.082$ liter-atm/g-mole-°K.

The molecular weight is nearest to

    (A) 18
    (B) 22
    (C) 25
    (D) 28
    (E) 31

Solution. To apply the equation $pV = \dfrac{m}{M}\bar{R}T$, we first must correct for the water vapor present:

$$p_{\text{total}} = p_{\text{gas}} + p_{\text{water}}$$

$$p_{\text{gas}} = p_t - p_w = 754 - 22.2 = 731.8 \text{ mm Hg} = \frac{731.8}{760} \text{ atm}$$

The number of moles of the gas is

$$M = \frac{m\bar{R}T}{pV} = \frac{1.251 \text{ g} \times 0.082(24 + 273)}{(731.8/760) \times 1.13} = 28 \text{ g/mole}$$

Thus the molecular weight of the gas is 28.

Answer is (D)

PROBLEM 8-42

A dry mixture of clay and barium sulfate has a density of 3.45 g/cm^3. The respective densities of the constituents are 2.67 g/cm^3 and 4.10 g/cm^3.
   The volume percent of clay in the mixture is nearest to

    (A) 25%
    (B) 35%
    (C) 45%
    (D) 55%
    (E) 65%

Solution. Basis: 100 cm^3 of mixture

Let x = volume clay and $100-x$ = volume of $BaSO_4$

Wt clay + wt $BaSO_4$ = total wt

$2.67x + 4.10(100-x) = 3.45(100)$

$2.67x + 410 - 4.10x = 345$

$-1.43x = -65$

Therefore

$x = 45.5 = \%$ vol. clay

$100 - x = 54.5 = \%$ vol. $BaSO_4$

Although not required in this problem, one could also compute the weight percent of the two constituents in the mixture as follows:

Weight = volume × density
Wt clay = (45.5)(2.67) = 121.5 g =   35.2% wt clay
Wt $BaSO_4$ = (54.5)(4.10) = 223.5 g =   64.8% wt $BaSO_4$

total wt = 345.0 g = 100.0%

Answer is (C)

PROBLEM 8-43

Which of the following metals has the slowest rate of combining with oxygen to form metallic oxide?

(A) tin
(B) copper
(C) iron
(D) magnesium
(E) zinc

Solution.   These metals when placed in order of activity as indicated by the emf scale would be

magnesium
zinc
iron
tin
copper

with the most active metal placed first.

Metals of the copper subgroup (Cu, Ag, and Au) are known for their inactivity as a result of their low oxidation potentials. These are the coinage metals. As the oxidation potential of Cu is low and nearest to that for the reduction potential of oxygen in the presence of water, it will thus have the slowest rate of reaction.

Answer is (B)

PROBLEM 8-44

Which of the following is the most chemically active (highest in electromotive series)?

(A) cesium
(B) potassium
(C) sodium
(D) strontium
(E) magnesium

Solution.

Electromotive Series

1. lithium	9. aluminum
2. potassium	10. manganese
3. cesium	11. zinc
4. barium	·
5. strontium	·
6. calcium	·
7. sodium	silver
8. magnesium	gold

Answer is (B)

PROBLEM 8-45

The concentration of H_2SO_4 is given in terms of moles per liter as 0.2 M—H_2SO_4. The normal concentration of this solution would be designated as

(A) 0.05 N—H_2SO_4
(B) 0.1 N—H_2SO_4

(C) $0.2\ N—H_2SO_4$
(D) $0.4\ N—H_2SO_4$
(E) $0.5\ N—H_2SO_4$

Solution. A molar solution contains a gram-molecular weight of the solute in 1 liter of solution. A normal solution contains a gram-equivalent weight of the solution in 1 liter of solution.

Since valence $= \dfrac{\text{atomic weight}}{\text{equivalent wt}}$ and in H_2SO_4 the radical is bivalent, it

follows that a $1\ M$ solution of H_2SO_4 is also a $2\ N$ solution. In this problem the $0.2\ M$ solution could also be called a $0.4\ N$ solution.

Answer is (D)

PROBLEM 8-46

A solution of $0.1\ N$ hydrochloric acid is required. The number of liters of water that should be added to 1 liter of $2.0\ N$ hydrochloric acid in order to obtain the desired solution is nearest

(A) 19
(B) 20
(C) 21
(D) 22
(E) 23

Atomic weights: $H = 1$ $O = 16$ $Cl = 35.5$

Solution. A normal solution contains a gram-equivalent weight of the solute in 1 liter of solution.

$$x = \text{total quantity}$$

$$2.0(1\ \text{liter}) = 0.1(x\ \text{liters})$$

$$x = \frac{2.0}{0.1} = 20\ \text{liters}$$

Therefore 1 liter of $2.0\ N$ HCl yields 20 liters of $0.1\ N$ HCl. So add $20 - 1 = 19$ liters of water.

Answer is (A)

PROBLEM 8-47

What volume of 2 N H_2SO_4 must be mixed with 6 N H_2SO_4 to yield 700 ml of a 3 N H_2SO_4 solution. Assume all volumes are additive. Volume of 2 N H_2SO_4 is nearest to

 (A) 175 ml
 (B) 250 ml
 (C) 350 ml
 (D) 425 ml
 (E) 525 ml

Solution.   Let x = ml of 2 N H_2SO_4 and y = ml of 6 N H_2SO_4. Now we can write two expressions:

$$\text{Total volume} \quad x + y = 700 \tag{1}$$
$$\text{Normality} \times \text{volume} \quad 2x + 6y = 3(700) \tag{2}$$

Substitute Eq. (1) into Eq. (2)

$$2(700 - y) + 6y = 2100$$
$$1400 - 2y + 6y = 2100$$
$$y = \tfrac{700}{4} = 175 \text{ ml of 6 } N \text{ } H_2SO_4$$
$$x = 700 - 175 = 525 \text{ ml of 2 } N \text{ } H_2SO_4$$

$$\text{Answer is (E)}$$

PROBLEM 8-48

Commercial preparation of magnesium metal is by electrolysis of a molten mixture of 70 wt % magnesium chloride and 30 wt % sodium chloride.

$$\text{Atomic weights: Mg} = 24.32 \quad \text{Cl} = 35.45 \quad \text{Na} = 23$$

Five kg of this mixture will yield an amount of magnesium closest to

 (A) 0.90 kg
 (B) 1.10 kg
 (C) 1.30 kg
 (D) 2.00 kg
 (E) 3.50 kg

Solution. Five kg of mixture contains $0.70 \times 5 = 3.5$ kg of $MgCl_2$. The weight of Mg in 3.5 kg of $MgCl_2 = \dfrac{24.32}{24.32 + 2(35.45)} \times 3.5 = 0.89$ kg

Answer is (A)

PROBLEM 8-49

It is desired to produce 100 g-moles of compound AD by reacting compounds A_2B and CD_2 at a constant temperature of 25°C.

$$A_2B + CD_2 \rightarrow BC + 2AD \pm Q \text{ kcal}$$

The heats of formation of these compounds from the elements A, B, C, and D at 25°C are as follows:

A_2B	100 kcal/g-mole
CD_2	200 kcal/g-mole
BC	300 kcal/g-mole
AD	150 kcal/g-mole

Assume that the reactants react completely to form the products. How much heat must be added or removed during the reaction to maintain the desired temperature of 25°C? Use − sign if heat is removed and + sign if heat is added.

The heat required is closest to

(A) −7500 kcal
(B) −15,000 kcal
(C) 0 kcal
(D) +7500 kcal
(E) +15,000 kcal

Solution.

$$50 \text{ g-moles } A_2B + 50 \text{ g-moles } CD_2 \rightarrow 50 \text{ g-moles } BC + 100 \text{ g-moles } AD$$

$$50(100 \text{ kcal}) + 50(200 \text{ kcal}) \rightarrow 50(300 \text{ kcal}) + 100(150 \text{ kcal}) + Q \text{ kcal}$$

$$5000 \text{ kcal} + 10,000 \text{ kcal} \rightarrow 15,000 \text{ kcal} + 15,000 \text{ kcal} + Q \text{ kcal}$$

For a heat balance

$$15,000 \text{ kcal} \rightarrow 30,000 \text{ kcal} - 15,000 \text{ kcal}$$

Therefore

$$Q = -15,000 \, \text{kcal}$$

Answer is (B)

PROBLEM 8-50

Naturally occurring boron consists of two isotopes, $_5B^{10}$ and $_5B^{11}$. It has an average atomic weight of 10.81. The mole % $_5B^{10}$ in the naturally occurring mixture is nearest to

(A) 7%
(B) 14%
(C) 19%
(D) 23%
(E) 37%

Solution. Let $x =$ mole fraction $_5B^{10}$

A mass balance yields $x(10) + (1-x)(11) = 1(10.81)$

$$10x + 11 - 11x = 10.81$$

$$x = 0.19 = 19\%$$

Answer is (C)

PROBLEM 8-51

If common salt, NaCl, is dissolved in water, which of the following statements is true about the behavior of the solution relative to the behavior of pure water?

(A) boiling point is lowered
(B) freezing point is decreased
(C) ionization of the water is decreased
(D) ionization of the water is increased
(E) none of the above is true

Solution. Adding solute (salt) to solvent (water) decreases the water freezing point and elevates its boiling point. Since there are no ions from NaCl common with those of water, ionization of the water is unaffected.

Answer is (B)

PROBLEM 8-52

The solubility product of MgF_2 at 18°C is 7.1×10^{-9}. Solubility of MgF_2 in grams per liter of water is nearest to

(A) 0.07
(B) 0.15
(C) 0.69
(D) 1.80
(E) 3.70

Atomic weights: $Mg = 24.3$ $F = 19.0$

Solution. The small amount of MgF_2 that dissolves in water produces Mg^{++} and F^- ions, according to $MgF_2 \rightarrow Mg^{++} + 2F^-$. The solubility product equals $[Mg^{++}][F^-]^2$, where the brackets indicate the molar concentration of the ions.

Let x = moles/liter Mg^{++}.

$$7.1 \times 10^{-9} = x(2x)^2$$
$$4x^3 = 7.1 \times 10^{-9}$$
$$x = 1.21 \times 10^{-3}$$

The solution contains 1.21×10^{-3} moles MgF_2 per liter.

Molecular weight $MgF_2 = 24.3 + 2(19.0) = 62.3$ g/mole

Solubility $= 62.3 \times 1.21 \times 10_0^{-3} = 0.075$ g/liter

Answer is (A)

PROBLEM 8-53

If K is the equilibrium constant for decomposition of nitrogen tetroxide according to $N_2O_4 = 2NO_2$, which is nitrogen dioxide, the expression in partial pressures for the equilibrium constant is

(A) $K = \dfrac{p_{NO_2}}{p_{N_2O_4}}$

(B) $K = p_{NO_2}p_{N_2O_4}$

(C) $K = \dfrac{p_{N_2O_4}}{p_{NO_2}}$

(D) $K = \dfrac{(p_{NO_2})^2}{p_{N_2O_4}}$

(E) $K = \dfrac{p_{NO_2}}{(p_{N_2O_4})^2}$

Solution. For a reaction of the form $aA + bB = cC + dD$, an expression for the equilibrium constant, expressed in partial pressures of each component in the system, is

$$K = \dfrac{(p_C)^c (p_D)^d}{(p_A)^a (p_B)^b}$$

Answer is (D)

PROBLEM 8-54

If the ionization constant of acetic acid, CH_3COOH, is 1.76×10^{-5} at 25°C and acetic acid ionizes according to $CH_3COOH = CH_3COO^- + H^+$, the H^+ concentration in moles/liter produced by $1\,M$ CH_3COOH in water is nearest to

(A) 1×10^{-7}

(B) 2×10^{-5}

(C) 4×10^{-3}

(D) 7×10^{-2}

(E) 1

Solution. $K = [CH_3COO^-][H^+]/[CH_3COOH]$, where brackets indicate concentrations in moles per liter.

Let x = moles/liter of CH_3COOH ionized, yielding x moles/liter of H^+ and CH_3COO^-.

$$1 - x = \text{moles/liter of acid remaining and not ionized}$$

$$K = \dfrac{(x)(x)}{(1-x)} = 1.76 \times 10^{-5}$$

Simplify by considering the denominator term $(1-x) \cong 1$, since ionization is small.

$$x = (1.76 \times 10^{-5})^{1/2} = 4.19 \times 10^{-3} = [H^+]$$

(An exact solution yields a comparable answer.)

Answer is (C)

PROBLEM 8-55

At 300°C the equilibrium constant is 45 for the water gas shift reaction: $CO + H_2O = CO_2 + H_2$. An equimolar mixture of CO and steam comes to equilibrium at 1 atm total pressure after being passed over a catalyst. The mole fraction of hydrogen in the resulting gas is

(A) 0.06
(B) 0.15
(C) 0.27
(D) 0.44
(E) 0.73

Solution.   The equilibrium constant for the shift reaction is

$$K = \frac{(p_{H_2})(p_{CO_2})}{(p_{CO})(p_{H_2O})} = 45$$

The partial pressure p of each component of the resulting mixture is total pressure p_t times mole fraction. Starting with one mole each of CO and H_2O, let $x =$ moles of H_2 produced at equilibrium. From the equation, total moles does not change from two. Mole fractions are

$$CO = \frac{(1-x)}{2} \qquad H_2O = \frac{(1-x)}{2} \qquad CO_2 = \frac{x}{2} \quad \text{and} \quad H_2 = \frac{x}{2}$$

$$\frac{(p_t x/2)(p_t x/2)}{[p_t(1-x)/2][p_t(1-x)/2]} = 45 = \frac{x^2}{(1-x)^2}$$

Take the square root of the entire equation and solve to get

$$\sqrt{45} = \frac{x}{1-x} \qquad x = \frac{\sqrt{45}}{1+\sqrt{45}} = 0.870$$

Mole fraction H_2 is $x/2 = 0.435$.

Answer is (D)

9 Electricity

The study and use of electrical energy in an ever-increasing variety of ways is the goal of an entire engineering discipline, that of electrical engineering. And as this field continues its rapid growth, the importance of a fundamental knowledge of electric circuits and electromechanical energy conversion grows with it. The problems collected here primarily emphasize a few basic principles of direct-current (D-C) and alternating-current (A-C) circuits and motors.

D-C CIRCUITS

In D-C circuitry a potential difference or voltage V across the terminals of some resistive device causes an electric current I, in amperes (A), to flow through it. According to Ohm's law,

$$V = IR \qquad (9\text{-}1)$$

where R is the resistance, in ohms (Ω), of the element to current flow, and the current is measured in amperes. Furthermore, power must be expended in sustaining this current flow in the amount

$$P = VI = I^2 R \qquad (9\text{-}2)$$

The units of this equation are volt-amperes or watts (W). The driving force for such current flows is called an electromotive force (emf). Typical sources of emf are batteries, generators, and motors.

For resistors of uniform geometric shape, the resistance R is a function of material properties and the size of the element. For a uniform wire, for example,

$$R = \rho \frac{l}{A} \qquad (9\text{-}3)$$

where ρ is the resistivity (a material property) and l and A are, respectively, the length and cross-sectional area of the wire.

Kirchhoff's two laws define the relations that exist when individual elements are combined to form an electric circuit:

1. The sum of all currents flowing into a junction must equal the sum of all currents flowing away from the junction.

2. The algebraic sum of the emf's and the voltage drops around a closed loop is equal to zero.

Rule 1 assures that charge is conserved at a point. In a large circuit or network containing n junctions, this rule must be applied $n-1$ times; the rule is then automatically satisfied at the last junction. Rule 2 is in a way a generalized application of Ohm's law. In the use of these rules care must be taken so that a chosen sign convention is used consistently.

In D-C circuits resistors and emf's may be connected in a seemingly bewildering variety of ways for different purposes. Most, but not all, resistor combinations are sequences of either the basic series or parallel arrangements of individual resistors. In these cases the single equivalent series resistance R_e is

$$R_e = \sum_{n=1}^{N} R_n \qquad (9\text{-}4)$$

for a series of N individual resistors and

$$\frac{1}{R_e} = \sum_{n=1}^{N} \frac{1}{R_n} \qquad (9\text{-}5)$$

for a parallel combination of N resistors. The Y-Δ transformation, an important exception, is treated in the problems.

Mesh analysis is normally·applied to the solution of larger networks or networks that are not easily simplified by use of equivalent resistances. The technique concentrates on writing loop equations or node equations for a network; only loop equations are considered here. If the common intersection of two or more lines in a network is called a junction J, and a line connecting two junctions and containing an electrical element is called a branch B, then the number of independent loop equations N required to solve the network is $N = 1 + B - J$. First the loops and the N independent unknown loop currents are identified. Then Kirchhoff's voltage law (rule 2) is written for each loop, and the resulting set of equations is solved. As example 2 will show, these loop equations could also be written directly by inspection.

Example 1

Solve for the five branch currents shown (Fig. 9-1) in this diagram of a D-C network.

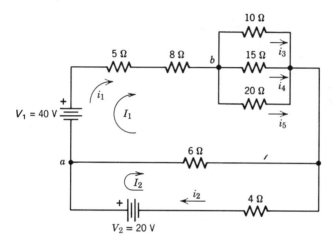

Figure 9-1

Solution. We begin by applying Kirchhoff's first rule. This network may be considered to have only two junctions if the parallel combination of resistors is replaced by one equivalent resistor R_e in series. Examining the junction marked a, the branch current i_6 through the 6-Ω resistor to the left is equal to the difference in loop currents I_1 and I_2

$$i_6 = I_1 - I_2$$

Next we replace the three resistors by the one equivalent series resistor R_e:

$$\frac{1}{R_e} = \tfrac{1}{10} + \tfrac{1}{15} + \tfrac{1}{20} = \tfrac{13}{60}$$

$$R_e = \tfrac{60}{13} \, \Omega$$

Now we can apply Kirchhoff's second rule to the two loops to obtain

$$40-(5+8+\tfrac{60}{13})I_1-6(I_1-I_2)=0 \tag{1}$$

$$20-6(I_2-I_1)-4I_2=0 \tag{2}$$

From Eq. (2), $I_2=0.1(20+6I_1)$. Substituting the relation into Eq. (1) gives $I_1=2.6$ A. Then we find $I_2=3.56$ A.

Returning to the parallel combination of resistors, the voltage drop across this combination is, by use of Ohm's law,

$$V_e = I_1 R_e = (2.6)(\tfrac{60}{13}) = 12 \text{ V}$$

This voltage drop occurs across each of these three resistors, whereas the use of Kirchhoff's first rule at point b shows that

$$I_1 = i_1 = i_3 + i_4 + i_5$$

Repeated use of Ohm's law gives

$$i_3 = \frac{V_e}{10} = 1.2 \text{ A}$$

$$i_4 = \frac{V_e}{15} = 0.8 \text{ A}$$

$$i_5 = \frac{V_e}{20} = 0.6 \text{ A}$$

Example 2

Find the three loop currents in the D-C network shown in Fig. 9-2.

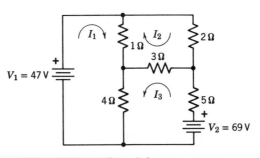

Figure 9-2

Solution.   This network has $B = 6$ branches and $J = 4$ junctions; hence the number of required loop equations is $N = 1 + B - J = 3$. Tracing each loop in the direction of the indicated current flow I and applying Kirchhoff's voltage law yields

$$1(I_1-I_2)+4(I_1+I_3)=47 \tag{1}$$

$$1(I_2-I_1)+2I_2+3(I_2+I_3)= 0 \tag{2}$$

$$5I_3+3(I_2+I_3)+4(I_1+I_3)=69 \tag{3}$$

which could be rearranged to the form

$$(1+4)I_1- \qquad 1I_2+ \qquad 4I_3=47 \tag{1'}$$

$$-1I_1+(1+2+3)I_2+ \qquad 3I_3= 0 \tag{2'}$$

$$4I_1+ \qquad 3I_2+(3+4+5)I_3=69 \tag{3'}$$

In the first equation (1') the coefficient of I_1 is the sum of the resistances in loop 1 and is called the self-resistance of the loop; the coefficients multiplying I_2 and I_3 are called mutual resistances and are the resistances common to loops 1 and 2, and 1 and 3, respectively. Equations (2') and (3') can be similarly interpreted. Modest practice will allow one to write these equations directly by inspection.

The method of substitution can be used to solve sets of equations, but here Cramer's method is used (see Chapter 2):

$$I_1=\frac{D_1}{D}, \qquad I_2=\frac{D_2}{D}, \qquad I_3=\frac{D_3}{D}$$

where

$$D = \begin{vmatrix} 5 & -1 & 4 \\ -1 & 6 & 3 \\ 4 & 3 & 12 \end{vmatrix} = 183, \qquad D_1 = \begin{vmatrix} 47 & -1 & 4 \\ 0 & 6 & 3 \\ 69 & 3 & 12 \end{vmatrix} = 1098$$

$$D_2 = \begin{vmatrix} 5 & 47 & 4 \\ -1 & 0 & 3 \\ 4 & 69 & 12 \end{vmatrix} = -183, \qquad D_3 = \begin{vmatrix} 5 & -1 & 47 \\ -1 & 6 & 0 \\ 4 & 3 & 69 \end{vmatrix} = 732$$

This leads directly to

$$I_1=6 \text{ A}, \qquad I_2=-1 \text{ A}, \qquad I_3=4 \text{ A}$$

A-C CIRCUITS

Ohm's law expresses only one of three possible relations between voltage and current, that resulting from the action of a resistive element. Two other relations exist and are characterized by two electrical elements: the inductance and the capacitance.

For an inductance

$$v = L\frac{di}{dt} \tag{9-6}$$

where L is the inductance in henries (H) when i is in amperes and v is in volts. For a sinusoidal alternating current the current lags 90° behind the voltage in a purely inductive circuit, as can be seen by assuming a sinusoidal form for i and computing v.

Capacitance, the third electrical element, directly relates the charge q to the voltage v. Thus the current-voltage relation is

$$i = \frac{dq}{dt} = C\frac{dv}{dt} \tag{9-7}$$

and C is measured in farads (F). In contrast to the inductive circuit, the current leads the voltage by 90° in a purely capacitive A-C circuit. Each phase relation contrasts with the purely resistive element, which causes no phase shift. The handy mnemonic "ELI the ICEman" can help one recall that voltage E leads current I for an inductor, whereas E lags behind I for a capacitor.

In A-C circuits we are usually interested in effective or rms values of the current or voltage rather than the maximum value of the quantity over one full cycle. For a sinusoidal current or voltage the effective values are

$$I = \frac{I_m}{\sqrt{2}} = 0.707I_m \qquad \text{and} \qquad V = \frac{V_m}{\sqrt{2}} = 0.707V_m$$

where the subscript m stands for the maximum value over a cycle. The effective values are unsubscripted since they are the commonly used quantities. In some instances the average value is desired over only a half cycle; for a sine wave over the first half cycle, for example, the results

$$I_h = \frac{2}{\pi}I_m = 0.637I_m \qquad \text{and} \qquad V_h = \frac{2}{\pi}V_m = 0.637V_m$$

are obtained.

Resistance, inductance, and capacitance are all generally present in A-C circuits. When elements are connected in series, a voltage drop between two points is the sum of the individual *phasor* voltage drops; for a parallel combination of elements, the total current at a junction is the *phasor* sum of individual currents through the parallel elements. Phasor algebra is the same as vector algebra since phasors have a magnitude and orientation, but we must remember that the angle is a phase angle between two sinusoidally varying quantities and not a definite direction in space.

The relation between current and voltage is conveniently given by

$$\mathbf{V} = \mathbf{ZI} \qquad (9\text{-}8)$$

where these quantities are all phasor quantities, and $\mathbf{Z}$ is the impedance. If the inductive and capacitive reactances X_L and X_C are defined in terms of frequency f, inductance L, and capacitance C as

$$X_L = 2\pi fL \qquad X_C = -\frac{1}{2\pi fC} \qquad (9\text{-}9)$$

then the relations between impedance Z, reactance X, and resistance R, as shown in Fig. 9-3 for $X > 0$, are $Z^2 = R^2 + X^2$, $\tan\theta = X/R$ with the reactance $X = X_L + X_R$. Note, however, that X may be either positive or negative.

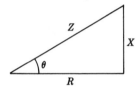

Figure 9-3

An alternative to the use of the impedance triangle is the volt-ampere method, which is based on A-C power considerations. Most A-C equipment has been rated in volt-amperes (VA), but the preferred practice now is simply to list power in watts. Because of the phase shift θ, however, the useful average power in an A-C circuit is $I^2R = P = VI\cos\theta$, where effective values are used for current and voltage. Cos θ is called the power factor. The reactive power is $Q = I^2X = VI\sin\theta$. These relations are summarized in Fig. 9-4, the volt-ampere triangle.

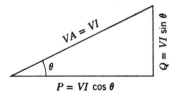

Figure 9-4

Equations (9-9) show that reactance is a function of frequency. By selecting this frequency so that $X = 0$, one can maximize the current in the system and at the same time maximize the power by operating with a power factor of unity. This is called the resonant frequency; at this frequency the response to a given voltage input will also be greatest.

Example 3

In a 440-V A-C circuit there is a resistive and inductive load that requires 30 A at 0.8 power factor. Calculate

(*a*) volt-amperes
(*b*) average power
(*c*) reactive volt-amperes
(*d*) resistance R
(*e*) impedance Z
(*f*) reactance X_L

Solution. We apply the volt-ampere method here.

(*a*) Volt-amperes $VA = VI = (440)(30) = 13,200$ W
(*b*) Average power $P = VI \cos \theta = (440)(30)(0.8) = 10,560$ W
(*c*) Reactive volt-amperes $Q = VI \sin \theta = (440)(30)(0.6) = 7920$ W
(*d*) The resistance R is found from $P = I^2 R$, or

$$R = \frac{P}{I^2} = \frac{10,560}{(30)^2} = 11.73 \ \Omega$$

(*e*) The impedance relation is $I^2 Z = VI$. The impedance magnitude is

$$Z = \frac{VI}{I^2} = \frac{V}{I} = \frac{440}{30} = 14.67 \ \Omega$$

with a phase angle $\theta = \cos^{-1} 0.8 = 37°$.

(*f*) The expression $I^2 X_L$ is an alternative expression for reactive volt-amperes. Hence

$$X_L = \frac{Q}{I^2} = \frac{7920}{(30)^2} = 8.80 \ \Omega$$

Example 4

A coil with an inductance of 1 H and a resistance of 50 Ω is connected in series with a 0.01 μF capacitor and a variable frequency supply, as in Fig. 9-5.

(*a*) At what frequency f_0 will the maximum voltage appear across the capacitor if the supply voltage is held constant as the frequency is varied?

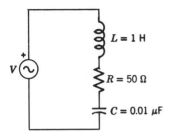

Figure 9-5

(b) Using the frequency f_0, what is the maximum supply voltage that may be used in this circuit if the capacitor voltage rating is 200 V?

Solution.

(a) The voltage is maximized at resonance when $X = 0$:

$$X = 2\pi f_0 L - \frac{1}{2\pi f_0 C} = 0$$

Solving,

$$2\pi f_0 = \frac{1}{(LC)^{1/2}} = \frac{1}{[1(10^{-8})]^{1/2}} = 10^4 \text{ rad/sec}$$

$$f_0 = \frac{10^4}{2\pi} = 1592 \text{ Hz}$$

(b) At the frequency f_0 the current $I_0 = V/R$. The maximum capacitor voltage $V_c = 200 = I_0 X_C$, where $X_C = -(2\pi f_0 C)^{-1}$. Combining these relations and solving for V yields

$$V = I_0 R = \frac{V_c}{X_c} R = 2\pi f_0 C V_c R$$

$$V = (10^4)(10^{-8})(200)(50) = 1.00 \text{ V}$$

ELECTRICAL MACHINERY

Both electric motors and generators utilize the same fundamental principles in their operation. For this reason the analysis of either type of machine is essentially the same. All these machines are basically usable as *either* a motor or a generator; it is only our optimization of the primary and secondary operating characteristics of a given machine so that it performs better in one role than in the other that has led us to separate the two.

Both A-C and D-C machines are widely used. By various means electric and magnetic fields, fields set up by the current flow through the machine, create a mechanical torque, or vice versa. However, there are sufficiently many different variations in the actual design of the major types of machinery that it is impractical to outline them here. One sample case is presented next; others may be found in the problems.

Example 5

A D-C shunt motor running at no load draws 5.0-A armature current and 1.0-A field current from a 120-V line. Its speed is 1190 rpm. When loaded, the armature current increases to 50 A without adjustment of the field. Neglecting the effect of armature reaction, what will be the speed under this load condition? The armature resistance is 0.2 Ω.

Solution.   One primary characteristic of this motor is the relative constancy of the armature voltage over a wide speed range.

Letting V_a = armature voltage, I_a = armature current, R_a = armature resistance, and V_t = terminal voltage, we can write

$$V_a = V_t - I_a R_a$$

At no load

$$V_a = 120 - (5.0)(0.2) = 119 \text{ V}$$

Under load

$$V_a = 120 - (50)(0.2) = 110 \text{ V}$$

For a particular machine the armature voltage V_a is proportional to the product of the field strength ϕ and the armature speed n in rpm, or

$$V_a = K\phi n$$

where K is a proportionality constant dependent on the machine. Here ϕ is also constant so that

$$\frac{V_{a1}}{V_{a2}} = \frac{n_1}{n_2}$$

$$n_2 = \frac{110}{119}(1190) = 1100 \text{ rpm}$$

PROBLEM 9-1

A 0.02 μF capacitor consists of two parallel plates of 100 in.^{2} each, with an air gap of 0.001 in. between them. If the air gap is increased to 0.0015 in., the resulting capacitance will be

(A) $0.0133\ \mu\text{F}$
(B) $0.0300\ \mu\text{F}$
(C) $0.0090\ \mu\text{F}$
(D) $0.0450\ \mu\text{F}$
(E) $0.0100\ \mu\text{F}$

Solution. Capacitance is proportional to A/d. Hence

$$\frac{C_2}{C_1}=\left(\frac{A_2}{d_2}\right)\left(\frac{d_1}{A_1}\right) \quad \text{and} \quad C_2=C_1\left(\frac{d_1}{d_2}\right)$$

since A is constant.

$$C_2=(0.02)\left(\frac{0.001}{0.0015}\right)=0.0133\ \mu\text{F}$$

Answer is (A)

PROBLEM 9-2

A capacitor is rated at $4000\ \text{F}$ (farads) and is given a charge of $80\ \text{C}$ (coulombs). The potential difference between the plates is

(A) 5 V
(B) 50 V
(C) 0.04 V
(D) 0.02 V
(E) 0.2 V

Solution.

$$C=\frac{Q}{V}$$

where $C=$ capacitance in farads, $Q=$ quantity of electricity in coulombs, and $V=$ difference in potential in volts. Thus

$$V=\frac{Q}{C}=\frac{80}{4000}=0.02\ \text{V}$$

Answer is (D)

PROBLEM 9-3

A copper rod 1 in. in diameter and 10 ft long is found to have a resistance of $10^{-4}\,\Omega$. If the rod is drawn into a wire with a uniform diameter of 0.05 in., the new resistance of the wire is nearest to

(A) $16\,\Omega$
(B) $1\,\Omega$
(C) $0.8\,\Omega$
(D) $0.04\,\Omega$
(E) $10^{-4}\,\Omega$

Solution.   Equation (9-3) applies here, and the volume of copper $V = lA$ is constant in the problem. For a circular cross section the area $A = \pi d^2/4$, and Eq. (9-3) now gives

$$R = \rho\frac{l}{A} = \rho\frac{V}{A^2} = \rho V\left(\frac{4}{\pi}\right)^2\frac{1}{d^4} = Kd^{-4}$$

where K is a constant. If subscript 1 refers to the original wire and subscript 2 to the new wire, we have

$$R_1 d_1^4 = R_2 d_2^4 = K$$

and

$$R_2 = R_1\left(\frac{d_1}{d_2}\right)^4 = (10^{-4})\left(\frac{1}{0.05}\right)^4 = 16\,\Omega$$

Answer is (A)

PROBLEM 9-4

A device that produces coherent light with predictable properties that can be controlled in a manner comparable to signals at radio and microwave frequencies is called a

(A) radar
(B) capacitor
(C) photoelectric cell
(D) reflector
(E) laser

Solution. The laser (*L*ight *A*mplification by *S*timulated *E*mission of *R*adiation) emits coherent light (a narrow, intense beam whose waves are all nearly parallel, in phase, and of the same wavelength).

Answer is (E)

PROBLEM 9-5

Cathode rays are

(A) high-energy X rays
(B) alpha particles
(C) protons
(D) electrons
(E) neutrons

Solution. Here we are dealing with a flow of electrons.

Answer is (D)

PROBLEM 9-6

A kilowatt-hour is a unit of

(A) momentum
(B) power
(C) acceleration
(D) energy
(E) impulse

Solution. A kilowatt-hour is a measure of work or energy. Energy = Power × Time. Power, on the other hand, is work per unit time; it might be measured in watts or kilowatts, not kilowatt-hours.

Answer is (D)

PROBLEM 9-7

A phonograph cartridge creates voltage by a process that is

(A) electromechanical
(B) electrochemical
(C) thermoelectric
(D) piezoelectric
(E) photovoltaic

Solution.   For completeness we cite an example of developing a voltage via each process.

A-C and D-C generators use the electromechanical process of rotating a coil through the field of a magnet.

D-C batteries use oxidation-reduction reactions in the cells in an electrochemical process.

A thermocouple uses the thermoelectric effect caused by heating a junction of dissimilar metals and completing the circuit at a different temperature to create a D-C voltage.

The phonograph cartridge employs the piezoelectric effect, whereby some crystals develop voltage differences when they are deformed by applied forces.

Photovoltaic cells are probably best known for converting sunlight into electricity in space vehicles.

Answer is (D)

PROBLEM 9-8

Permittivity is expressed by the Greek letter ε, and its defining equation is Coulomb's law

$$\varepsilon = \frac{QQ'}{4\pi FS^2}$$

where Q and $Q' =$ two point charges, $F =$ force between these charges, and $S =$ distance between these charges.

Permittivity in the MKS system is

    (A) $\text{coulombs}^2/\text{newton-meter}^2$
    (B) $\text{coulombs}^2/\text{dyne-meter}^2$
    (C) $\text{statcoulombs}^2/\text{dyne-meters}^2$
    (D) $\text{coulombs}^2/\text{newton-centimeter}^2$
    (E) $\text{statcoulombs}^2/\text{gram-centimeter}^2$

Solution.   Permittivity in the MKS system is $\text{coulombs}^2/\text{newton-meter}^2$

Answer is (A)

PROBLEM 9-9

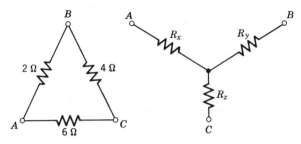

Figure 9-6

The Δ and Y circuits are intended to be equivalent. The resistance R_y is

(A) 0.50 Ω
(B) 0.67 Ω
(C) 1.00 Ω
(D) 1.67 Ω
(E) 2.67 Ω

Solution. First determine the line-to-line resistance for the Δ circuit:

$$AB: \quad \frac{1}{R_{AB}} = \frac{1}{2} + \frac{1}{6+4} = 0.6 \qquad R_{AB} = \frac{1}{0.6} = 1.67 \ \Omega$$

$$BC: \quad \frac{1}{R_{BC}} = \frac{1}{4} + \frac{1}{2+6} = 0.375 \qquad R_{BC} = \frac{1}{0.375} = 2.67 \ \Omega$$

$$CA: \quad \frac{1}{R_{CA}} = \frac{1}{6} + \frac{1}{4+2} = 0.333 \qquad R_{CA} = \frac{1}{0.333} = 3.0 \ \Omega$$

Similarly, determine the line-to-line resistance for the Y circuit:

$$AB: \quad R_{AB} = R_x + R_y$$
$$BC: \quad R_{BC} = R_y + R_z$$
$$CA: \quad R_{CA} = R_z + R_x$$

For the circuits to be equivalent, the line-to-line resistances must be equal. Therefore

$$R_{AB} = R_x + R_y \qquad = 1.67 \qquad (1)$$
$$R_{BC} = \qquad R_y + R_z = 2.67 \qquad (2)$$
$$R_{CA} = R_x \qquad + R_z = 3.0 \qquad (3)$$

Solving these three equations simultaneously,

$$(1) \qquad R_x + R_y \qquad = \quad 1.67$$
$$-(3) \qquad -R_x \qquad -R_z = -3.0$$

$$\overline{\qquad\qquad R_y - R_z = -1.33}$$

$$(2) \qquad\qquad R_y + R_z = \quad 2.67$$

$$\overline{\qquad 2R_y \qquad = \quad 1.34 \quad R_y = \frac{1.34}{2} = 0.67 \ \Omega}$$

Completing the solution would yield

$$(2) \qquad\qquad 0.67 + R_z = 2.67 \qquad R_z = 2.00 \ \Omega$$
$$(1) \qquad R_x + 0.67 \qquad = 1.67 \qquad R_x = 1.00 \ \Omega$$

Answer is (B)

PROBLEM 9-10

In a Y-connected circuit the line current equals

(A) the phase current

(B) $\dfrac{1}{\sqrt{3}}$ times the phase current

(C) $\sqrt{3}$ times the phase current

(D) $\dfrac{\sqrt{3}}{2}$ times the phase current

(E) $\dfrac{2}{\sqrt{3}}$ times the phase current

Solution. For the balanced three-phase Y connection, line current equals phase current.

Answer is (A)

PROBLEM 9-11

A bank of three transformers is to be connected to reduce the voltage from a three-phase, 12,000-volt (line-to-line) distribution line to supply power for a small irrigation pump driven by a 440-V induction motor. A Y connection will be used for the primary and a Δ connection for the secondary.

(1) The primary voltage rating of each transformer is nearest to

    (A) 4000 V
    (B) 7000 V
    (C) 12,000 V
    (D) 21,000 V
    (E) 36,000 V

Solution. In a Y connection, $V_{line} = \sqrt{3}\, V_{phase}$. Hence

$$V_{phase} = \frac{12,000}{\sqrt{3}} = 6930\ V = \text{primary rating}$$

Answer is (B)

(2) The secondary voltage rating of each transformer is nearest to

    (A) 150 V
    (B) 250 V
    (C) 440 V
    (D) 760 V
    (E) 1320 V

Solution. In a Δ connection, $V_{line} = V_{phase} = 440\ V = \text{secondary rating}$.

Answer is (C)

PROBLEM 9-12

One has five capacitors that are each rated at 5 μF.

(1) When the five are connected in series, the equivalent capacitance is nearest to

    (A) 1 μF
    (B) 2 μF
    (C) 5 μF
    (D) 11 μF
    (E) 25 μF

Solution. Capacitors in series add reciprocally:

$$\frac{1}{C_{eq}} = \frac{1}{C_1} + \frac{1}{C_2} + \frac{1}{C_3} + \frac{1}{C_4} + \frac{1}{C_5}$$

$$\frac{1}{C_{eq}} = 5\left(\frac{1}{5\,\mu F}\right) \qquad C_{eq} = 1\,\mu F$$

Answer is (A)

(2) When the five are connected in parallel, the equivalent capacitance is nearest to

(A) $1\,\mu F$
(B) $2\,\mu F$
(C) $5\,\mu F$
(D) $11\,\mu F$
(E) $25\,\mu F$

Solution. Capacitors in parallel add directly:

$$C_{eq} = C_1 + C_2 + C_3 + C_4 + C_5 = 5(5\,\mu F) = 25\,\mu F$$

Answer is (E)

PROBLEM 9-13

One has n identical cells that each have emf V and internal resistance r. When they are connected to an external resistance R, they will produce a current I.

(1) If the cells are connected in series, the current I is

(A) nV/R
(B) V/r
(C) $V/(R+r)$
(D) $nV/(R+nr)$
(E) $nV/(nR+r)$

Solution. Figure 9-7 shows the series circuit for $n = 3$ cells:

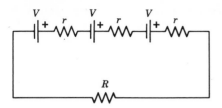

Figure 9-7

Each cell internal resistance r has been shown separately from the emf. Direct application of Kirchhoff's voltage law around the closed loop yields

$$nV - I(nr) - IR = 0$$

$$I = \frac{nV}{R + nr}$$

Answer is (D)

(2) If the cells are connected in parallel, the current I is

(A) V/nR
(B) V/R
(C) $V/(R + r)$
(D) $nV/(R + nr)$
(E) $nV/(nR + r)$

Solution. Figure 9-8 depicts the parallel circuit for $n = 3$ cells:

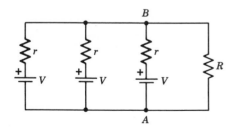

Figure 9-8

The current through R is I, but the current from A to B can take any one of n identical paths through a cell; by Kirchhoff's current law the current through a cell is I/n. Application of the voltage law around a loop containing A, B and R gives

$$V - \left(\frac{I}{n}\right)r - IR = 0$$

$$I = \frac{V}{R + r/n} = \frac{nV}{nR + r}$$

Answer is (E)

PROBLEM 9-14

The total equivalent resistance between A and B is nearest to

(A) 3.2 Ω
(B) 6.5 Ω
(C) 7.5 Ω
(D) 10.2 Ω
(E) 36 Ω

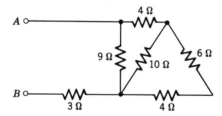

Figure 9-9

Solution.   The circuit is first redrawn in a more conventional form:

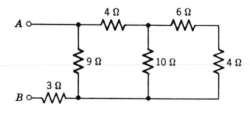

Figure 9-10

The 6-Ω and 4–Ω resistors in series may be added:

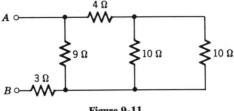

Figure 9-11

The equivalent resistance of the two 10-Ω resistors in parallel is

$$\frac{1}{R_e} = \frac{1}{10} + \frac{1}{10} \qquad R_e = 5\,\Omega$$

This 5-Ω resistor can now be added to the 4-Ω resistor in series to produce

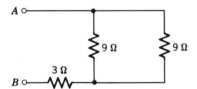

Figure 9-12

The equivalent resistance of the two parallel 9-Ω resistors is

$$\frac{1}{R_e} = \frac{1}{9} + \frac{1}{9} \qquad R_e = 4.5\,\Omega$$

When this resistor is added in series to the remaining 3-Ω resistor, we find the total equivalent resistance to be 7.5 Ω.

Answer is (C)

PROBLEM 9-15

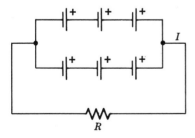

Figure 9-13

When Kirchhoff's laws are applied to three similar cells that are connected in series, and then they are connected in parallel with three other similar cells, as in the figure, the total current is

(A) $I = \dfrac{V}{\dfrac{r}{3} + R}$

(B) $I = \dfrac{3V}{\dfrac{3r}{2} + R}$

(C) $I = \dfrac{V}{\dfrac{3r}{2} + R}$

(D) $I = \dfrac{3V}{3r + R}$

(E) $I = \dfrac{1}{\dfrac{6}{r} + R}$

V is the emf of each cell, r is the internal resistance of each cell, and R is the resistance of the external circuit.

Solution.   Voltage and resistance are each directly additive in simple series. When this is done within each parallel limb, the equivalent circuit with the internal resistances shown explicitly is

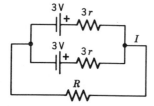

Figure 9-14

By symmetry the current in each parallel limb is $I/2$. By applying Kirchhoff's voltage law around the main circuit one obtains

$$3V - \frac{I}{2}(3r) - IR = 0$$

$$I = \frac{3V}{\dfrac{3r}{2} + R}$$

Answer is (B)

PROBLEM 9-16

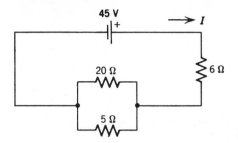

Figure 9-15

The current I in this circuit is nearest to

(A) 1.5 A

(B) 4.5 A

(C) 7.2 A

(D) 9.3 A

(E) 18.8 A

Solution. The equivalent resistance for the two parallel resistors is

$$\frac{1}{R_e} = \frac{1}{5} + \frac{1}{20} \qquad R_e = 4\,\Omega$$

The total resistance of the circuit is $4\,\Omega + 6\,\Omega = 10\,\Omega$. Applying Ohm's law gives

$$I = \frac{V}{R} = \frac{45}{10} = 4.5\text{ A}$$

Answer is (B)

PROBLEM 9-17

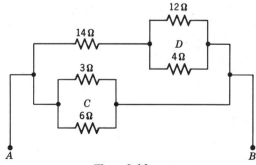

Figure 9-16

The resistance between A and B is nearest to

(A) 1.1 Ω
(B) 1.8 Ω
(C) 6.9 Ω
(D) 16.5 Ω
(E) 39.0 Ω

Solution. The two parallel pairs of resistors at C and D can each be replaced by an equivalent resistor:

$$\frac{1}{R_C}=\frac{1}{3}+\frac{1}{6} \quad \text{and} \quad \frac{1}{R_D}=\frac{1}{12}+\frac{1}{4}$$

$$R_C = 2 \,\Omega \qquad R_D = 3 \,\Omega$$

The total resistance in the upper branch is $14\,\Omega + 3\,\Omega = 17\,\Omega$. The circuit now reduces to

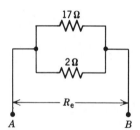

A   *B*   **Figure 9-17**

$$\frac{1}{R_e}=\frac{1}{17}+\frac{1}{2} \qquad R_e = \frac{34}{19}\,\Omega = 1.79\,\Omega$$

Answer is (B)

PROBLEM 9-18

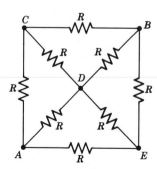

Figure 9-18

The circuit shown is composed of resistors that have equal resistances R. The resistance between points A and B is nearest to

(A) $8R \, \Omega$
(B) $6R \, \Omega$
(C) $2R \, \Omega$
(D) $2R/3 \, \Omega$
(E) $R/2 \, \Omega$

Solution. By examining Fig. 9-18 we find that the resistance encountered along paths ACB, ADB, and AEB between A and B are identical ($2R$) and furthermore that the voltage at points C, D, and E must be the same. In other words, no current flows along the diagonal wire connecting C, D, and E, and therefore the resistors between C and D, and between D and E, have no effect on the circuit and can be removed. A simplified diagram can now be drawn:

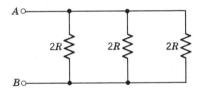

Figure 9-19

For parallel resistors the equation for equivalent resistance is

$$\frac{1}{R_e} = \frac{1}{R_1} + \frac{1}{R_2} + \frac{1}{R_3} \qquad \frac{1}{R_{AB}} = \frac{1}{2R} + \frac{1}{2R} + \frac{1}{2R} = \frac{3}{2R}$$

Therefore, the resistance R_{AB} between A and B is

$$R_{AB} = \frac{2R}{3} \, \Omega$$

Answer is (D)

PROBLEM 9-19

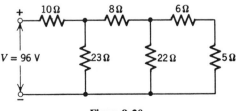

Figure 9-20

The current input to the circuit in Fig. 9-20 is nearest to

    (A) 1.3 A
    (B) 3.5 A
    (C) 5.0 A
    (D) 12.9 A
    (E) 14.8 A

Solution.   Add the resistors in series to get $6+5=11\ \Omega$:

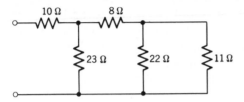

Figure 9-21

Now add the 11-Ω and 22-Ω resistors in parallel,

$$\frac{1}{R_e}=\frac{1}{11}+\frac{1}{22} \qquad R_e = 7\tfrac{1}{3}\,\Omega$$

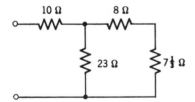

Figure 9-22

Again add the series resistors to obtain $7\tfrac{1}{3}\,\Omega+8\ \Omega=15\tfrac{1}{3}\,\Omega$, and find the equivalent resistance for the last pair of parallel resistors:

$$\frac{1}{R_e}=\frac{1}{15\tfrac{1}{3}}+\frac{1}{23} \qquad R_e = 9.2\,\Omega$$

Finally, we add the last resistors in series to find $9.2+10=19.2\ \Omega$ is the resistance of the entire combination. By Ohm's law

$$I=\frac{V}{R}=\frac{96}{19.2}=5\ \text{A}$$

Answer is (C)

PROBLEM 9-20

The resistance of the circuit shown in the figure is

(A) 0.43 Ω
(B) 0.80 Ω
(C) 2.28 Ω
(D) 5.50 Ω
(E) 14.00 Ω

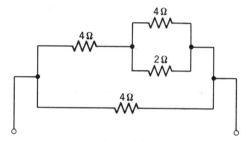

Figure 9-23

Solution. The 2- and 4-Ω resistors in parallel can be replaced by R_e given by

$$\frac{1}{R_e} = \frac{1}{2} + \frac{1}{4} \qquad R_e = \frac{4}{3}\,\Omega$$

Adding this value to the 4-Ω resistor in series with it leads to

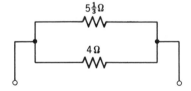

Figure 9-24

The remaining two resistors are in parallel and can be combined to yield

$$\frac{1}{R} = \frac{1}{4} + \frac{1}{5\frac{1}{3}} = 0.250 + 0.188 = 0.438$$

$$R = \frac{1}{0.438} = = 2.28\ \Omega$$

Answer is (C)

PROBLEM 9-21

The value of the resistor R is nearest to

    (A) $1.9\ \Omega$

    (B) $4.0\ \Omega$

    (C) $5.0\ \Omega$

    (D) $8.0\ \Omega$

    (E) $9.8\ \Omega$

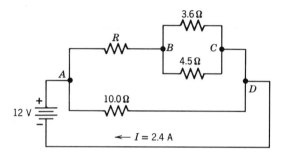

Figure 9–25

Solution. If we apply Ohm's law directly to the exterior branch of the circuit, we find the equivalent resistance R_{AD} to be given by

$$R_{AD} = \frac{V}{I} = \frac{12.0}{2.4} = 5.0\ \Omega$$

Now we must find R_{AD} in terms of R. The resistance R_{BC} is found from

$$\frac{1}{R_{BC}} = \frac{1}{3.6} + \frac{1}{4.5} \qquad R_{BC} = 2.0\ \Omega$$

and in the same general manner

$$\frac{1}{R_{AD}} = \frac{1}{10.0} + \frac{1}{R+2.0}$$

and also $1/R_{AD} = 1/5.0$. We find that

$$\frac{1}{5.0} - \frac{1}{10.0} = \frac{1}{10.0} = \frac{1}{R+2.0}$$

and $R = 8.0\ \Omega$.

$$\text{Answer is (D)}$$

PROBLEM 9-22

A two-loop D-C circuit is shown below.

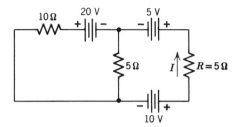

Figure 9-26

(1) The current I through the resistor R is closest to

(A) 1.8 A

(B) 1.4 A

(C) 1.0 A

(D) 0.4 A

(E) 0.2 A

Solution. This problem will be solved by using mesh equations. Let us assume a counterclockwise direction for the current.

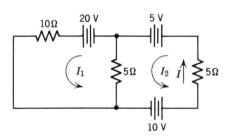

Figure 9-27

Around the first loop we obtain

$$(10+5)I_1 - 5I_2 = 20 \tag{1}$$

and for the second loop

$$-5I_1 + (5+5)I_2 = -5+10 \tag{2}$$

Add $3 \times$ Eq. (2) to Eq. (1) to obtain

$$25I_2 = 35 \qquad I_2 = I = \frac{35}{25} = 1.4 \text{ A}$$

Answer is (B)

(2) The power dissipated in resistor R is nearest to

    (A) 25.0 W
    (B) 16.2 W
    (C) 9.8 W
    (D) 5.0 W
    (E) 0.8 W

Solution.   The power dissipated is $P = I^2 R = (1.4)^2(5) = 9.8$ W

Answer is (C)

PROBLEM 9-23

For the circuit shown, assume the inductance has negligible resistance.

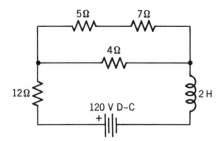

Figure 9-28

    The voltage drop across the 7-Ω resistor during steady-state operation is nearest to

    (A) 12 V
    (B) 14 V
    (C) 26 V
    (D) 30 V
    (E) 31 V

Solution.   Since this is a steady-state problem, the inductance plays no role, for it responds only to changes in the current. Rather than use the equivalent resistance approach, let us select loop currents and write two mesh equations:

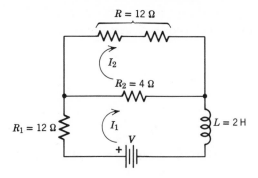

Figure 9-29

Loop 1: $(R_1 + R_2)I_1 -$ $R_2I_2 = V$

Loop 2: $-R_2I_1 + (R + R_2)I_2 = 0$

or

$$16I_1 - 4I_2 = 120$$

$$-4I_1 + 16I_2 = 0$$

From the second equation $I_1 = 4I_2$ so that the first equation gives $16(4I_2) - 4I_2 = 120$, $60I_2 = 120$ and $I_2 = 2A$. The voltage drop across the 7-Ω resistor is then

$$V_7 = I_2 R_7 = 2(7) = 14 \text{ V}$$

Answer is (B)

PROBLEM 9-24

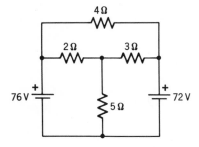

Figure 9-30

If it is known that the current through the 4-Ω resistor is 1 A, then the current through the 5-Ω resistor is closest to

(A) 1 A
(B) 3 A
(C) 6 A
(D) 9 A
(E) 12 A

Solution. Let us set up three loop equations for the loop currents I_1, I_2, and I_3:

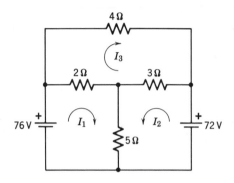

$$\text{Figure 9-31}$$

$$(2+5)I_1 + \quad 5I_2 - \quad 2I_3 = 76 \text{ V} \tag{1}$$
$$5I_1 + (5+3)I_2 + \quad 3I_3 = 72 \text{ V} \tag{2}$$
$$-2I_1 + \quad 3I_2 + (2+3+4)I_3 = \quad 0 \text{ V} \tag{3}$$

It is given that $I_3 = 1 \, A$; Eq. (3) requires that

$$I_1 = \frac{9+3I_2}{2}$$

Then from one of the other equations, say Eq. (1), we find

$$\tfrac{7}{2}(9+3I_2)+5I_2 = 76+2 \qquad \text{or} \qquad I_2 = 3 \text{ A}$$

and

$$I_1 = \frac{9+3(3)}{2} = 9 \text{ A}$$

The current through the 5-Ω resistor is $I_1 + I_2 = 12 \, A$.

$$\text{Answer is (E)}$$

PROBLEM 9-25

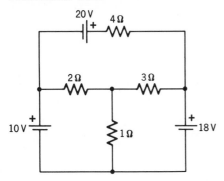

Figure 9-32

For the circuit shown, the current through the 1-Ω resistor is nearest to

(A) 2 A

(B) 4 A

(C) 5 A

(D) 6 A

(E) 10 A

Solution.   We will write three loop equations. The loops are chosen so that only I_1 passes through the 1-Ω resistor; the third loop traverses the entire outer boundary.

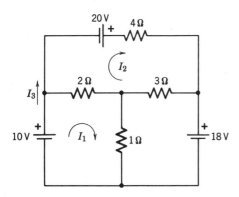

Figure 9-33

The loop equations are

$$(2+1)I_1 - \qquad\qquad 2I_2 \qquad = 10 \tag{1}$$

$$-2I_1 + (2+4+3)I_2 + 4I_3 = 20 \tag{2}$$

$$4I_2 + 4I_3 = 10 + 20 - 18 \tag{3}$$

From Eq. (3) $4I_3 = 12 - 4I_2$. Substitution of this result into Eq. (2) yields

$$-2I_1 + 5I_2 = 8 \qquad (4)$$

Now multiply Eq. (1) by 5, multiply Eq. (4) by 2 and add the two results to find

$$11I_1 = 66$$

$$I_1 = 6 \text{ A}$$

Answer is (D)

PROBLEM 9-26

A radio circuit oscillates at a frequency $f = 1/[2\pi(LC)^{1/2}]$. A given circuit has an inductance L of 0.1 H, but its capacitance is not exact, being given as $0.1 \, \mu\text{F}$ with a 10% tolerance.

(1) The resonant frequency is most nearly

   (A) 1.6 Hz
   (B) 160 Hz
   (C) 1600 Hz
   (D) 10,000 Hz
   (E) 62,800 Hz

Solution.

$$f = \frac{1}{2\pi(LC)^{1/2}} = \frac{1}{2\pi(0.1 \times 0.1 \times 10^{-6})^{1/2}} = \frac{1}{2\pi(1 \times 10^{-8})^{1/2}} = \frac{10^4}{2\pi}$$

$$f = 1592 \text{ Hz}$$

Answer is (C)

(2) The variation in the resonant frequency caused by the capacitor tolerance is most nearly

   (A) 0.2 Hz
   (B) 10 Hz
   (C) 20 Hz
   (D) 80 Hz
   (E) 160 Hz

Solution.

$$f = \frac{1}{2\pi(L)^{1/2}} C^{-1/2}$$

Hence

$$\frac{df}{dC} = \frac{1}{2\pi L^{1/2}}(-\tfrac{1}{2}C^{-3/2})$$

$$= \frac{-1}{2C} \frac{1}{2\pi(LC)^{1/2}} = -\frac{1}{2C}f$$

Using finite increments,

$$\left|\frac{\Delta f}{\Delta C}\right| = \frac{1}{2C}f \quad \text{or} \quad |\Delta f| = \frac{f}{2}\frac{\Delta C}{C}$$

The variation in frequency is therefore

$$\Delta f = \left(\frac{1592}{2}\right)(0.1) = 80 \text{ Hz}$$

Answer is (D)

PROBLEM 9-27

In a series circuit containing resistance, capacitance, and inductance, to increase the resonant frequency it is necessary to

(A) increase the resistance
(B) decrease the inductance
(C) increase the capacitance
(D) decrease the resistance
(E) increase the voltage

Solution.

At resonance, $2\pi f L = \dfrac{1}{2\pi f C}$

$$2\pi f = \frac{1}{(LC)^{1/2}} \quad \text{and} \quad f = \frac{1}{2\pi(LC)^{1/2}}$$

Therefore to increase f, L or C must be decreased.

Answer is (B)

PROBLEM 9-28

A 12-V car battery, with an internal resistance of 4 Ω, supplies a bank of 25 lights that are connected in parallel. Each light has an effective resistance of 500 Ω.

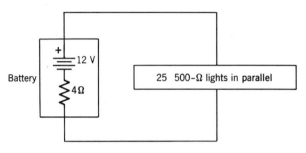

Figure 9-34

The power supplied to the bank of lights by the battery is nearest to

(A) 7.0 W
(B) 6.0 W
(C) 5.0 W
(D) 0.01 W
(E) 0.001 W

Solution. The resistance equivalent to the 25 lights in parallel is

$$\frac{1}{R_e} = 25\left(\frac{1}{500}\right) \qquad R_e = 20\ \Omega$$

The circuit current is then

$$I = \frac{V}{R} = \frac{12}{20+4} = 0.5\ \text{A}$$

and the power supplied *to the bank* (not the total power supplied) is

$$P = I^2 R_e = (0.5)^2(20) = 5.0\ \text{W}$$

Answer is (C)

PROBLEM 9-29

A "black box" has within it several batteries and resistors. An external resistor R and an ammeter are placed in series between the terminals A and B of the box as shown. When $R = 3.0 \, \Omega$, the current flow indicated by the meter is 1.0 A. When $R = 9.0 \, \Omega$, the indicated flow of current is 0.5 A.

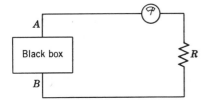

Figure 9-35

The resistance between terminals A and B is most closely

(A) $2.0 \, \Omega$
(B) $3.0 \, \Omega$
(C) $5.0 \, \Omega$
(D) $6.0 \, \Omega$
(E) $9.0 \, \Omega$

Solution. We can draw the circuit diagram:

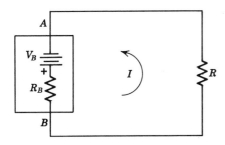

Figure 9-36

Using Kirchhoff's voltage law, we can write the loop equation for the circuit:

$$IR_B + IR = V_B$$

When $R = 3.0 \, \Omega$ and $I = 1.0$ A,

$$1.0 \, R_B + 1.0(3.0) = V_B \qquad (1)$$

When $R = 9.0\ \Omega$ and $I = 0.5$ A,

$$0.5R_B + 0.5(9.0) = V_B \qquad (2)$$

Subtracting Eq. (2) from Eq. (1) results in

$$0.5R_B - 1.5 = 0 \qquad R_B = 3.0\ \Omega$$

Answer is (B)

PROBLEM 9-30

For a D-C shunt motor operating from a constant potential supply, one of the following will happen:

(A) Increasing the shunt field resistance will cause the motor speed to increase.

(B) Increasing the shunt field resistance will cause the motor speed to decrease.

(C) Increasing the shunt field resistance will cause the shunt field current to increase.

(D) Decreasing the shunt field resistance will cause the shunt field current to decrease.

(E) Increasing the shunt field resistance will cause the armature current to decrease.

Solution. The practical way to control the speed of a D-C shunt motor is to place a rheostat in series with the shunt field. Increasing the resistance will decrease the field current I_f, which is proportional to the flux ϕ_f. This can be seen in the equation for motor speed,

$$\text{Speed} = \frac{V - I_a R_a}{K \phi_f}$$

Answer is (A)

PROBLEM 9-31

A four-pole synchronous motor operating from a 50-Hz supply will have a synchronous speed of

(A) 3600 rpm

(B) 3000 rpm

(C) 1800 rpm

(D) 1500 rpm

(E) 1200 rpm

Solution.

$$\text{Frequency } f = \frac{\text{no. of poles}}{2} \times \frac{\text{rpm}}{60} \qquad 50 = \frac{4}{2} \times \frac{\text{rpm}}{60}$$

$$\text{rpm} = 1500$$

Answer is (D)

PROBLEM 9-32

Two wattmeters are properly connected in a three-phase circuit to read the power supplied to a 220-volt, 15-hp, Δ-connected, three-phase induction motor running at no load. Voltmeters and ammeters are also properly connected in the circuit to supply lines A, B, and C. The readings are

Voltmeter (volts)	Ammeter (amperes)	Wattmeter (watts)
$V_{AB} = 221$	$I_A = 4.04$	810
$V_{BC} = 221$	$I_C = 4.04$	-200

(1) The power input to the motor is nearest to

(A) 200 W
(B) 610 W
(C) 810 W
(D) 893 W
(E) 1010 W

Solution. The power in a three-phase circuit can be determined by the use of two wattmeters. The total power is the algebraic sum of the two meter readings.

$$P_{\text{total}} = W_1 + W_2 = 810 + (-200) = 610 \text{ W}$$

Answer is (B)

(2) The power factor of the motor, with current lagging, is nearest to

    (A) 0.394
    (B) 0.523
    (C) 0.577
    (D) 0.683
    (E) 0.907

Solution. Current lags for the induction motor. The relation for total power is $P_t = (3)^{1/2} VI \cos \theta$. Hence

$$\cos \theta = \frac{P_t}{(3)^{1/2} VI} = \frac{610}{(3)^{1/2} \times 221 \times 4.04} = 0.394$$

$$\text{Power factor} = 0.394 \text{ current lagging}$$

$$\text{Answer is (A)}$$

PROBLEM 9-33

The armature resistance of a 50-hp, 550-V, D-C shunt wound motor is 0.35 Ω. The full-load armature current of this motor is 76 A.

(1) So that the initial starting current is 150% of the full-load value, the resistance of the starting coil should be closest to

    (A) 7.59 Ω
    (B) 7.24 Ω
    (C) 6.89 Ω
    (D) 4.82 Ω
    (E) 4.47 Ω

Solution. The starting current is to be limited to $1.50(76) = 114$ A. The total resistance is then

$$R = \frac{V}{I_s} = \frac{550}{114} = 4.82 \ \Omega$$

and the resistance of the starter is $4.82 - 0.35 = 4.47 \ \Omega$.

$$\text{Answer is (E)}$$

(2) If the field current under full load is 3 A, the overall efficiency of the motor is nearest to

(A) 81%
(B) 84%
(C) 86%
(D) 89%
(E) 91%

Solution. Line current = armature current + field current = $76 + 3 = 79$ A.

$$\text{Efficiency} = \frac{\text{output}}{\text{input}} = \frac{50 \times 746}{79 \times 550} = 0.859 = 85.9\%$$

Answer is (C)

PROBLEMS 9-34

The rpm of an A-C electric motor

(A) varies directly as the number of poles
(B) varies inversely as the number of poles
(C) is independent of the number of poles
(D) is independent of the frequency
(E) is directly proportional to the square of the frequency

Solution. The speed varies directly as the frequency and inversely as the number of poles.

Answer is (B)

PROBLEM 9-35

A metal transport plane has a wing spread of 88 ft and flies horizontally with a speed of 150 miles/hr. At the elevation of the plane the vertical component of the earth's magnetic field is 0.65×10^4 tesla.

The difference in potential that exists between the ends of the wings is most nearly

(A) 10^7 V
(B) 10^3 V

(C) 10^1 V

(D) 10^{-1} V

(E) 10^{-3} V

Solution.   An emf is induced in a circuit whenever any of its conductors cuts magnetic flux. The equation is

$$e_i = BLv$$

where e_i = induced voltage (in units of 10^{-8} V)

B = flux density in teslas

L = conductor length in meters

v = velocity of conductor in m/sec

$$e_i = 0.65 \times 10^4 \left(\frac{88 \times 12 \times 2.54}{100} \right) \left(150 \times \frac{5280}{3600} \times \frac{12 \times 2.54}{100} \right)$$

$$e_i = 0.117 \text{ V}$$

Answer is (D)

PROBLEM 9-36

A circuit consisting of $R = 3 \, \Omega$ and $X_L = 4 \, \Omega$ in series is connected to a 100-V, 60-Hz source as shown.

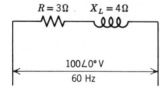

$R = 3\Omega$ $X_L = 4\Omega$

$100\angle 0°$ V

60 Hz

Figure 9-37

(1) The average power is nearest to

(A) 1200 W

(B) −1200 W

(C) 1600 W

(D) −1600 W

(E) 2000 W

Solution.

$$\text{Impedance } \bar{Z} = R + jX_L = 3 + j4$$

In polar form

$$\bar{Z} = (3^2 + 4^2)^{1/2} \tan^{-1}\left(\tfrac{4}{3}\right) = 5\angle 53.1° \ \Omega$$

$$\text{Current } \bar{I} = \frac{\bar{V}}{\bar{Z}} = \frac{100\angle 0°}{5\angle 53.1°} = 20\angle -53.1°$$

The average power or real volt-amperes is

$$P = \bar{V} \times \bar{I} = 100\angle 0° \times 20\angle -53.1° = 100 \times 20 \cos(-53.1°)$$
$$P = 2000 \times 0.6 = 1200 \text{ W}$$

Answer is (A)

(2) The reactive power is nearest to

(A) 1200 W
(B) −1200 W
(C) 1600 W
(D) −1600 W
(E) 2000 W

Solution. The reactive power or reactive volt-amperes is

$$Q = 1200 \tan(-53.1°) = 1200 \times (-\tfrac{4}{3}) = -1600 \text{ W}$$

Answer is (D)

(3) The total volt-amperes is nearest to

(A) 1150 W
(B) 1200 W
(C) 1600 W
(D) 2000 W
(E) 3500 W

Solution. The apparent or total volt-amperes is

$$VI = 100 \times 20 = 2000 \text{ W}$$

Answer is (D)

The relations between these results are conveniently summarized by the volt-ampere triangle for this problem:

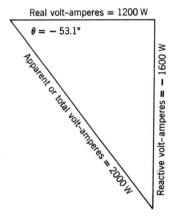

Real volt-amperes = 1200 W
$\theta = -53.1°$
Apparent or total volt-amperes = 2000 W
Reactive volt-amperes = $-$ 1600 W

Figure 9-38

PROBLEM 9-37

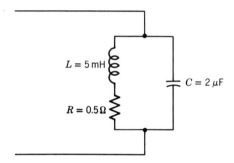

$L = 5\,\text{mH}$

$R = 0.5\,\Omega$

$C = 2\,\mu\text{F}$

Figure 9-39

Consider the circuit shown in Fig. 9-39.

(1) The resonant frequency f_0 is nearest to

    (A) 50,000 Hz
    (B) 10,000 Hz
    (C) 8000 Hz
    (D) 4000 Hz
    (E) 1590 Hz

Solution. The resonant frequency occurs when $X_L = X_C$, assuming $X_L \gg R$.

$$2\pi f_0 L = \frac{1}{2\pi f_0 C} \qquad 2\pi f_0 \times 0.005 = \frac{1}{2\pi f_0 \times 2 \times 10^{-6}}$$

$$f_0 = \left(\frac{10^6}{8\pi^2 \times 0.005}\right)^{1/2} = (2.525 \times 10^6)^{1/2} = 1590 \text{ Hz}$$

Answer is (E)

(2) The figure of merit Q of the circuit at f_0 is nearest to

    (A) 1
    (B) 10
    (C) 25
    (D) 100
    (E) 200

Solution.

$$X_C = X_L = 2\pi \times 1590 \times 0.005 = 50 \ \Omega$$

The figure of merit is therefore

$$Q = \frac{X}{R} = \frac{50}{0.5} = 100$$

Answer is (D)

(3) The wavelength at the resonant frequency is nearest to

    (A) 2.0×10^7 m
    (B) 2.0×10^5 m
    (C) 7.5×10^4 m
    (D) 4.0×10^4 m
    (E) 3.0×10^3 m

Solution. Using the value of c = velocity of light = 300×10^6 m/sec, the wavelength is

$$\lambda = \frac{c}{f_0} = \frac{300 \times 10^6}{1590} = 1.89 \times 10^5 \text{ m}$$

Answer is (B)

PROBLEM 9-38

A choke coil having an inductance of 150 mH is connected across a 60-Hz source.

(1) To have the same reactance as the choke coil, the value of the capacitance C that must be connected across a 50-Hz source is most nearly

    (A) $5.6 \, \mu F$
    (B) $17.6 \, \mu F$
    (C) $39.2 \, \mu F$
    (D) $56.3 \, \mu F$
    (E) $354 \, \mu F$

Solution. Inductive reactance $X_L = 2\pi f L = 2\pi \times 60 \times 0.150 = 56.5 \, \Omega$

$$\text{Capacitive reactance } X_C = \frac{1}{2\pi f C} = \frac{1}{2\pi \times 50 C}$$

For equal reactance,

$$\frac{1}{2\pi \times 50 C} = 56.5 \, \Omega$$

$$C = \frac{1}{2\pi \times 50 \times 56.5} = \frac{1}{17,750} = 56.3 \times 10^{-6} \, \text{farad} = 56.3 \, \mu F$$

Answer is (D)

(2) Assuming the capacitor has negligible resistance, the series resonant frequency of L and C (part 1) is nearest to

    (A) 50.0 Hz
    (B) 54.8 Hz
    (C) 55.0 Hz
    (D) 56.3 Hz
    (E) 60.0 Hz

Solution. The series resonant frequency can be found by equating X_L and X_C.

$$2\pi f L = \frac{1}{2\pi f C} \qquad 4\pi^2 f^2 = \frac{1}{LC} \qquad f = \frac{1}{2\pi (LC)^{1/2}}$$

$$f = \frac{1}{2\pi (0.15 \times 56.3 \times 10^{-6})^{1/2}} = \frac{10^3}{2\pi (8.45)^{1/2}}$$

$$f = 54.8 \text{ Hz}$$

Answer is (B)

PROBLEM 9-39

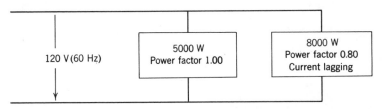

Figure 9-40

A circuit is shown in Fig. 9-40.

(1) The power is most nearly

    (A) 4800 W

    (B) 5000 W

    (C) 8000 W

    (D) 11,400 W

    (E) 13,000 W

Solution. For each load, power $P = VI \cos \theta$

$$P = 5000(1.00) + 8000(0.80) = 11,400 \text{ W}$$

Answer is (D)

(2) The reactive power is most nearly

    (A) 4800 W

    (B) 6400 W

    (C) 8000 W

    (D) 8400 W

    (E) 11,400 W

Solution. For each load, reactive power $Q = VI \sin \theta$

$$Q = 5000(0.00) + 8000(0.60) = 4800 \text{ W}$$

Answer is (A)

(3) The total volt-amperes input is most nearly

  (A) 8000 W
  (B) 9800 W
  (C) 11,400 W
  (D) 12,400 W
  (E) 13,000 W

Solution.

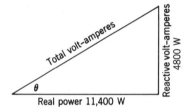

Real power 11,400 W   **Figure 9-41**

As can be seen from Fig. 9-41, the total volt-amperes is equal to

$$[P^2 + Q^2]^{1/2} = [(11,400)^2 + (4800)^2]^{1/2} = 12,400 \text{ W}$$

Answer is (D)

(4) The power factor for the entire circuit is most nearly

  (A) 0.80
  (B) 0.88
  (C) 0.92
  (D) 0.95
  (E) 1.00

Solution.

$$\text{Power factor} = \cos \theta = \frac{11,400}{12,400} = 0.92 \text{ lagging}$$

Answer is (C)

PROBLEM 9-40

A rectifier circuit connected to a 60-Hz A-C supply is shown below. Each secondary winding of the transformer has a voltage of 120 V rms. Selenium

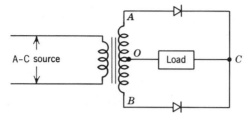

Figure 9-42

disks are used as the rectifying element; their voltage drop is negligible. The average or D-C voltage across the load is most nearly

(A) 170 V

(B) 120 V

(C) 108 V

(D) 85 V

(E) 60 V

Solution. From the problem statement $V_{AO} = V_{OB} = 120$ V rms. The instantaneous voltage is therefore

$$V = 120\sqrt{2}\sin 2\pi(60)t \qquad 0 < t < \tfrac{1}{120}$$

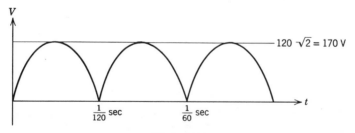

Figure 9-43

For a half cycle the average or effective voltage is

$$V_h = \frac{2}{\pi}V_m = \frac{2}{\pi}(120\sqrt{2}) = 108 \text{ V}$$

Answer is (C)

PROBLEM 9-41

In a series A-C circuit with a lagging power factor, to increase the power factor you should

  (A) increase the current
  (B) increase the voltage
  (C) increase the frequency
  (D) add inductance to the circuit
  (E) add capacitance to the circuit

Solution. A lagging power factor indicates that the current is behind the voltage by a phase angle θ. This is caused by the inductive load in the circuit, which causes the inductive reactance X_L to be larger than the capacitive reactance X_C. The power factor, $\cos \theta$, can be increased by adding capacitance to the circuit in such a way that X_C increases and the reactance X thereby decreases, as shown in Fig. 9-44.

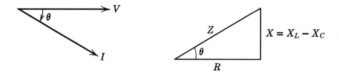

Figure 9-44

Answer is (E)

PROBLEM 9-42

Given the circuit shown in Fig. 9-45:

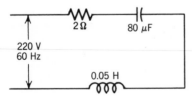

Figure 9-45

(1) The impedance is nearest to

    (A) $2\ \Omega\ \angle 0°$

    (B) $14.4\ \Omega\ \angle -82°$

    (C) $16.3\ \Omega\ \angle 2°$

    (D) $19.0\ \Omega\ \angle 84°$

    (E) $33.3\ \Omega\ \angle -87°$

Solution.

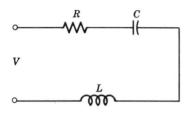

Figure 9-46

The impedance Z is

$$Z = R + jX_L - jX_C$$

where R = resistance (ohms)

      $X_L = 2\pi f L$ (ohms) (inductive reactance)

      $X_C = \dfrac{1}{2\pi f C}$(ohms) (capacitive reactance)

$$Z = 2 + j(2\pi 60)(0.05) - j\left[\frac{1}{(2\pi 60)(80 \times 10^{-6})}\right]$$

$$= 2 + j(18.85) - j(33.2)$$

$$\cong 2 - j14.3\ \Omega \text{ (in rectangular form)}$$

$$\cong 14.4\angle -82.0°$$

                               Answer is (B)

(2) The line current is closest to

    (A) 110 A

    (B) 15.2 A

    (C) 13.5 A

    (D) 11.6 A

    (E) 6.6 A

Solution.

$$\bar{I} = \frac{V\angle 0°}{\bar{Z}} = \frac{220\angle 0°}{14.4\angle -82.0°} = 15.22\angle +82.0°$$

Hence $|I| \approx 15.2$ A

Answer is (B)

(3) The power factor is closest to

    (A) 0.0
    (B) 0.052
    (C) 0.105
    (D) 0.138
    (E) 1.0

Solution. Power factor $= \cos 82.0° = 0.138$ (leading)

Answer is (D)

PROBLEM 9-43

Given the circuit and voltage measurements shown in Fig. 9-47:

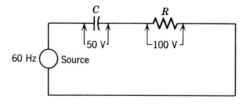

Figure 9-47

(1) The voltage of the 60-Hz source is nearest to

    (A) 50 V
    (B) 100 V
    (C) 112 V
    (D) 150 V
    (E) 212 V

Solution.

$$IR = 100 \text{ V} \qquad \text{and} \qquad I(-jX_C) = 50 \text{ V}$$

Therefore

$$\bar{V} = IR - jIX_C = 100 - j50 \text{ V (rectangular form)}$$
$$\bar{V} = 112 \angle -26.6°$$

$$|\bar{V}| = 112 \text{ V}$$

Answer is (C)

(2) The power factor of the circuit is nearest to

    (A) 0.45
    (B) 0.67
    (C) 0.75
    (D) 0.89
    (E) 1.00

Solution.

$$\text{Power factor} = \cos 26.6° = 0.89 \text{ (leading)}$$

Answer is (D)

PROBLEM 9-44

It is desired, by discharging a capacitance, to send an electric impulse through a 1000-Ω resistor so that the initial current will be 1.00 A. At the end of 0.010 sec (10 msec), the voltage across the resistance is to be 368 V.
  The correct value of the required capacitance is most nearly

    (A) $1 \mu\text{F}$
    (B) $3.7 \mu\text{F}$
    (C) $6.3 \mu\text{F}$
    (D) $10 \mu\text{F}$
    (E) $37 \mu\text{F}$

Solution.   The time constant T is the time in seconds needed for a capacitor to lose 63% of its voltage, that is, for the voltage to drop to 37% of its initial value. In this case $T = 0.01$ sec. For a capacitor the time constant is

$$T = RC \qquad \text{or} \qquad C = \frac{T}{R} = \frac{10 \times 10^{-3}}{10^3} = 10 \times 10^{-6}\,\text{F} = 10\,\mu\text{F}$$

Answer is (D)

10   Economic Analysis

Engineering economic analysis (often called engineering economy) is the title given to a group of techniques for the systematic examination of alternative courses of action. Being money based, the analysis gives guidance for economically efficient decision making.

CASH FLOW

In examining alternative ways of solving a problem we recognize the need to resolve the various consequences (both favorable and unfavorable) of each alternative into some common unit. One convenient unit, and the one typically used in economic analysis, is money. Thus an initial step in resolving economic analysis problems is to convert the various consequences of an alternative into a table of year-by-year cash flows.

For example, a simple problem might be to portray the consequences of purchasing a new car as follows.

	Year	*Cash flow*	
Beginning of first year	0	−$4500	Car purchased "now" for $4500 cash. The minus sign indicates a disbursement.
End of year	1	−350	Maintenance costs are $350 per year.
End of year	2	−350	
End of year	3	−350	
End of year	4	−350	
		+2000	The car is sold at the end of the fourth year for $2000.
			The plus sign represents a receipt for money.

This cash flow is represented graphically in Fig. 10-1. The upward arrow represents a receipt of money, and the downward arrows represent disbursements. The horizontal axis represents the passage of time.

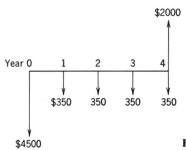

Figure 10-1   Graphical representation of a cash flow

TIME VALUE OF MONEY

When the money consequences of an alternative occur in a short period of time, say less than one year, we might simply algebraically add up the various sums of money and obtain the net result. But we cannot treat money this same way over longer periods of time. This is because money today is not the same as money at some future time. The general rule is: *A dollar now is worth more than the prospect of a dollar at any future time.* Thus $100 today is more valuable to us than someone's guarantee to give us $100 a year from now. The reason is simple; we could deposit our present $100 in a bank at 5% interest, and at the end of the year the bank would return our $100 together with $5 interest. So $100 today *is* worth more than $100 one year hence, and at 5% interest we know that $100 today is really equivalent to $105 one year hence.

EQUIVALENCE

If we are unconcerned whether we have money now or a year from now, and if we believe that 5% is a suitable interest rate, we would say that $100 today is *equivalent* to $105 one year hence. Equivalence is an essential element in engineering economic analysis. Suppose we wish to select the better of two alternatives. First, we must compute their cash flows. An example would be

Alternative

Year	A	B
0	−$2000	−$2800
1	+800	+1100
2	+800	+1100
3	+800	+1100

The larger investment in alternative B results in larger subsequent benefits, but we have no direct way of knowing if alternative B is better than alternative A. Therefore we do not know which alternative should be selected. To make a decision we must resolve the alternatives into equivalent sums so they may be compared accurately.

COMPOUND INTEREST FORMULAS

To facilitate equivalence computations a series of compound interest formulas will be derived and used. The generally accepted notation is:

$i =$ interest rate per interest period. In the formulas the interest rate is stated as a decimal. For example, 5% interest is entered in formulas as 0.05.

$n =$ number of interest periods. Frequently the interest period is 1 year, but it might be 3 months, 6 months, or some other time period.

$P =$ a present sum of money

$F =$ a sum of money n periods from the present time that is equivalent to P with interest rate i.

$A =$ the end-of-period payment or disbursement in a uniform series continuing for n periods, the entire series being equivalent to P at interest rate i.

$G =$ uniform period-by-period increase in cash flows; the arithmetic gradient.

Single Payment Formulas

Assume a present sum of money P is invested for one period at interest rate i. At the end of the period we would receive back our initial investment P together with interest equal to Pi or a total of $P + Pi$. This equals $P(1 + i)$. If the investment is continued for subsequent periods, the complete progression is as follows:

	Amount at beginning of period	+	Interest for the period	=	Amount at end of the period
1st period	P	+	Pi	=	$P(1+i)$
2nd period	$P(1+i)$	+	$Pi(1+i)$	=	$P(1+i)^2$
3rd period	$P(1+i)^2$	+	$Pi(1+i)^2$	=	$P(1+i)^3$
nth period	$P(1+i)^{n-1}$	+	$Pi(1+i)^{n-1}$	=	$P(1+i)^n$

The present sum P increases in n periods to $P(1+i)^n$. This gives a relationship between a present sum P and its equivalent future sum F.

$$\text{Future sum} = (\text{present sum})(1+i)^n$$

$$F = P(1+i)^n$$

This is the *single payment compound amount factor*. In functional notation* it is

$$F = P(F/P, i\%, n)$$

Solving for P, the equation is rewritten as

$$P = F(1+i)^{-n}$$

This is the *single payment present worth factor*. It is written

$$P = F(P/F, i\%, n)$$

Example 1

An individual wishes to deposit a certain quantity of money now so that at the end of five years he will have $500. With interest at 6% compounded annually, how much must he deposit now?

Solution.

$$n = 5 \quad i = 0.06 \quad F = \$500 \quad P = ?$$

$$P = F(1+i)^{-n} = 500(1+0.06)^{-5} = \$373.63$$

---

* In a notation now considered obsolete, the single payment compound amount factor was written $(\text{caf}' - i\% - n)$.

Instead of direct calculations, tables of compound interest factors are widely used. The problem may be written

$$P = F(P/F, i\%, n) = 500(P/F, 6\%, 5)$$

The compound interest factor is read from a table (See Appendix C). Find the 6% table, and read from the line for $n = 5$, the value of $(P/F, 6\%, 5)$ as 0.7473.

$$P = 500(0.7473) = \$373.65$$

Example 2

If you were to deposit \$2000 in a bank whose interest policy is "4% interest, compounded quarterly," how much would be in your account at the end of two years?

Solution. We must compute the number of interest periods and the interest rate per interest period.

$$n = 8 \qquad i = 0.01 \qquad P = \$2000 \qquad F = ?$$
$$F = P(1+i)^n \qquad \text{or} \qquad F = P(F/P, i\%, n)$$

From the compound interest tables in Appendix C we find $(F/P, 1\%, 8)$ equals 1.083.

$$F = 2000(1.083) = \$2166$$

Uniform Series Formulas

Consider the following situation.

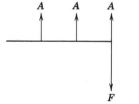

F   **Figure 10-2**

Using the single payment compound amount factor, we can write an equation for F in terms of A.

$$F = A + A(1+i) + A(1+i)^2 \tag{1}$$

In our situation, with $n = 3$, Eq. (1) may be written in a more general form:

Multiply
$$F = A + A(1+i) + A(1+i)^{n-1} \tag{2}$$

Eq. (2)
by $(1+i)$
$$(1+i)F = A(1+i) + A(1+i)^{n-1} + A(1+i)^n \tag{3}$$

Write
Eq. (2)
$$F = A + A(1+i) + A(1+i)^{n-1} \tag{2}$$

Subtract:
(3)−(2)
$$iF = -A + A(1+i)^n$$

$$F = A\left[\frac{(1+i)^n - 1}{i}\right] \qquad \textbf{Uniform series}$$
$$\textbf{Compound amount factor}$$

Solving this equation for A,

$$A = F\left[\frac{i}{(1+i)^n - 1}\right] \qquad \begin{array}{l}\textbf{Uniform series}\\ \textbf{Sinking fund factor}\end{array}$$

Since $F = P(1+i)^n$, we can substitute this expression for F in the equation and obtain

$$A = P\left[\frac{i(1+i)^n}{(1+i)^n - 1}\right] \qquad \begin{array}{l}\textbf{Uniform series}\\ \textbf{Capital recovery factor}\end{array}$$

Solving the equation for P,

$$P = A\left[\frac{(1+i)^n - 1}{i(1+i)^n}\right] \qquad \begin{array}{l}\textbf{Uniform series}\\ \textbf{Present worth factor}\end{array}$$

In functional notation the uniform series factors are

Compound Amount	$(F/A, i\%, n)$
Sinking Fund	$(A/F, i\%, n)$
Capital Recovery	$(A/P, i\%, n)$
Present Worth	$(P/A, i\%, n)$

Example 3

A man on January 1 deposits $1000 in a bank that pays 5% interest, compounded annually. He wishes to make five equal end-of-year withdrawals beginning December 31 of the first year. How much should he withdraw each year?

Solution.
$$n = 5 \qquad i = 0.05 \qquad P = \$1000 \qquad A = ?$$

$$A = P(A/P, i\%, n) = 1000(A/P, 5\%, 5) = 1000(0.2310) = \$231$$

Gradient Formulas

The gradient present worth factor is represented by Fig. 10-3. One should note carefully that the first G occurs at the end of the *second* interest period.

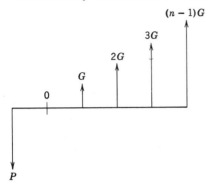

Figure 10-3 Gradient present worth diagram

As a result, in n interest periods there are $n-1$ cash flows. In the situation illustrated in Fig. 10-3, for example, there are four cash flows (G, $2G$, $3G$, and $4G$), and n equals five. The equation for the relationship between P and G is

$$P = \frac{G}{i}\left[\frac{(1+i)^n - 1}{i} - n\right]\left[\frac{1}{(1+i)^n}\right]$$

$$P = G(P/G, i\%, n)$$

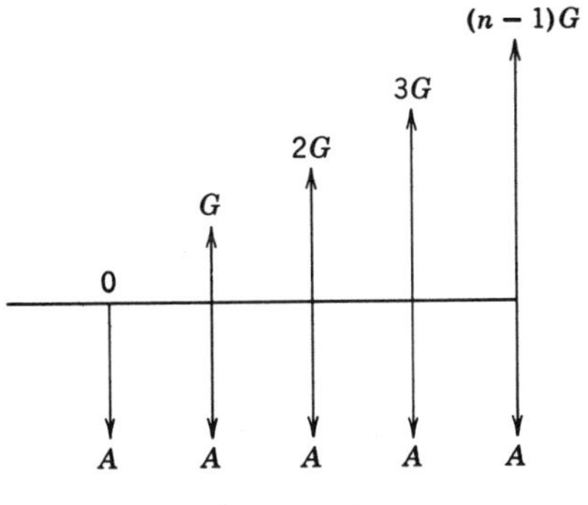

Figure 10-4 for the gradient uniform series has a form similar to Fig. 10-3. There is an A at the end of each of the n interest periods, but only $n-1$ values of G.

$$A = G\left[\frac{1}{i} - \frac{n}{(1+i)^n - 1}\right]$$

$$A = G(A/G, i\%, n)$$

Figure 10-4 Gradient uniform series diagram

ECONOMY STUDIES

Several fundamental elements are involved in economy studies:

1. The economy study is usually made from the viewpoint of the owners of the enterprise.

2. The study is a comparison of alternatives and deals with prospective differences between the alternatives.

3. As far as possible, the differences are reduced to differences in money receipts and disbursements.

4. A minimum attractive rate of return or suitable interest rate is selected or will be calculated.

5. The decision concerning the best alternative considers not only the monetary comparison but also the prospective differences between alternatives that are not reduced to money terms.

ANNUAL COST

To compare nonuniform series of cash flows it is necessary to make them comparable. One way to do this is by reducing each to an equivalent uniform annual series of payments (or receipts).

Factors in the annual cost method are:

1. A uniform rate of interest is charged on all money, no matter whether borrowed or not.

2. If a net money receipt from salvage value is in prospect at the end of the life of the asset, this reduces the annual cost.

3. Differences in the estimated lives of alternatives need not create a special problem when making an annual cost comparison. The underlying assumption made in this situation is that when the shorter lived alternative has reached the end of its useful life, it can be replaced with an identical item with identical cost and so forth. Therefore the equivalent uniform annual cost (EUAC) of the initial alternative for a relatively

short period of time is equal to the EUAC for the continuing series of replacements.

4. The conversion of variable annual disbursements to equivalent uniform annual disbursements may be accomplished by calculating the present worth of each of the variable disbursements. Then the present worth sum is multiplied by a suitable capital recovery factor to obtain the equivalent uniform annual cost.

Criteria

Economic analysis problems inevitably fall into one of three categories.

1. *Fixed input*:

The amount of money or other input resources is fixed.

Example: A project engineer has a budget of $450,000 to overhaul a petroleum process plant.

2. *Fixed output*:

There is a fixed task or other output to be accomplished.

Example: A contractor has been awarded a fixed price contract to build a building.

3. *Neither input nor output fixed*:

This is the general situation where neither the amount of money or other inputs nor the amount of benefits or other outputs is fixed.

Example: A consulting engineering firm has more work available than it can handle. It is considering paying the staff for working evenings to increase the amount of design work it can perform.

In each of these categories we can determine what the criterion should be for economic efficiency. For annual cost analysis the proper criteria are

Category	Annual Cost Criterion
Fixed input	Maximize the equivalent uniform annual benefits; that is, maximize EUAB.
Fixed output	Minimize the equivalent uniform annual cost. Simply stated, minimize EUAC.
Neither input nor output fixed	Maximize [EUAB – EUAC].

Example 4

A manufacturing firm is considering replacing an old machine with a new machine which would perform the same task. The old machine cost $5000 five years ago and has an estimated useful life of an additional ten years. The salvage value at the present time is $2400, but no salvage value will remain after ten years. The annual operating costs, excluding depreciation and interest, are $6500. The proposed new machine would cost $7000, and the salvage value at the end of its useful life, estimated to be ten years, would be zero. The annual operating costs, excluding depreciation and interest, are $5400. If the firm uses a 10% interest rate, what should it do?

Solution.

	Old machine	*New machine*
Current value	$2400	$7000
Useful life	10 more years	10 years
Salvage value	0	0
Annual operating cost	$6500	$5400

The problem will be solved by the annual cost method.

Old machine
$2400(A/P, 10\%, 10) = 2400(0.1627) = \quad 390$
Annual operating cost $\qquad = \;6500$

Equivalent Uniform Annual Cost $\quad = \$6890$

New machine
$7000(A/P, 10\%, 10) = 7000(0.1627) = \;1140$
Annual operating cost $\qquad = \;5400$

Equivalent Uniform Annual Cost $\quad = \$6540$

Based on the available data, we would choose the alternative with the smaller equivalent uniform annual cost.

PRESENT WORTH

Present worth analysis is used to determine the present value of future money receipts and disbursements. One might want to know, for example, the present worth of an income-producing property like an oil well. The computation would provide an estimate of the price at which the property might be bought or sold.

Criteria

Category	Present Worth Criterion
Fixed input	Maximize the present worth of benefits or other outputs.
Fixed output	Minimize the present worth of costs or other inputs.
Neither input nor output fixed	Maximize [present worth of benefits minus present worth of costs]. Simply stated, maximize net present worth.

Analysis period

An important restriction in the use of present worth calculations is that there must be a common analysis period when comparing alternatives. It would be incorrect, for example, to compare the present worth (PW) of cost of machine A, expected to last five years, with the PW of cost of machine B that will last for 10 years. In situations like this the solution is either to use some other analysis technique or to restructure the problem so there is a common analysis period. In the example of Fig. 10-5 a customary assumption would

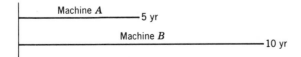

Figure 10-5   Improper present worth comparison

be that there is a need for the machine for 10 years and that machine *A* will be replaced by an identical machine *A* at the end of five years. This gives a 10-year common analysis period. This approach is easy to use when the different lives of the alternatives have a practical least common multiple life.

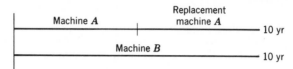

Figure 10-6   Correct present worth comparison

When this is not true (for example, machine *A* has a seven-year life and machine *B* has an 11-year life), some assumptions must be made to select a suitable common analysis period or the present worth method should not be used.

Example 5

Three alternatives are being considered to solve a material handling problem in a plant. Any of the three alternatives will provide the desired result.

	A	*B*	*C*
Initial cost	$5000	$8000	$13,000
Annual operation and maintenance	1000	500	500
Useful life	2 yr	3 yr	6 yr
End-of-useful-life salvage value	$0	$2000	$0

If a 10% interest rate is used in the economic analysis, which alternative should be selected?

Solution.   A common analysis period must be selected. Unless there are other facts provided, the normal assumption is to choose the six-year least common multiple life.

Present worth (PW) of six years of alternative *A*:

$$\text{PW of cost} = 5000 + 5000(P/F, 10\%, 2) + 5000(P/F, 10\%, 4)$$
$$+ 1000(P/A, 10\%, 6)$$
$$= 5000(1 + 0.8264 + 0.6830) + 1000(4.355)$$
$$= \$16,902$$

PW of six years of alternative B:

PW of cost $= 8000 + (8000 - 2000)(P/F, 10\%, 3) + 500(P/A, 10\%, 6)$
$$- 2000(P/F, 10\%, 6)$$

$$= 8000 + 6000(0.7513) + 500(4.355) - 2000(0.5645)$$

$$= \$13,556$$

PW of six years of alternative C:

PW of cost $= 13,000 + 500(P/A, 10\%, 6) = \$15,178$

Choose the alternative with the least PW of cost. Choose alternative B.

CAPITALIZED COST

In the special situation where the analysis period is infinite $(n = \infty)$, an analysis of the present worth of cost is called *capitalized cost*. A present sum P will accrue interest of Pi for every interest period. For the principal sum P to continue undiminished (an essential requirement for n equal to infinity), the end-of-period sum A that can be disbursed is Pi.

$$A = Pi$$

Some form of this equation is used whenever there is a problem with an infinite analysis period.

Example 6

What present investment is necessary to secure a perpetual income of $1800 a year if interest is 5%?

Solution.

$$\text{Capitalized cost} \quad P = \frac{A}{i} = \frac{1800}{0.05} = \$36,000$$

RATE OF RETURN

Present worth and annual cost calculations start with the assumption of an interest rate or a minimum attractive rate of return. Sometimes we want to know the prospective rate of return on an investment rather than simply to determine whether or not it meets some fixed standard of attractiveness. In the rate of return method the typical situation is where there is a cash flow representing both costs and benefits. The rate of return may be defined as the interest rate where

$$PW \text{ of cost} = PW \text{ of benefit}$$

or

$$EUAC = EUAB$$

These calculations frequently require trial and error solution.

Example 7

A man invested $10,000 in a project. At the end of three years he received $9000, and at the end of six years he received $6000. What rate of return did he receive on the project?

Solution.

$$PW \text{ of cost} = PW \text{ of benefits}$$

$$10,000 = 9000(P/F, i\%, 3) + 6000(P/F, i\%, 6)$$

Try $i = 12\%$.

$$10,000 = 9000(0.7118) + 6000(0.5066)$$

$$= 6406 + 3040 = 9446$$

The PW of benefits is lower than the PW of cost, indicating that the interest rate i is too high.

Try $i = 10\%$.

$$10,000 = 9000(0.7513) + 6000(0.5645)$$

$$= 6762 + 3387 = 10,149$$

Using linear interpolation between 10% and 12%,

$$\text{Rate of return } i = 10\% + (2\%)\left(\frac{10,149 - 10,000}{10,149 - 9446}\right) = 10.42\%$$

RATE OF RETURN ANALYSIS

Earlier in this chapter two alternatives were described in terms of their cash flows.

	Alternative	
Year	A	B
0	−$2000	−$2800
1	+800	+1100
2	+800	+1100
3	+800	+1100

If one considers 5% the minimum attractive rate of return (MARR), which alternative should be selected? A computation of the rate of return for each alternative shows that both alternatives have a rate of return in excess of 5%.

A conventional assumption in economic analysis is that we will select the larger investment (rather than the smaller one) if the rate of return on the *additional* investment is greater than or equal to the minimum attractive rate of return. This means that there will be situations where the alternative selected is *not* the alternative with the largest rate of return. To decide which alternative in the example to select, compute the cash flow that represents the difference between the alternatives. Then compute the rate of return on this increment.

	Alternative		Difference between the alternatives
Year	A	B	B − A
0	−$2000	−$2800	−$800
1	+800	+1100	+300
2	+800	+1100	+300
3	+800	+1100	+300
Computed rate of return	9.7%	8.7%	6.1%

Since the rate of return on the difference between the alternatives (6.1%) exceeds the 5% MARR, the increment of additional investment is desirable. We would select alternative *B*.

When there are three or more mutually exclusive (doing one precludes doing the rest) alternatives, one may proceed following the same general logic presented for two alternatives. The components of incremental analysis are

1. Compute the rate of return for each alternative. Reject any alternative where the rate of return is less than the given MARR. This step is not essential but helps to immediately eliminate unacceptable alternatives. One must ensure, however, that the lowest cost alternative has a rate of return $\geq$ MARR.

2. Rank the remaining alternatives in their order of increasing cost.

3. Examine the difference between the two lowest cost alternatives as described for the two alternative problem. Select the better of the two alternatives and reject the other one.

4. Take the preferred alternative from step 3. Consider the next higher cost alternative and proceed with another two-alternative comparison.

5. Continue until all alternatives have been examined and the best of the multiple alternatives has been identified.

FUTURE WORTH

In present worth analysis the comparison is made in terms of the equivalent *present* costs and benefits. The selection of the present time as the point for the computations is simply a matter of convenience. The computations may be made as of any date: past, present, or future. Although the numerical calculations may look different, the decision is unaffected by the choice of a time about which to make the computations. When a future time is used, we compute the equivalent *future* costs and benefits and call the computations future worth analysis.

Example 8

Solve Example 5 by future worth analysis. At a 10% interest rate which of the three alternatives should be selected?

	A	B	C
Initial cost	$5000	$8000	$13,000
Annual operation and maintenance	1000	500	500
Useful life	2 yr	3 yr	6 yr
End-of-useful-life salvage value	$0	$2000	$0

Solution.

FW of six years of alternative A:

$$FW \text{ of cost} = 5000(F/P, 10\%, 6) + 5000(F/P, 10\%, 4)$$
$$+ 5000(F/P, 10\%, 2) + 1000(F/A, 10\%, 6)$$

$$= 5000(1.772 + 1.464 + 1.210) + 1000(7.716)$$

$$= \$29,946$$

FW of six years of alternative B:

$$FW \text{ of cost} = 8000(F/P, 10\%, 6) + (8000 - 2000)(F/P, 10\%, 3)$$
$$+ 500(F/A, 10\%, 6) - 2000$$

$$= 8000(1.772) + 6000(1.331) + 500(7.716) - 2000$$

$$= \$24,020$$

FW of six years of alternative C:

$$FW \text{ of cost} = 13,000(F/P, 10\%, 6) + 500(F/A, 10\%, 6) = \$26,894$$

Choose the alternative with the least FW of cost. Choose alternative B.

BREAKEVEN ANALYSIS

Breakeven is the point where two alternatives are equivalent. This method is used to find the value of a single parameter with all other values held fixed.

Example 9

A new snow removal machine costs $50,000. The new machine will operate at a reported saving of $400 per day over the present equipment in terms of time and efficiency. If interest is 5% and the machine's life is assumed to be 10 years with zero salvage, how many days per year must the machine be used to make the investment economical?

Solution. The problem will be solved by the annual cost method. Set the annual saving with the snow removal machine equal to its annual cost and solve for the required utilization.

Let x = annual operation in days per year.

$$\text{Annual saving} = \text{annual cost}$$

$$400x = 50{,}000(A/P, 5\%, 10)$$

$$= 50{,}000(0.1295) = 6475$$

$$x = \frac{6475}{400} = 16.2 \text{ days per year}$$

DEPRECIATION

Depreciation of equipment and other capital assets is an important component of many after-tax economic analyses. Depreciation is defined, in its accounting sense, as the systematic allocation of the cost of a capital asset over its useful life. In computing a schedule of depreciation charges, three items may be considered.

1. Cost of the property P

2. Useful life in years n

3. Salvage value of the property at the end of its useful life F

For *straight-line depreciation,*

$$\text{Depreciation charge in any year} = \frac{P - F}{n}$$

For *sum-of-years digits depreciation,*

$$\text{Depreciation charge in any year} = \frac{\substack{\text{Remaining useful life} \\ \text{at beginning of year}}}{\substack{\text{Sum-of-years digits} \\ \text{for total useful life}}}(P - F)$$

$$\text{where sum-of-years digits} = 1 + 2 + 3 + \cdots + n = \frac{n}{2}(n + 1).$$

For *double declining balance depreciation,*

$$\text{Depreciation charge in any year} = \frac{2}{n}(P - \text{Depreciation charges to date})$$

INCOME TAXES

Income taxes represent another of the various kinds of disbursements encountered in an economic analysis. The starting point in an after-tax computation is the before-tax cash flow. The next step is to compute the depreciation schedule for any capital assets. Taxable income is the taxable component of the before-tax cash flow, minus the depreciation. The income tax disbursement is the taxable income times the appropriate tax rate. Finally, the after-tax cash flow is the before-tax cash flow minus income taxes.

Example 10

A firm is investing $150,000 in equipment from which it expects to receive $50,000 a year in benefits for five years. There is no expected salvage value at the end of the five years. Based on straight-line depreciation and a 48% income tax rate, what is the after-tax rate of return?

Solution.

$$\text{Straight-line depreciation} = \frac{P-F}{n} = \frac{150,000-0}{5} = \$30,000 \text{ per year}$$

Year	Before-tax cash flow	Straight-line depreciation	Taxable income	48% Income taxes	After-tax cash flow
0	−$150,000				−$150,000
1	+50,000	$30,000	+$20,000	−$9600[a]	+40,400
2	+50,000	30,000	+20,000	−9600	+40,400
3	+50,000	30,000	+20,000	−9600	+40,400
4	+50,000	30,000	+20,000	−9600	+40,400
5	+50,000	30,000	+20,000	−9600	+40,400

[a] The minus sign represents the payment of taxes.

The rate of return based on the after-tax cash flow is

$$\text{PW of cost} = \text{PW of benefits}$$

$$150,000 = 40,400(P/A, i\%, 5)$$

$$(P/A, i\%, 5) = \frac{150,000}{40,400} = 3.71$$

From interest tables, $i = 10.8\%$.

ANNUAL COST COMPUTED BY STRAIGHT-LINE DEPRECIATION PLUS AVERAGE INTEREST

At times an average annual cost is computed rather than an equivalent uniform annual cost (EUAC). The principal advantage of average annual cost is that it may be computed without compound interest tables.

In this approximate method the decline in value of the asset, over its useful life, is distributed equally to each year of asset life. This amounts to straight-line depreciation, $(P-F)/n$. In addition, there must be a charge to reflect the investment in the unrecovered cost of the asset. This is an interest charge on the money still invested in the asset. In the first year the interest is $(P-F)i+Fi$, and in the last year of the useful life it is $(P-F)i/n+Fi$. The average interest charge is half the sum of the first and last year's interest charges, or

$$\frac{(P-F)i + Fi + \left(\dfrac{P-F}{n}\right)i + Fi}{2} = (P-F)\left(\frac{i}{2}\right)\left(\frac{n+1}{n}\right) + Fi$$

Thus the straight-line depreciation plus average interest method is based on two equations:

$$\text{Straight-line depreciation} = \frac{P-F}{n}$$

$$\text{Average interest} = (P-F)\left(\frac{i}{2}\right)\left(\frac{n+1}{n}\right) + Fi$$

This method should *not* be used in the engineering fundamentals examination unless the problem specifically asks for a computation of annual cost by the method of straight-line depreciation plus average interest.

ADDITIONAL ECONOMIC ANALYSIS INFORMATION

For an expanded discussion of economic analysis a textbook written by one of the authors should be helpful: Newnan, *Engineering Economic Analysis* (Engineering Press, P.O. Box 5, San Jose, CA 95103). A 30-page set of compound interest tables (from $\frac{1}{4}$% to 60%) is contained in the appendix of that book.

PROBLEM 10-1

How long will it take for $10,000 invested in a bank savings account to double in value? If the bank pays 6% interest, compounded semiannually, the time for the $10,000 to double is nearest to

(A) 8 yr

(B) 12 yr

(C) 14 yr

(D) 16 yr

(E) 24 yr

Solution.

$$F = P(F/P, 3\%, n) \qquad 2 = 1(F/P, 3\%, n) \qquad (F/P, 3\%, n) = 2$$

From the 3% compound interest tables we see that n is between 23 and 24 for $(F/P, 3\%, n)$ equal to 2. Thus it will take 24 six-month periods or 12 years for money to double.

Answer is (B)

PROBLEM 10-2

The capital recovery factor equals the

(A) present worth factor plus the sinking fund factor

(B) sinking fund factor plus the interest rate

(C) single payment compound amount factor minus the sinking fund factor

(D) uniform series compound amount factor plus the single payment present worth factor

(E) gradient present worth factor minus one

Solution. The capital recovery factor equals the sinking fund factor plus the interest rate.

Answer is (B)

PROBLEM 10-3

The uniform annual end-of-year payment to repay a debt in n years, with an interest rate i, is determined by multiplying the capital recovery factor by the

(A) average debt

(B) initial debt plus total interest

(C) average debt plus interest

(D) initial debt plus the first year's interest

(E) initial debt

Solution. The correct relation is the following: Annual payment equals the capital recovery factor multiplied by the initial debt, or $A = P(A/P, i\%, n)$.

Answer is (E)

PROBLEM 10-4

If $100,000 is invested at 12% interest, compounded monthly, the annual interest is nearest to

(A) $1000
(B) $12,000
(C) $12,350
(D) $12,700
(E) $144,000

Solution.

$$\text{Effective interest rate} = (1+i)^m - 1$$

where i is the interest rate per interest period and m is the number of compoundings per year.

In this problem

$$\text{Effective interest rate} = (1+0.01)^{12} - 1 = 0.1268 = 12.68\%$$

$$\text{Annual interest} = 100,000(0.1268) = \$12,680$$

Answer is (D)

PROBLEM 10-5

A 12% annual interest rate, compounded annually, is equivalent to what annual interest rate, compounded quarterly?

(A) 2.9%
(B) 3.0%
(C) 11.5%
(D) 12.0%
(E) 12.6%

Solution.

$$\text{Effective interest rate} = (1+i)^m - 1 = 0.12 = 12\%$$

$$(1+i)^4 = 1.12 \qquad 1.12^{0.25} = 1+i \qquad 1.0287 = 1+i$$

$$i \text{ per quarter year} = 0.0287 = 2.87\%$$

$$\text{Annual interest rate} = 4(2.87) = 11.5\%$$

$$\text{Answer is (C)}$$

PROBLEM 10-6

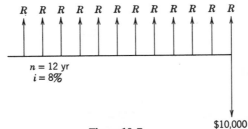

$$n = 12 \text{ yr}$$
$$i = 8\%$$

$10,000

Figure 10-7

The value of R in Fig. 10-7 is closest to

(A) $527
(B) $833
(C) $900
(D) $1327
(E) $1633

Solution.

$$R = 10,000(A/F, 8\%, 12) = 10,000(0.0527) = \$527$$

$$\text{Answer is (A)}$$

PROBLEM 10-7

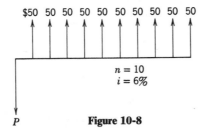

$50 50 50 50 50 50 50 50 50 50

$$n = 10$$
$$i = 6\%$$

P **Figure 10-8**

The value of P in Fig. 10-8 is closest to

(A) $200
(B) $400
(C) $450
(D) $500
(E) $700

Solution.

$$P = 50(P/A, 6\%, 10) = 50(7.360) = \$368$$

Answer is (B)

PROBLEM 10-8

A company is about to purchase three trucks valued at a total of $78,000. The truck dealer offers terms of $5000 downpayment with 12 equal end-of-month payments at an interest rate of 1% per month on the unpaid balance. The monthly payment required by the dealer is closest to

(A) $6000
(B) $6500
(C) $6800
(D) $6900
(E). $9800

Solution.

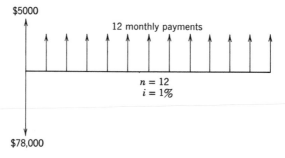

$n = 12$
$i = 1\%$

Figure 10-9

$$A = (78,000 - 5000)(A/P, 1\%, 12) = 73,000(0.0888) = \$6482$$

Answer is (B)

PROBLEM 10-9

An industrial firm is purchasing a machine for $10,000. It estimates there will be a $3000 salvage value at the end of the machine's 12-year useful life. The annual maintenance cost is estimated to be $150. The equivalent uniform annual cost (EUAC) for the machine, based on a 10% interest rate, is nearest to

(A) $800
(B) $1000
(C) $1300
(D) $1500
(E) $1600

Solution.

$$\text{EUAC} = 10,000(A/P, 10\%, 12) + 150 - 3000(A/F, 10\%, 12)$$

$$= 10,000(0.1468) + 150 - 3000(0.0468)$$

$$= \$1478$$

Answer is (D)

PROBLEM 10-10

A building costs $500,000. Its estimated life is 20 years. At 12% interest the annual amount that could be spent for extra maintenance, if this would extend the life of the building to 30 years, is closest to

(A) $5000
(B) $10,000
(C) $30,000
(D) $62,000
(E) $67,000

Solution. One should be willing to spend money on extra annual mainte-nance until the point is reached where the equivalent uniform annual cost

(EUAC) for the building with its life extended to 30 years equals the equivalent uniform annual cost (EUAC) for the building with a 20-year life.

$$\text{EUAC for 30-year life} = \text{EUAC for 20-year life}$$

$$500,000(A/P, 12\%, 30) + \text{extra annual maintenance} =$$
$$500,000(A/P, 12\%, 20)$$

$$500,000(0.1241) + \text{extra annual maintenance} = 500,000(0.1339)$$

$$\text{Extra annual maintenance} = 500,000(0.1339 - 0.1241) = \$4900$$

Answer is (A)

PROBLEM 10-11

A contractor wishes to set up a special fund by making uniform semiannual end-of-period deposits for 20 years. The fund is to provide $10,000 at the end of each of the last five years of the 20-year period. If interest if 8%, compounded semiannually, the required semiannual deposit is closest to

(A) $600
(B) $750
(C) $1000
(D) $1250
(E) $2500

Solution.

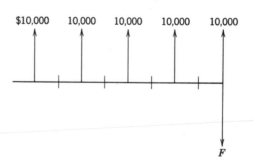

Figure 10-10   Withdrawals from the special fund

$$F = 10{,}000(F/P, 4\%, 8) + 10{,}000(F/P, 4\%, 6) + 10{,}000(F/P, 4\%, 4)$$
$$+ 10{,}000(F/P, 4\%, 2) + 10{,}000$$
$$= 10{,}000(1.369 + 1.265 + 1.170 + 1.082 + 1)$$
$$= \$58{,}860$$

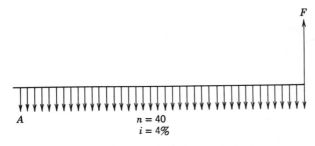

Figure 10-11   Deposits to the special fund

$$A = F(A/F, 4\%, 40) = 58{,}860(0.0105) = \$618$$

Answer is (A)

PROBLEM 10-12

A woman wishes to accumulate a total of $10,000 in a savings account at the end of 10 years. If the bank pays 4%, compounded quarterly, the sum that should be deposited now is closest to

(A) $3350
(B) $5000
(C) $6000
(D) $6700
(E) $6760

Solution.

$$P = F(P/F, 1\%, 40) = 10{,}000(P/F, 1\%, 40) = 10{,}000(0.6717) = \$6717$$

Answer is (D)

PROBLEM 10-13

An engineer has been asked if he would like to purchase a mortgage on an expensive home. The holder of the mortgage will receive $840 interest at the end of each year for seven years, and a lump sum payment of $12,000 at the end of seven years. How much would the engineer be willing to pay for the mortgage if he considers 10% to be the proper interest rate? The amount is closest to

(A) $8000
(B) $10,000
(C) $12,000
(D) $14,000
(E) $16,000

Solution.

$$P = 840(P/A, 10\%, 7) + 12,000(P/F, 10\%, 7)$$

$$= 840(4.868) + 12,000(0.5132) = \$10,248$$

Answer is (B)

PROBLEM 10-14

How much money would need to be deposited in a bank now at 5% interest to provide $1000 per year for the next 50 years? The amount is closest to

(A) $16,000
(B) $18,000
(C) $20,000
(D) $25,000
(E) $50,000

Solution.

$$P = 1000(P/A, 5\%, 50) = 1000(18.256) = \$18,256$$

Answer is (B)

PROBLEM 10-15

The federal government has loaned $10 million to one of the states. The loan is at 5% interest and is for an infinite period. What annual payment must the state make to the federal government?

(A) $0
(B) $50,000
(C) $500,000
(D) $504,000
(E) $600,000

Solution. When n equals infinity, the general equation is $A = Pi$.

Required annual payment $A = Pi = 10,000,000(0.05) = \$500,000$

Answer is (C)

PROBLEM 10-16

A city has been offered land for a park and enough money to pay $10,000 per year maintenance forever. Assuming a 6% interest rate, the amount of money needed to provide the perpetual maintenance is closest to

(A) $6000
(B) $10,000
(C) $166,667
(D) $600,000
(E) $1,000,000

Solution. When n equals infinity, the capitalized cost P equals A/i.

$$P = \frac{A}{i} = \frac{10,000}{0.06} = \$166,667$$

Answer is (C)

PROBLEM 10-17

A trust fund is to be established to pay $100,000 per year operating costs of a laboratory, and in addition to provide $75,000 of replacement equipment every four years, beginning four years from now. At 8% interest the amount of money now required in the perpetual trust fund is nearest

(A) $115,000
(B) $210,000
(C) $1,250,000
(D) $1,450,000
(E) $1,600,000

Solution.   For $100,000 per year operating costs

$$\text{Required amount in the trust fund} = \frac{A}{i} = \frac{100,000}{0.08} = \$1,250,000$$

For $75,000 replacement equipment every four years the perpetual series is

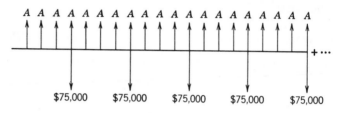

$75,000 $75,000 $75,000 $75,000 $75,000

Figure 10-12

If the equivalent amount A is computed, then the $A = Pi$ equation may be used to compute the necessary amount in the trust fund, P. We can solve one portion of the perpetual series for A

A A A A

$75,000   **Figure 10-13**

$$A = 75,000(A/F, 8\%, 4) = 75,000(0.2219) = \$16,642.50$$

This value for A for the four-year period is the same as the value of A for the perpetual series.

$$\text{Required amount in trust fund} = \frac{A}{i} = \frac{16,642.50}{0.08} = \$208,000$$

For annual maintenance plus periodic equipment replacement the total needed in the trust fund is $1,250,000 + 208,000 = \$1,458,000$

Answer is (D)

PROBLEM 10-18

A subdivider offers lots for sale for $15,000. A downpayment of $1500 is required and $1500 is paid at the end of each year for nine years with no interest charged. Further negotiation reveals that the lots may also be purchased for $10,000 cash. The actual interest rate the buyer will pay if he chooses the installment payment plan is closest to

(A) 0%
(B) 4%
(C) 6%
(D) 8%
(E) 10%

Solution.

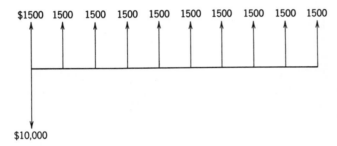

Figure 10-14

$$10,000 = 1500 + 1500(P/A, i\%, 9)$$

$$(P/A, i\%, 9) = \frac{8500}{1500} = 5.67$$

From the compound interest tables:

i	$(P/A, i\%, 9)$
10%	5.759
12%	5.328

Linear interpolation:

$$i = 10\% + (2\%)\left(\frac{5.759 - 5.67}{5.759 - 5.328}\right) = 10\% + (2\%)\left(\frac{0.09}{0.431}\right) = 10.4\%$$

Answer is (E)

PROBLEM 10-19

A machine is to be purchased for $15,500; it has an estimated life of eight years and a salvage value of $600. A sinking fund is to be established so money will be available to purchase a replacement when the first machine wears out at the end of eight years. An amount of $1303 is to be deposited at the end of each year during the lifetime of the first machine into this sinking fund. The interest rate this fund must earn to produce sufficient funds to purchase the replacement machine at the end of eight years is closest to

(A) 8%
(B) 9%
(C) 10%
(D) 11%
(E) 12%

Solution. The amount of money that must be accumulated in the sinking fund by the end of eight years equals the cost of the new machine minus the salvage value of the original machine, or $15,500 - 600 = $14,900$.

$$(A/F, i\%, 8) = \frac{A}{F} = \frac{1303}{14,900} = 0.0874$$

From the interest tables we see that this is exactly the value of the sinking fund factor when the interest rate i is 10%.

Answer is (C)

PROBLEM 10-20

An investor purchased 10 shares of Omega Company stock for $9000. He held the stock for nine years. For the first four years he received annual end-of-year dividends of $800. For the next four years he received annual dividends of $400. He received no dividend for the ninth year. At the end of the ninth year he sold his stock for $8000. The rate of return he received on his investment is closest to

(A) 4%
(B) 5%
(C) 6%
(D) 8%
(E) 10%

Solution. Tabulate the cash flow for the problem.

Year	Cash flow
0	−$9000
1	+800
2	+800
3	+800
4	+800
5	+400
6	+400
7	+400
8	+400
9	+8000

Write one equation for the cash flow with the rate of return i as the only unknown. There are several ways of writing this equation. One way is

$$\text{Net Present Worth} = \text{PW of benefits} - \text{PW of cost} = 0$$
$$800(P/A, i\%, 4) + 400(P/A, i\%, 4)(P/F, i\%, 4)$$
$$+ 8000(P/F, i\%, 9) - 9000 = 0$$

Solve the equation by trial and error to determine the rate of return i.
Try $i = 5\%$.

$$\text{NPW} = 800(3.546) + 400(3.546)(0.8227) + 8000(0.6446) - 9000$$

$$= +161$$

The rate of return is too low.
Try $i = 6\%$.

$$\text{NPW} = 800(3.465) + 400(3.465)(0.7921) + 8000(0.5919) - 9000$$

$$= -395$$

This time the rate of return is too high. Interpolate between 5% and 6%.

$$\text{Rate of return } i = 5\% + (1\%)\left(\frac{161}{395 + 161}\right) = 5.29\%$$

Answer is (B)

PROBLEM 10-21

A firm whose minimum attractive rate of return is 8% is considering which one of four alternatives it should select. The alternatives and their computed rates of return are as follows.

Alternative

	1	2	3	4
Initial cost	$100	$130	$200	$330
Uniform annual benefit	26.38	38.78	47.48	91.55
Useful life	5 yr	5 yr	5 yr	5 yr
Computed rate of return	10%	15%	6%	12%

Any money not spent on one of the alternatives can be invested elsewhere at an 8% rate of return. The alternative the firm should select is

(A) 1
(B) 2
(C) 3
(D) 4
(E) None of them

Solution.   By inspection, two of the alternatives may be rejected. Alternative 2 costs more and has a greater rate of return than alternative 1. This means the incremental rate of return of doing alternative 2 rather than alternative 1 exceeds the rate of return of alternative 2, or 15%. Thus alternative 2 dominates alternative 1 and is preferred over alternative 1. Reject alternative 1.

Following this same logic, alternative 4 dominates alternative 3. Reject alternative 3. There is also another basis for rejecting alternative 3. Since alternative 3 has a rate of return less than the 8% minimum attractive rate of return, and there are other alternatives that exceed the minimum attractive rate of return, alternative 3 will always be eliminated.

The problem reduces to a choice between alternative 2 and alternative 4. Compute the incremental rate of return ($\triangle$ROR) on the difference between alternatives 4 and 2.

	Increment
	4 − 2
$\triangle$ Initial cost	$200
$\triangle$ Uniform annual benefit	52.77
Useful life	5 yr
$\triangle$ Rate of return	10%

Since the 10% incremental rate of return exceeds the 8% minimum attractive rate of return, the increment of investment is desirable. Choose alternative 4. Note particularly that the alternative with the highest rate of return is *not* the selected alternative.

Answer is (D)

PROBLEM 10-22

A manufacturer who produces a single item has a maximum production capacity of 40,000 units per year. The overall and unit costs for different levels of operation are as follows:

Output (units)	Total cost	Total cost per unit
0	$60,000	
5,000	85,000	$17.00
10,000	109,000	10.90
20,000	155,000	7.75
40,000	243,000	6.08

Early in the year orders are received for 10,000 units at $12 each. Owing to depressed business conditions it is realized that no further domestic orders can be expected in the current year. The manufacturer has an opportunity to sell 10,000 units overseas to a foreign buyer. The lowest selling price per unit the manufacturer is willing to accept on this overseas order is closest to

(A) $4.40
(B) $4.70
(C) $7.75
(D) $10.90
(E) $12.00

Solution.

The manufacturer should accept any order where the additional revenue exceeds the cost of supplying the order. The table on the next page shows the incremental cost per unit is $4.60.

Output (units)	Total cost	Incremental cost	Incremental cost per unit
0	$ 60,000		
		$25,000	$5.00
5,000	85,000		
		24,000	4.80
10,000	109,000		
		46,000	4.60
20,000	155,000		
		88,000	4.40
40,000	243,000		

Answer is (B)

PROBLEM 10-23

Three hundred dollars is deposited in a bank savings account at the beginning of each of 15 years. If the account pays 8% annual interest, the amount in the account at the end of 15 years is nearest to

(A) $2800
(B) $4500
(C) $7600
(D) $8100
(E) $8800

Solution.

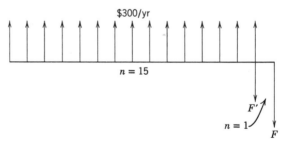

Figure 10-15

$$F' = 300(F/A, 8\%, 15) = 300(27.152) = \$8146$$

$$F = 8146(F/P, 8\%, 1) = 8146(1.080) = \$8798$$

Answer is (E)

PROBLEM 10-24

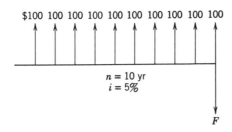

$100 100 100 100 100 100 100 100 100 100

$n = 10$ yr
$i = 5\%$

F **Figure 10-16**

The value of F is closest to

(A) $800
(B) $1000
(C) $1050
(D) $1300
(E) $1500

Solution.

$$F = 100(F/A, 5\%, 10) = 100(12.578) = \$1258$$

Answer is (D)

PROBLEM 10-25

A firm believes that it will need land for another warehouse 20 years hence. A suitable piece of land can now be bought for $95,000. If purchased, the firm will pay $3000 a year taxes on the property. A 10% interest rate is to be used. To justify its purchase now, 20 years from now the land must be worth approximately

(A) $155,000
(B) $170,000
(C) $640,000
(D) $700,000
(E) $810,000

Solution.

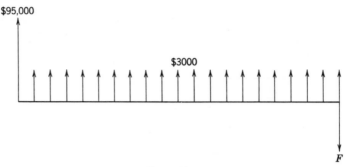

Figure 10-17

$$F = 95{,}000(F/P, 10\%, 20) + 3000(F/A, 10\%, 20)$$

$$= 95{,}000(6.727) + 3000(57.275) = \$811{,}000$$

Answer is (E)

PROBLEM 10-26

A contractor can purchase a dump truck for $18,000. Its estimated salvage value is $4000 at the end of its seven-year useful life. Daily operating expenses are $125, including maintenance and the cost of the driver. The contractor can hire a similar truck and its driver for $160 a day. At 8% interest the number of days per year the dump truck must be needed to justify its purchase is closest to

(A) 20
(B) 40
(C) 60
(D) 80
(E) 100

Solution.   Set the annual cost of hiring a truck equal to the annual cost of truck ownership. Let x = truck utilization in days per year.

$$160x = 18{,}000(A/P, 8\%, 7) + 125x - 4000(A/F, 8\%, 7)$$

$$35x = 18{,}000(0.1921) - 4000(0.1121)$$

$$x = 86 \text{ days per year}$$

Answer is (D)

PROBLEM 10-27

A company which manufactures electric motors has a production capacity of 200 motors a month. The variable costs are $150 per motor. The average selling price of the motors is $275. Fixed costs of the company amount to $20,000 per month, which includes all taxes. The number of motors that must be sold each month to break even is closest to

(A) 40
(B) 80
(C) 120
(D) 160
(E) 200

Solution. Let $x =$ number of motor sales per month to break even.

$$275x = 150x + 20,000$$

$$x = 160 \text{ motors}$$

Answer is (D)

PROBLEM 10-28

A 10-hp electric motor is required and two different models are being considered. Motor A has a first cost of $350 and an overall efficiency of 90%. Motor B has a first cost of $250 and an efficiency of 80%. Both motors have identical lives and will be loaded to rated capacity when in use. The cost of electric energy is $0.05/kW-hr. The annual investment charge is 20% of the first cost. The annual breakeven operating time in hours at which the annual cost of operation and ownership of motor A equals that of B is closest to

(A) 200 hr
(B) 400 hr
(C) 600 hr
(D) 800 hr
(E) 1000 hr

Solution.

$$\text{Efficiency} = \frac{\text{output}}{\text{input}} = \frac{10 \text{ hp}}{\text{input}}$$

$$\text{Difference in input} = \frac{10 \text{ hp}}{0.80} - \frac{10 \text{ hp}}{0.90} = 12.5 - 11.1 = 1.4 \text{ hp}$$

$$= 1.4 \text{ hp} \times 0.746 = 1.0 \text{ kW}$$

Let the annual incremental investment charge equal the annual incremental power charge:

$$0.20(\$350 - \$250) = 1.0(\$0.05)(\text{hours of operation})$$

$$\text{Hours of operation} = \frac{0.20(100)}{0.05} = 400$$

Answer is (B)

PROBLEM 10-29

Given the following data for two capacitors:

	Ferrous	Copper
Installed cost	$800	$1200
Useful life	10 yr	unknown
Annual operating cost	$63	$40

The useful life of the ferrous capacitor is 10 years, but the useful life of the copper alloy capacitor is not known. Either capacitor may be replaced by an identical replacement at the end of its useful life. Neither capacitor has any salvage value. The life for the copper capacitor to be as economical as the ferrous one, assuming 10% interest, is closest to

(A) 12 yr
(B) 14 yr
(C) 16 yr
(D) 18 yr
(E) 20 yr

Solution. Set the annual cost of the ferrous capacitor equal to the annual cost of the copper capacitor.

$$800(A/P, 10\%, 10) + 63 = 1200(A/P, 10\%, n) + 40$$

$$(A/P, 10\%, n) = \frac{800(0.1627) + 23}{1200} = 0.1276$$

$(A/P, 10\%, n)$ equals 0.1276 when n is very close to 16 years.

Answer is (C)

PROBLEM 10-30

A machine costs $9000 and has an estimated salvage value of $800 at the end of its 10-year life. Depreciation is computed by the straight-line method. When the machine is three years old, its book value will be nearest to

    (A) $800
    (B) $3260
    (C) $6300
    (D) $6540
    (E) $9000

Solution.

$$\text{Straight-line depreciation in any year} = \frac{P - F}{n} = \frac{9000 - 800}{10} = \$820$$

$$\text{Book value} = \text{cost} - \text{depreciation}$$

After three years,

$$\text{Book value} = 9000 - 3(820) = \$6540$$

Answer is (D)

PROBLEM 10-31

The value of P in Fig. 10-18 is closest to

    (A) $750
    (B) $800
    (C) $850
    (D) $900
    (E) $950

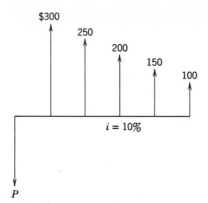

$i = 10\%$

P

Figure 10-18

Solution. Using the single payment present worth factor

$$P = 300(P/F, 10\%, 1) + 250(P/F, 10\%, 2) + 200(P/F, 10\%, 3)$$
$$+ 150(P/F, 10\%, 4) + 100(P/F, 10\%, 5)$$
$$= 300(0.9091) + 250(0.8264) + 200(0.7513) + 150(0.6830)$$
$$+ 100(0.6209)$$

$$= \$794$$

The problem may be solved more easily using the gradient present worth factor.

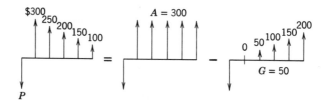

Figure 10-19

$$P = 300(P/A, 10\%, 5) - 50(P/G, 10\%, 5) = 300(3.791) - 50(6.862)$$
$$= \$794$$

Answer is (B)

PROBLEM 10-32

A firm has invested $14,000 in machinery with a seven-year useful life. The machinery has no salvage value. The uniform annual benefits from the machinery are $3600. For a 48% income tax rate, and sum-of-years digits depreciation, the after-tax rate of return is nearest

    (A) 0%

    (B) 5%

    (C) 11%

    (D) 18%

    (E) 26%

Solution.

Year	Before-tax cash flow	SOYD depreciation	Taxable income	48% Income taxes	After-tax cash flow
0	−$14,000				−$14,000
1	+3,600	$3,500	$100	−$48	+3,552
2	+3,600	3,000	600	−288	+3,312
3	+3,600	2,500	1,100	−528	+3,072
4	+3,600	2,000	1,600	−768	+2,832
5	+3,600	1,500	2,100	−1,008	+2,592
6	+3,600	1,000	2,600	−1,248	+2,352
7	+3,600	500	3,100	−1,488	+2,112
		14,000			

Write one equation for the after-tax cash flow with i the only unknown. One can see there is a $240 declining gradient in the cash flow for years 1 through 7. This fact allows us to solve easily what would otherwise be a lengthy computation.

$$NPW = PW \text{ of benefits} - PW \text{ of cost} = 0$$

$$NPW = 3552(P/A, i\%, 7) - 240(P/G, i\%, 7) - 14,000$$

Try $i = 12\%$.

$$NPW = 3552(4.564) - 240(11.644) - 14,000 = -583$$

The interest rate is too high. Try $i = 10\%$.

$$NPW = 3552(4.868) - 240(12.763) - 14,000 = +228$$

Linear interpolation:

$$\text{After-tax rate of return} = 10\% + (2\%)\left(\frac{228}{583 + 228}\right) = 10.6\%$$

Answer is (C)

PROBLEM 10-33

A piece of machinery costs $8300 and has a projected $2000 salvage value at the end of its six-year useful life. Sum-Of-Years Digits depreciation for the third year is closest to

(A) $900
(B) $1000
(C) $1100
(D) $1200
(E) $1300

Solution.

$$\text{Sum-Of-Years Digits depreciation in any year} = \frac{\text{Remaining useful life at beginning of year}}{\frac{n}{2}(n+1)}(P-F)$$

$$\text{SOYD depreciation for the third year} = \frac{4}{\frac{6}{2}(6+1)}(8300-2000) = \$1200$$

Answer is (D)

PROBLEM 10-34

The formula for determining average annual interest is

(A) $(P-F)\left(\dfrac{i}{2}\right)\left(\dfrac{n+1}{n}\right)+Fi$

(B) $(P-F)\left[\dfrac{i}{2}\left(\dfrac{n+1}{1}\right)\right]$

(C) $\left(\dfrac{P+F}{2}\right)i$

(D) $(P-F)\left[\dfrac{1}{(1+i)^n-1}\right]+Fi$

(E) $(P-F)\left[\dfrac{1}{(1+i)^n-1}\right]$

Solution.   The formula for average annual interest is given in (A) above, where $P=$ initial cost, $F=$ salvage value, $n=$ life in years, and $i=$ annual interest rate stated as a decimal.

Answer is (A)

PROBLEM 10-35

An engineer purchased an automobile for $5000. She believes it will have a $600 resale value at the end of five years. Fixed costs are $180 per year. Fuel and oil are $0.03/km. The engineer drives 15,000 kilometers per year and considers 8% to be a suitable interest rate. Her average annual cost, computed by the method of straight-line depreciation plus average interest, is closest to

(A) $900
(B) $1200
(C) $1500
(D) $1800
(E) $1900

Solution.

$$\text{Straight-line depreciation} = \frac{5000-600}{5} = \ \$880$$

$$\text{Average interest} = (5000-600)\left(\frac{0.08}{2}\right)\left(\frac{6}{5}\right)+600(0.08) = \ 259$$

$$\text{Fixed costs plus fuel} = 180+0.03(15,000) = \ \underline{630}$$

$$\text{Average annual cost} = \$1769$$

Answer is (D)

PROBLEM 10-36

This is a problem set containing 10 questions. Two machines are being compared.

	Machine A	Machine B
Initial cost	$50,000	$120,000
Uniform annual maintenance	4000	2000
Useful life	10 yr	20 yr
End of useful life salvage value	$10,000	0

Interest rate: 8%

Question 1.   The equivalent uniform annual cost for 10 years of machine A is closest to

   (A) $4000
   (B) $7000
   (C) $8900
   (D) $10,800
   (E) $11,500

Solution.

$$EUAC = 50,000(A/P, 8\%, 10) + 4000 - 10,000(A/F, 8\%, 10)$$

$$= 50,000(0.1490) + 4000 - 10,000(0.0690) = \$10,760$$

Answer is (D)

Question 2.   At the end of 10 years the original machine A will be replaced by another machine A with identical initial cost, annual maintenance, useful life and salvage value. The equivalent uniform annual cost for the 20-year period (10 years of machine A plus 10 years of the replacement machine A) is closest to

   (A) $4000
   (B) $7000
   (C) $8900
   (D) $10,800
   (E) $11,500

Solution.   In question 1 the equivalent uniform annual cost of 10 years of machine A was computed to be $10,760. During the second 10 years the EUAC will be identical. Therefore the EUAC for the 20-year period is the same as the EUAC for the initial 10-year period.

Answer is (D)

Question 3.   The capitalized cost of machine A is nearest

   (A) $50,000
   (B) $65,000
   (C) $80,000
   (D) $90,000
   (E) $135,000

Solution. Capitalized cost is the present worth of cost for an infinite analysis period. For n equal to infinity, capitalized cost equals A divided by i. From question 1, the equivalent uniform annual cost for machine A is $10,760.

$$\text{Capitalized cost } P = \frac{A}{i} = \frac{10,760}{0.08} = \$134,500$$

Answer is (E)

Question 4. The present worth of cost for 20 years of machine B is nearest

(A) $8000
(B) $12,000
(C) $120,000
(D) $140,000
(E) $160,000

Solution.

$$\text{Present worth of cost} = 120,000 + 2000(P/A, 8\%, 20)$$
$$= 120,000 + 2000(9.818) = \$139,640$$

Answer is (D)

Question 5. To have sufficient money to replace machine A with an identical machine A at the end of 10 years, annual end-of-year deposits will be made into a sinking fund. If the sinking fund earns 8% interest, the annual deposit is nearest

(A) $2800
(B) $3500
(C) $6800
(D) $12,200
(E) $18,500

Solution. To have sufficient money to replace machine A one must accumulate $40,000 in the fund. The balance of the needed $50,000 will come from the salvage value of the original machine A.

$$\text{Annual deposit} = 40,000(A/F, 8\%, 10) = 40,000(0.0690) = \$2760$$

Answer is (A)

Question 6. To provide a 10% before-tax rate of return, machine B must have annual benefits nearest to

(A) $8000
(B) $10,000
(C) $12,000
(D) $14,000
(E) $16,000

Solution. Set the equivalent uniform annual benefits equal to the equivalent uniform annual cost at a 10% interest rate.

$$EAUB = EUAC$$
$$= 120,000(A/P, 10\%, 20) + 2000$$
$$= 120,000(0.1175) + 2000 = \$16,100$$

Answer is (E)

Question 7. The uniform annual production benefits from installing machine B are estimated at $12,460 per year. If installed, machine B would have a rate of return nearest

(A) 4%
(B) 5%
(C) 6%
(D) 8%
(E) 10%

Solution.

$$PW \text{ of cost} = PW \text{ of benefits}$$
$$120,000 + 2000(P/A, i\%, 20) = 12,460(P/A, i\%, 20)$$
$$(P/A, i\%, 20) = 11.47$$

From compound interest tables, $i = 6\%$.

Answer is (C)

Question 8. The annual straight-line depreciation for machine A is nearest

(A) $4000
(B) $5000
(C) $6000
(D) $8000
(E) $10,000

Solution.

$$\text{Straight-line depreciation in any year} = \frac{P-F}{n} = \frac{50,000-10,000}{10} = \$4000$$

Answer is (A)

Question 9. The Sum-Of-Years Digits depreciation for the fourth year of machine A is closest to

(A) $2000
(B) $4000
(C) $5000
(D) $7000
(E) $28,000

Solution.

$$\text{SOYD depreciation charge for any year} = \frac{\substack{\text{Remaining useful life} \\ \text{at beginning of year}}}{\substack{\text{Sum-of-years digits} \\ \text{for total useful life}}}(P-F)$$

where

$$\text{Sum-Of-Years Digits} = \frac{n}{2}(n+1) = \frac{10}{2}(10+1) = 55$$

$$\text{Fourth-year SOYD depreciation} = \frac{7}{55}(50,000-10,000) = \$5091$$

Answer is (C)

Question 10. Based on double declining balance depreciation, the book value of machine A at the end of three years is nearest

(A) $20,000
(B) $26,000
(C) $30,000
(D) $38,000
(E) $40,000

Solution. The important thing to note in double declining balance (DDB) depreciation is that the salvage value is not a component of the calculation.

$$\text{First year DDB depreciation} = \frac{2}{10}(50{,}000 - 0) = \$10{,}000$$

$$\text{Second year DDB depreciation} = \frac{2}{10}(50{,}000 - 10{,}000) = \quad 8{,}000$$

$$\text{Third year DDB depreciation} = \frac{2}{10}(50{,}000 - 18{,}000) = \quad \underline{6{,}400}$$

$$\$24{,}400$$

Book value at the end of three years $= 50{,}000 - 24{,}400 = \$25{,}600.$

Answer is (B)

PROBLEM 10-37

This is a problem set containing 10 questions. A firm is trying to determine which of three different machines it should install in the plant to perform a given task.

	Machine A	Machine B	Machine C
Initial cost	$3000	$4300	$5000
Annual maintenance	1500	1000	1000
Annual operating cost	1500	1500	1100
End of useful life			
salvage value	800	500	0
Useful life	5 yr	5 yr	5 yr

Question 1. Using the method of straight-line depreciation plus average interest, which machine is the best economic choice at a 10% interest rate?

(A) Machine A
(B) Machine B
(C) Machine C
(D) Machines A and B are equally good
(E) Machines B and C are equally good

Solution.

Straight-line depreciation plus average interest

$$= \frac{P-F}{n} + (P-F)\left(\frac{i}{2}\right)\left(\frac{n+1}{n}\right) + Fi$$

Machine A:

$$\text{Annual cost} = \frac{3000-800}{5} + (3000-800)\left(\frac{0.10}{2}\right)\left(\frac{6}{5}\right) + 800(0.10)$$

$$+1500+1500 = \$3652$$

Machine B:

$$\text{Annual cost} = \frac{4300-500}{5} + (4300-500)\left(\frac{0.10}{2}\right)\left(\frac{6}{5}\right)$$

$$+500(0.10)+1000+1500 = \$3538$$

Machine C:

$$\text{Annual cost} = \frac{5000-0}{5} + (5000-0)\left(\frac{0.10}{2}\right)\left(\frac{6}{5}\right) + 1000+1100$$

$$= \$3400$$

Answer is (C)

Question 2. If machine B had a useful life of seven years, rather than five years, which machine would now be the best economic choice based on straight-line depreciation plus average interest?

(A) Machine A
(B) Machine B
(C) Machine C
(D) Machines A and B are equally good
(E) Machines B and C are equally good

Solution.

Machine B:

$$\text{Annual cost} = \frac{4300 - 500}{7} + (4300 - 500)\left(\frac{0.10}{2}\right)\left(\frac{8}{7}\right) + 500(0.10)$$

$$+ 1000 + 1500 = \$3310$$

Machine *B* now has the least annual cost, as computed by the method of straight-line depreciation plus average interest.

Answer is (B)

Question 3. The equivalent uniform annual cost of machine *A*, using a 10% interest rate, is closest to

(A) $3550
(B) $3600
(C) $3650
(D) $3700
(E) $3750

Solution.

$$\text{EUAC} = (3000 - 800)(A/P, 10\%, 5) + 800(0.10) + 1500 + 1500$$

$$= 2200(0.2638) + 80 + 3000 = \$3660$$

Answer is (C)

Question 4. If machine *C* produces production benefits of $3487 per year, what is its rate of return?

(A) 0%
(B) 8%
(C) 10%
(D) 12%
(E) 15%

Solution. Set the present worth of cost equal to the present worth of benefits and solve for the unknown rate of return i.

$$\text{PW of cost} = \text{PW of benefits}$$

$$5000 + (1000 + 1100)(P/A, i\%, 5) = 3487(P/A, i\%, 5)$$

$$(P/A, i\%, 5) = \frac{5000}{3487 - 2100} = 3.605$$

From compound interest tables, $(P/A, 12\%, 5)$ equals 3.605. The rate of return is exactly 12%.

Answer is (D)

Question 5. If machine A is depreciated by the sum-of-years digits method, its book value at the end of three years is closest to

(A) $440
(B) $1240
(C) $1760
(D) $2020
(E) $2200

Solution.

$$\text{Sum} = 1 + 2 + 3 + 4 + 5 = 15$$

$$\text{First year SOYD depreciation} = \frac{5}{15}(3000 - 800) = \ \ \$733.33$$

$$\text{Second year SOYD depreciation} = \frac{4}{15}(3000 - 800) = \ \ \ 586.67$$

$$\text{Third year SOYD depreciation} = \frac{3}{15}(3000 - 800) = \ \ \ \underline{440.00}$$

$$\$1760.00$$

$$\text{Book value} = \text{cost} - \text{depreciation} = 3000 - 1760 = \$1240$$

Answer is (B)

Question 6. A sinking fund will be set up to replace machine A if it is the selected alternative. The sinking fund earns 5% annual interest. If the cost of machine A will be 7% higher than the prior year, each year for the five-year period, what uniform annual deposit would need to be made to the sinking fund to replace machine A at the end of five years? The amount is closest to

(A) $440
(B) $530
(C) $620
(D) $710
(E) $800

Solution.

$$\text{Replacement cost of machine } A = 3000(F/P, 7\%, 5)$$
$$= 3000(1.403) = \$4209$$
$$\text{Annual sinking fund deposit} = (4209 - 800)(A/F, 5\%, 5)$$
$$= 3409(0.1810) = \$617$$

Answer is (C)

Question 7. If machine A is depreciated by the double declining balance method, the depreciation for year 2 would be nearest to

(A) $600
(B) $720
(C) $880
(D) $1000
(E) $1240

Solution.

$$\text{First year DDB depreciation} = \tfrac{2}{5}(3000 - 0) = \$1200$$
$$\text{Second year DDB depreciation} = \tfrac{2}{5}(3000 - 1200) = \$720$$

Answer is (B)

Question 8. Assume that if machine B is installed the firm could reduce its labor cost by $3404 per year. What rate of return would the firm obtain on this investment? The rate of return is nearest to

(A) 6%
(B) 8%
(C) 10%
(D) 12%
(E) 15%

Solution.

$$EUAC = EUAB$$

$$4300(A/P, i\%, 5) + 1000 + 1500 = 3404 + 500(A/F, i\%, 5)$$

The equation will be solved by trial and error for the unknown i.
Try $i = 5\%$.

$$4300(A/P, 5\%, 5) + 1000 + 1500 = 3404 + 500(A/F, 5\%, 5)$$

$$4300(0.2310) + 2500 = 3404 + 500(0.1810)$$

$$3493 = 3494$$

The rate of return i is 5%.

Answer is (A)

Question 9. If machine A is installed, the firm would be able to reduce its labor costs by $3800 per year. Based on a 10% interest rate, compute the net present worth of machine A for the five-year period. Net present worth is closest to

(A) 0
(B) +$50
(C) +$500
(D) +$5000
(E) +$10,000

Solution.

NPW = PW of benefits − PW of cost

$$= 3800(P/A, 10\%, 5) + 800(P/F, 10\%, 5)$$
$$- (1500 + 1500)(P/A, 10\%, 5) - 3000$$

$$= 3800(3.791) + 800(0.6209) - 3000(3.791) - 3000 = +530$$

Answer is (C)

Question 10. Some specialized attachments could be purchased for use with one of the machines. The attachments would increase the productivity of the machine. If this increased productivity is valued at $3000 per year, how much could the firm afford to pay for the attachments? Assume a five-year useful life, a $1000 salvage value at the end of five years, and a 10% interest rate. The amount the firm could afford to pay is closest to

(A) $4000
(B) $10,000
(C) $11,000
(D) $12,000
(E) $15,000

Solution. The firm could pay an amount equal to the present worth of the benefits of having the attachments.

$$PW \text{ of benefits} = 3000(P/A, 10\%, 5) + 1000(P/F, 10\%, 5)$$

$$= 3000(3.791) + 1000(0.6209) = \$11,994$$

Answer is (D)

A SI Units and Conversion Factors

The international system of units is a modernized version of the metric system established by international agreement. Officially abbreviated SI, the system is built upon a foundation of seven base units, plus two supplementary units.

Quantity	Unit	Symbol
Length	meter	m
Mass	kilogram	kg
Time	second	s
Electric current	ampere	A
Thermodynamic temperature	kelvin	K
Amount of substance	mole	mol
Luminous intensity	candela	cd
Plane angle	radian	rad
Solid angle	steradian	sr

This appendix is based on *Standard for Metric Practice*, ANSI/ASTM E 380-76 and is reprinted by permission of the American Society for Testing and Materials.

In addition, there are derived SI units which have special names and symbols.

Quantity	Unit	Symbol	Formula
Frequency (of a periodic phenomenon)	hertz	Hz	$1/s$
Force	newton	N	$kg \cdot m/s^2$
Pressure, stress	pascal	Pa	N/m^2
Energy, work, quantity of heat	joule	J	$N \cdot m$
Power, radiant flux	watt	W	J/s
Quantity of electricity, electric charge	coulomb	C	$A \cdot s$
Electric potential, potential difference, electromotive force	volt	V	W/A
Capacitance	farad	F	C/V
Electric resistance	ohm	Ω	V/A
Conductance	siemens	S	A/V
Magnetic flux	weber	Wb	$V \cdot s$
Magnetic flux density	tesla	T	Wb/m^2
Inductance	henry	H	Wb/A
Luminous flux	lumen	lm	$cd \cdot sr$
Illuminance	lux	lx	lm/m^2
Activity (of radionuclides)	becquerel	Bq	$1/s$

Listed below are some frequently used SI units.

Quantity	Unit	Symbol
Acceleration	meter per second squared	m/s^2
Angular acceleration	radian per second squared	rad/s^2
Angular velocity	radian per second	rad/s
Area	square meter	m^2
Concentration (of amount of substance)	mole per cubic meter	mol/m^3
Current density	ampere per square meter	A/m^2
Density, mass	kilogram per cubic meter	kg/m^3
Electric charge density	coulomb per cubic meter	C/m^3
Electric field strength	volt per meter	V/m
Electric flux density	coulomb per square meter	C/m^2
Energy density	joule per cubic meter	J/m^3
Entropy	joule per kelvin	J/K
Heat capacity	joule per kelvin	J/K
Heat flux density⎫ Irradiance ⎭	watt per square meter	W/m^2
Luminance	candela per square meter	cd/m^2
Magnetic field strength	ampere per meter	A/m
Molar energy	joule per mole	J/mol
Molar entropy	joule per mole kelvin	$J/(mol \cdot K)$
Molar heat capacity	joule per mole kelvin	$J/(mol \cdot K)$
Moment of force	newton meter	$N \cdot m$
Permeability	henry per meter	H/m
Permittivity	farad per meter	F/m
Radiance	watt per square meter steradian	$W/(m^2 \cdot sr)$
Radiant intensity	watt per steradian	W/sr
Specific heat capacity	joule per kilogram kelvin	$J/(kg \cdot K)$
Specific energy	joule per kilogram	J/kg
Specific entropy	joule per kilogram kelvin	$J/(kg \cdot K)$
Specific volume	cubic meter per kilogram	m^3/kg
Surface tension	newton per meter	N/m
Thermal conductivity	watt per meter kelvin	$W/(m \cdot K)$
Velocity	meter per second	m/s
Viscosity, dynamic	pascal second	$Pa \cdot s$
Viscosity, kinematic	square meter per second	m^2/s
Volume	cubic meter	m^3

The factors given on the next page allow one to convert easily *from* English *to* SI units; to convert *from* the given SI unit *to* the stated English unit, one divides by the factor.

SI CONVERSION FACTORS

To convert from	to	multiply by
Acceleration		
ft/s^2	meter per second^2 (m/s^2)	0.3048
free fall, standard (g)	meter per second^2 (m/s^2)	9.8066
Area		
acre (U.S. survey)	meter^2 (m^2)	4.0469×10^3
ft^2	meter^2 (m^2)	9.2903×10^{-2}
in.^{2}	meter^2 (m^2)	6.4516×10^{-4}
yd^2	meter^2 (m^2)	0.8361
Bending Moment or Torque		
dyne · cm	newton meter (N · m)	1.0000×10^{-7}
kgf · m	newton meter (N · m)	9.8066
lbf · ft	newton meter (N · m)	1.3558
Energy (includes work)		
Btu (International Table)	joule (J)	1.0551×10^3
calorie (mean)	joule (J)	4.1900
erg	joule (J)	1.0000×10^{-7}
ft · lbf	joule (J)	1.3558
kW · h	joule (J)	3.6000×10^6
Force		
dyne	newton (N)	1.0000×10^{-5}
kilogram-force	newton (N)	9.8066
kip (1000 lbf)	newton (N)	4.4482×10^3
pound-force (lbf)	newton (N)	4.4482
Heat		
Btu (International Table)/lb	joule per kilogram (J/kg)	2.3260×10^3

SI CONVERSION FACTORS—continued

To convert from	to	multiply by
Length		
inch	meter (m)	2.5400×10^{-2}
foot	meter (m)	0.3048
mile (U.S. survey)	meter (m)	1.6093×10^{3}
Mass		
pound (lbm)	kilogram (kg)	0.4536
Mass per Unit Volume		
g/cm^3	kilogram per meter^3 (kg/m^3)	1.0000×10^{3}
lb/ft^3	kilogram per meter^3 (kg/m^3)	1.6018×10^{1}
Power		
ft · lbf/min	watt (W)	2.2597×10^{-2}
horsepower (550 ft · lbf/s)	watt (W)	7.4570×10^{2}
Pressure or Stress		
atmosphere (standard)	pascal (Pa)	1.0132×10^{5}
lbf/ft^2	pascal (Pa)	4.7880×10^{1}
psi	pascal (Pa)	6.8948×10^{3}
Velocity		
ft/s	meter per second (m/s)	0.3048
mi/h	meter per second (m/s)	0.4470
mi/h	kilometers per hour (km/h)	1.6093
Volume		
ft^3	meter^3 (m^3)	2.8317×10^{-2}
gallon (U.S. liquid)	meter^3 (m^3)	3.7854×10^{-3}
yd^3	meter^3 (m^3)	0.7646

B Centroidal Coordinates and Moments of Inertia for Common Shapes

	Centroid	Moment of Inertia

Triangle

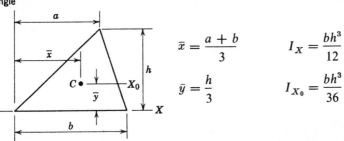

$$\bar{x} = \frac{a+b}{3}$$

$$\bar{y} = \frac{h}{3}$$

$$I_X = \frac{bh^3}{12}$$

$$I_{X_0} = \frac{bh^3}{36}$$

Rectangle

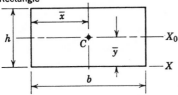

$$\bar{x} = \frac{b}{2}$$

$$\bar{y} = \frac{h}{2}$$

$$I_X = \frac{bh^3}{3}$$

$$I_{X_0} = \frac{bh^3}{12}$$

Quarter circle

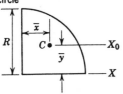

$$\bar{x} = \bar{y} = \frac{4R}{3\pi}$$

$$I_X = \frac{\pi R^4}{16}$$

$$I_{X_0} = R^4\left(\frac{\pi}{16} + \frac{4}{9\pi}\right)$$

$$= 0.0549R^4$$

Semicircle

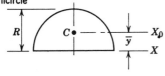

$$\bar{y} = \frac{4R}{3\pi}$$

$$I_X = \frac{\pi R^4}{8}$$

$$I_{X_0} = R^4\left(\frac{\pi}{8} - \frac{8}{9\pi}\right)$$

$$= 0.1098R^4$$

Circle

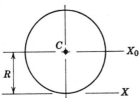

$$I_X = \frac{5\pi R^4}{4}$$

$$I_{X_0} = \frac{\pi R^4}{4}$$

C Compound Interest Tables

The tables are from *Engineering Economic Analysis* by Donald G. Newnan and are reprinted by permission from Engineering Press, the copyright owner.

	SINGLE PAYMENT		UNIFORM PAYMENT SERIES				GRADIENT SERIES	
	Compound Amount Factor	Present Worth Factor	Sinking Fund Factor	Capital Recovery Factor	Compound Amount Factor	Present Worth Factor	Gradient Uniform Series	Gradient Present Worth
	Find F Given P	Find P Given F	Find A Given F	Find A Given P	Find F Given A	Find P Given A	Find A Given G	Find P Given G
n	F/P	P/F	A/F	A/P	F/A	P/A	A/G	P/G
1	1.010	.9901	1.0000	1.0100	1.000	.990	0	0
2	1.020	.9803	.4975	.5075	2.010	1.970	.498	.980
3	1.030	.9706	.3300	.3400	3.030	2.941	.993	2.921
4	1.041	.9610	.2463	.2563	4.060	3.902	1.488	5.804
5	1.051	.9515	.1960	.2060	5.101	4.853	1.980	9.610
6	1.062	.9420	.1625	.1725	6.152	5.795	2.471	14.321
7	1.072	.9327	.1386	.1486	7.214	6.728	2.960	19.917
8	1.083	.9235	.1207	.1307	8.286	7.652	3.448	26.381
9	1.094	.9143	.1067	.1167	9.369	8.566	3.934	33.696
10	1.105	.9053	.0956	.1056	10.462	9.471	4.418	41.843
11	1.116	.8963	.0865	.0965	11.567	10.368	4.901	50.807
12	1.127	.8874	.0788	.0888	12.683	11.255	5.381	60.569
13	1.138	.8787	.0724	.0824	13.809	12.134	5.861	71.113
14	1.149	.8700	.0669	.0769	14.947	13.004	6.338	82.422
15	1.161	.8613	.0621	.0721	16.097	13.865	6.814	94.481
16	1.173	.8528	.0579	.0679	17.258	14.718	7.289	107.273
17	1.184	.8444	.0543	.0643	18.430	15.562	7.761	120.783
18	1.196	.8360	.0510	.0610	19.615	16.398	8.232	134.996
19	1.208	.8277	.0481	.0581	20.811	17.226	8.702	149.895
20	1.220	.8195	.0454	.0554	22.019	18.046	9.169	165.466
21	1.232	.8114	.0430	.0530	23.239	18.857	9.635	181.695
22	1.245	.8034	.0409	.0509	24.472	19.660	10.100	198.566
23	1.257	.7954	.0389	.0489	25.716	20.456	10.563	216.066
24	1.270	.7876	.0371	.0471	26.973	21.243	11.024	234.180
25	1.282	.7798	.0354	.0454	28.243	22.023	11.483	252.894
26	1.295	.7720	.0339	.0439	29.526	22.795	11.941	272.196
27	1.308	.7644	.0324	.0424	30.821	23.560	12.397	292.070
28	1.321	.7568	.0311	.0411	32.129	24.316	12.852	312.505
29	1.335	.7493	.0299	.0399	33.450	25.066	13.304	333.486
30	1.348	.7419	.0287	.0387	34.785	25.808	13.756	355.002
36	1.431	.6989	.0232	.0332	43.077	30.108	16.428	494.621
40	1.489	.6717	.0205	.0305	48.886	32.835	18.178	596.856
48	1.612	.6203	.0163	.0263	61.223	37.974	21.598	820.146
50	1.645	.6080	.0155	.0255	64.463	39.196	22.436	879.418
52	1.678	.5961	.0148	.0248	67.769	40.394	23.269	939.918
60	1.817	.5504	.0122	.0222	81.670	44.955	26.533	1192.806
70	2.007	.4983	.0099	.0199	100.676	50.169	30.470	1528.647
72	2.047	.4885	.0096	.0196	104.710	51.150	31.239	1597.867
80	2.217	.4511	.0082	.0182	121.672	54.888	34.249	1879.877
84	2.307	.4335	.0077	.0177	130.672	56.648	35.717	2023.315
90	2.449	.4084	.0069	.0169	144.863	59.161	37.872	2240.567
96	2.599	.3847	.0063	.0163	159.927	61.528	39.973	2459.430
100	2.705	.3697	.0059	.0159	170.481	63.029	41.343	2605.776
104	2.815	.3553	.0055	.0155	181.464	64.471	42.688	2752.182
120	3.300	.3030	.0043	.0143	230.039	69.701	47.835	3334.115
240	10.893	.0918	.0010	.0110	989.255	90.819	75.739	6878.602
360	35.950	.0278	.0003	.0103	3494.694	97.218	89.699	8720.432
480	118.648	.0084	.0001	.0101	11764.773	99.157	95.920	9511.158

Compound Interest Factors

	SINGLE PAYMENT		UNIFORM PAYMENT SERIES				GRADIENT SERIES	
	Compound Amount Factor	Present Worth Factor	Sinking Fund Factor	Capital Recovery Factor	Compound Amount Factor	Present Worth Factor	Gradient Uniform Series	Gradient Present Worth
	Find F Given P	Find P Given F	Find A Given F	Find A Given P	Find F Given A	Find P Given A	Find A Given G	Find P Given G
n	F/P	P/F	A/F	A/P	F/A	P/A	A/G	P/G
1	1.030	.9709	1.0000	1.0300	1.000	.971	0	0
2	1.061	.9426	.4926	.5226	2.030	1.913	.493	.943
3	1.093	.9151	.3235	.3535	3.091	2.829	.980	2.773
4	1.126	.8885	.2390	.2690	4.184	3.717	1.463	5.438
5	1.159	.8626	.1884	.2184	5.309	4.580	1.941	8.889
6	1.194	.8375	.1546	.1846	6.468	5.417	2.414	13.076
7	1.230	.8131	.1305	.1605	7.662	6.230	2.882	17.955
8	1.267	.7894	.1125	.1425	8.892	7.020	3.345	23.481
9	1.305	.7664	.0984	.1284	10.159	7.786	3.803	29.612
10	1.344	.7441	.0872	.1172	11.464	8.530	4.256	36.309
11	1.384	.7224	.0781	.1081	12.808	9.253	4.705	43.533
12	1.426	.7014	.0705	.1005	14.192	9.954	5.148	51.248
13	1.469	.6810	.0640	.0940	15.618	10.635	5.587	59.420
14	1.513	.6611	.0585	.0885	17.086	11.296	6.021	68.014
15	1.558	.6419	.0538	.0838	18.599	11.938	6.450	77.000
16	1.605	.6232	.0496	.0796	20.157	12.561	6.874	86.348
17	1.653	.6050	.0460	.0760	21.762	13.166	7.294	96.028
18	1.702	.5874	.0427	.0727	23.414	13.754	7.708	106.014
19	1.754	.5703	.0398	.0698	25.117	14.324	8.118	116.279
20	1.806	.5537	.0372	.0672	26.870	14.877	8.523	126.799
21	1.860	.5375	.0349	.0649	28.676	15.415	8.923	137.550
22	1.916	.5219	.0327	.0627	30.537	15.937	9.319	148.509
23	1.974	.5067	.0308	.0608	32.453	16.444	9.709	159.657
24	2.033	.4919	.0290	.0590	34.426	16.936	10.095	170.971
25	2.094	.4776	.0274	.0574	36.459	17.413	10.477	182.434
26	2.157	.4637	.0259	.0559	38.553	17.877	10.853	194.026
27	2.221	.4502	.0246	.0546	40.710	18.327	11.226	205.731
28	2.288	.4371	.0233	.0533	42.931	18.764	11.593	217.532
29	2.357	.4243	.0221	.0521	45.219	19.188	11.956	229.414
30	2.427	.4120	.0210	.0510	47.575	19.600	12.314	241.361
31	2.500	.4000	.0200	.0500	50.003	20.000	12.668	253.361
32	2.575	.3883	.0190	.0490	52.503	20.389	13.017	265.399
33	2.652	.3770	.0182	.0482	55.078	20.766	13.362	277.464
34	2.732	.3660	.0173	.0473	57.730	21.132	13.702	289.544
35	2.814	.3554	.0165	.0465	60.462	21.487	14.037	301.627
40	3.262	.3066	.0133	.0433	75.401	23.115	15.650	361.750
45	3.782	.2644	.0108	.0408	92.720	24.519	17.156	420.632
50	4.384	.2281	.0089	.0389	112.797	25.730	18.558	477.480
55	5.082	.1968	.0073	.0373	136.072	26.774	19.860	531.741
60	5.892	.1697	.0061	.0361	163.053	27.676	21.067	583.053
65	6.830	.1464	.0051	.0351	194.333	28.453	22.184	631.201
70	7.918	.1263	.0043	.0343	230.594	29.123	23.215	676.087
75	9.179	.1089	.0037	.0337	272.631	29.702	24.163	717.698
80	10.641	.0940	.0031	.0331	321.363	30.201	25.035	756.087
85	12.336	.0811	.0026	.0326	377.857	30.631	25.835	791.353
90	14.300	.0699	.0023	.0323	443.349	31.002	26.567	823.630
95	16.578	.0603	.0019	.0319	519.272	31.323	27.235	853.074
100	19.219	.0520	.0016	.0316	607.288	31.599	27.844	879.854

Compound Interest Factors

	SINGLE PAYMENT		UNIFORM PAYMENT SERIES				GRADIENT SERIES	
	Compound Amount Factor	Present Worth Factor	Sinking Fund Factor	Capital Recovery Factor	Compound Amount Factor	Present Worth Factor	Gradient Uniform Series	Gradient Present Worth
	Find F Given P F/P	Find P Given F P/F	Find A Given F A/F	Find A Given P A/P	Find F Given A F/A	Find P Given A P/A	Find A Given G A/G	Find P Given G P/G
n								
1	1.040	.9615	1.0000	1.0400	1.000	.962	0	0
2	1.082	.9246	.4902	.5302	2.040	1.886	.490	.925
3	1.125	.8890	.3203	.3603	3.122	2.775	.974	2.703
4	1.170	.8548	.2355	.2755	4.246	3.630	1.451	5.267
5	1.217	.8219	.1846	.2246	5.416	4.452	1.922	8.555
6	1.265	.7903	.1508	.1908	6.633	5.242	2.386	12.506
7	1.316	.7599	.1266	.1666	7.898	6.002	2.843	17.066
8	1.369	.7307	.1085	.1485	9.214	6.733	3.294	22.181
9	1.423	.7026	.0945	.1345	10.583	7.435	3.739	27.801
10	1.480	.6756	.0833	.1233	12.006	8.111	4.177	33.881
11	1.539	.6496	.0741	.1141	13.486	8.760	4.609	40.377
12	1.601	.6246	.0666	.1066	15.026	9.385	5.034	47.248
13	1.665	.6006	.0601	.1001	16.627	9.986	5.453	54.455
14	1.732	.5775	.0547	.0947	18.292	10.563	5.866	61.962
15	1.801	.5553	.0499	.0899	20.024	11.118	6.272	69.735
16	1.873	.5339	.0458	.0858	21.825	11.652	6.672	77.744
17	1.948	.5134	.0422	.0822	23.698	12.166	7.066	85.958
18	2.026	.4936	.0390	.0790	25.645	12.659	7.453	94.350
19	2.107	.4746	.0361	.0761	27.671	13.134	7.834	102.893
20	2.191	.4564	.0336	.0736	29.778	13.590	8.209	111.565
21	2.279	.4388	.0313	.0713	31.969	14.029	8.578	120.341
22	2.370	.4220	.0292	.0692	34.248	14.451	8.941	129.202
23	2.465	.4057	.0273	.0673	36.618	14.857	9.297	138.128
24	2.563	.3901	.0256	.0656	39.083	15.247	9.648	147.101
25	2.666	.3751	.0240	.0640	41.646	15.622	9.993	156.104
26	2.772	.3607	.0226	.0626	44.312	15.983	10.331	165.121
27	2.883	.3468	.0212	.0612	47.084	16.330	10.664	174.138
28	2.999	.3335	.0200	.0600	49.968	16.663	10.991	183.142
29	3.119	.3207	.0189	.0589	52.966	16.984	11.312	192.121
30	3.243	.3083	.0178	.0578	56.085	17.292	11.627	201.062
31	3.373	.2965	.0169	.0569	59.328	17.588	11.937	209.956
32	3.508	.2851	.0159	.0559	62.701	17.874	12.241	218.792
33	3.648	.2741	.0151	.0551	66.210	18.148	12.540	227.563
34	3.794	.2636	.0143	.0543	69.858	18.411	12.832	236.261
35	3.946	.2534	.0136	.0536	73.652	18.665	13.120	244.877
40	4.801	.2083	.0105	.0505	95.026	19.793	14.477	286.530
45	5.841	.1712	.0083	.0483	121.029	20.720	15.705	325.403
50	7.107	.1407	.0066	.0466	152.667	21.482	16.812	361.164
55	8.646	.1157	.0052	.0452	191.159	22.109	17.807	393.689
60	10.520	.0951	.0042	.0442	237.991	22.623	18.697	422.997
65	12.799	.0781	.0034	.0434	294.968	23.047	19.491	449.201
70	15.572	.0642	.0027	.0427	364.290	23.395	20.196	472.479
75	18.945	.0528	.0022	.0422	448.631	23.680	20.821	493.041
80	23.050	.0434	.0018	.0418	551.245	23.915	21.372	511.116
85	28.044	.0357	.0015	.0415	676.090	24.109	21.857	526.938
90	34.119	.0293	.0012	.0412	827.983	24.267	22.283	540.737
95	41.511	.0241	.0010	.0410	1012.785	24.398	22.655	552.731
100	50.505	.0198	.0008	.0408	1237.624	24.505	22.980	563.125

	SINGLE PAYMENT		UNIFORM PAYMENT SERIES				GRADIENT SERIES	
	Compound Amount Factor	Present Worth Factor	Sinking Fund Factor	Capital Recovery Factor	Compound Amount Factor	Present Worth Factor	Gradient Uniform Series	Gradient Present Worth
	Find F Given P	Find P Given F	Find A Given F	Find A Given P	Find F Given A	Find P Given A	Find A Given G	Find P Given G
n	F/P	P/F	A/F	A/P	F/A	P/A	A/G	P/G
1	1.050	.9524	1.0000	1.0500	1.000	.952	0	0
2	1.102	.9070	.4878	.5378	2.050	1.859	.488	.907
3	1.158	.8638	.3172	.3672	3.152	2.723	.967	2.635
4	1.216	.8227	.2320	.2820	4.310	3.546	1.439	5.103
5	1.276	.7835	.1810	.2310	5.526	4.329	1.903	8.237
6	1.340	.7462	.1470	.1970	6.802	5.076	2.358	11.968
7	1.407	.7107	.1228	.1728	8.142	5.786	2.805	16.232
8	1.477	.6768	.1047	.1547	9.549	6.463	3.245	20.970
9	1.551	.6446	.0907	.1407	11.027	7.108	3.676	26.127
10	1.629	.6139	.0795	.1295	12.578	7.722	4.099	31.652
11	1.710	.5847	.0704	.1204	14.207	8.306	4.514	37.499
12	1.796	.5568	.0628	.1128	15.917	8.863	4.922	43.624
13	1.886	.5303	.0565	.1065	17.713	9.394	5.322	49.988
14	1.980	.5051	.0510	.1010	19.599	9.899	5.713	56.554
15	2.079	.4810	.0463	.0963	21.579	10.380	6.097	63.288
16	2.183	.4581	.0423	.0923	23.657	10.838	6.474	70.160
17	2.292	.4363	.0387	.0887	25.840	11.274	6.842	77.140
18	2.407	.4155	.0355	.0855	28.132	11.690	7.203	84.204
19	2.527	.3957	.0327	.0827	30.539	12.085	7.557	91.328
20	2.653	.3769	.0302	.0802	33.066	12.462	7.903	98.488
21	2.786	.3589	.0280	.0780	35.719	12.821	8.242	105.667
22	2.925	.3418	.0260	.0760	38.505	13.163	8.573	112.846
23	3.072	.3256	.0241	.0741	41.430	13.489	8.897	120.009
24	3.225	.3101	.0225	.0725	44.502	13.799	9.214	127.140
25	3.386	.2953	.0210	.0710	47.727	14.094	9.524	134.228
26	3.556	.2812	.0196	.0696	51.113	14.375	9.827	141.259
27	3.733	.2678	.0183	.0683	54.669	14.643	10.122	148.223
28	3.920	.2551	.0171	.0671	58.403	14.898	10.411	155.110
29	4.116	.2429	.0160	.0660	62.323	15.141	10.694	161.913
30	4.322	.2314	.0151	.0651	66.439	15.372	10.969	168.623
31	4.538	.2204	.0141	.0641	70.761	15.593	11.238	175.233
32	4.765	.2099	.0133	.0633	75.299	15.803	11.501	181.739
33	5.003	.1999	.0125	.0625	80.064	16.003	11.757	188.135
34	5.253	.1904	.0118	.0618	85.067	16.193	12.006	194.417
35	5.516	.1813	.0111	.0611	90.320	16.374	12.250	200.581
40	7.040	.1420	.0083	.0583	120.800	17.159	13.377	229.545
45	8.985	.Ν13	.0063	.0563	159.700	17.774	14.364	255.315
50	11.467	.0872	.0048	.0548	209.348	18.256	15.223	277.915
55	14.636	.0683	.0037	.0537	272.713	18.633	15.966	297.510
60	18.679	.0535	.0028	.0528	353.584	18.929	16.606	314.343
65	23.840	.0419	.0022	.0522	456.798	19.161	17.154	328.691
70	30.426	.0329	.0017	.0517	588.529	19.343	17.621	340.841
75	38.833	.0258	.0013	.0513	756.654	19.485	18.018	351.072
80	49.561	.0202	.0010	.0510	971.229	19.596	18.353	359.646
85	63.254	.0158	.0008	.0508	1245.087	19.684	18.635	366.801
90	80.730	.0124	.0006	.0506	1594.607	19.752	18.871	372.749
95	103.035	.0097	.0005	.0505	2040.694	19.806	19.069	377.677
100	131.501	.0076	.0004	.0504	2610.025	19.848	19.234	381.749

	SINGLE PAYMENT		UNIFORM PAYMENT SERIES				GRADIENT SERIES	
	Compound Amount Factor	Present Worth Factor	Sinking Fund Factor	Capital Recovery Factor	Compound Amount Factor	Present Worth Factor	Gradient Uniform Series	Gradient Present Worth
	Find F Given P F/P	Find P Given F P/F	Find A Given F A/F	Find A Given P A/P	Find F Given A F/A	Find P Given A P/A	Find A Given G A/G	Find P Given G P/G
n								
1	1.060	.9434	1.0000	1.0600	1.000	.943	0	0
2	1.124	.8900	.4854	.5454	2.060	1.833	.485	.890
3	1.191	.8396	.3141	.3741	3.184	2.673	.961	2.569
4	1.262	.7921	.2286	.2886	4.375	3.465	1.427	4.946
5	1.338	.7473	.1774	.2374	5.637	4.212	1.884	7.935
6	1.419	.7050	.1434	.2034	6.975	4.917	2.330	11.459
7	1.504	.6651	.1191	.1791	8.394	5.582	2.768	15.450
8	1.594	.6274	.1010	.1610	9.897	6.210	3.195	19.842
9	1.689	.5919	.0870	.1470	11.491	6.802	3.613	24.577
10	1.791	.5584	.0759	.1359	13.181	7.360	4.022	29.602
11	1.898	.5268	.0668	.1268	14.972	7.887	4.421	34.870
12	2.012	.4970	.0593	.1193	16.870	8.384	4.811	40.337
13	2.133	.4688	.0530	.1130	18.882	8.853	5.192	45.963
14	2.261	.4423	.0476	.1076	21.015	9.295	5.564	51.713
15	2.397	.4173	.0430	.1030	23.276	9.712	5.926	57.555
16	2.540	.3936	.0390	.0990	25.673	10.106	6.279	63.459
17	2.693	.3714	.0354	.0954	28.213	10.477	6.624	69.401
18	2.854	.3503	.0324	.0924	30.906	10.828	6.960	75.357
19	3.026	.3305	.0296	.0896	33.760	11.158	7.287	81.306
20	3.207	.3118	.0272	.0872	36.786	11.470	7.605	87.230
21	3.400	.2942	.0250	.0850	39.993	11.764	7.915	93.114
22	3.604	.2775	.0230	.0830	43.392	12.042	8.217	98.941
23	3.820	.2618	.0213	.0813	46.996	12.303	8.510	104.701
24	4.049	.2470	.0197	.0797	50.816	12.550	8.795	110.381
25	4.292	.2330	.0182	.0782	54.865	12.783	9.072	115.973
26	4.549	.2198	.0169	.0769	59.156	13.003	9.341	121.468
27	4.822	.2074	.0157	.0757	63.706	13.211	9.603	126.860
28	5.112	.1956	.0146	.0746	68.528	13.406	9.857	132.142
29	5.418	.1846	.0136	.0736	73.640	13.591	10.103	137.310
30	5.743	.1741	.0126	.0726	79.058	13.765	10.342	142.359
31	6.088	.1643	.0118	.0718	84.802	13.929	10.547	147.286
32	6.453	.1550	.0110	.0710	90.890	14.084	10.799	152.090
33	6.841	.1462	.0103	.0703	97.343	14.230	11.017	156.768
34	7.251	.1379	.0096	.0696	104.184	14.368	11.228	161.319
35	7.686	.1301	.0090	.0690	111.435	14.498	11.432	165.743
40	10.286	.0972	.0065	.0665	154.762	15.046	12.359	185.957
45	13.765	.0727	.0047	.0647	212.744	15.456	13.141	203.110
50	18.420	.0543	.0034	.0634	290.336	15.762	13.796	217.457
55	24.650	.0406	.0025	.0625	394.172	15.991	14.341	229.322
60	32.988	.0303	.0019	.0619	533.128	16.161	14.791	239.043
65	44.145	.0227	.0014	.0614	719.083	16.289	15.160	246.945
70	59.076	.0169	.0010	.0610	967.932	16.385	15.461	253.327
75	79.057	.0126	.0008	.0608	1300.949	16.456	15.706	258.453
80	105.796	.0095	.0006	.0606	1746.600	16.509	15.903	262.549
85	141.579	.0071	.0004	.0604	2342.982	16.549	16.062	265.810
90	189.465	.0053	.0003	.0603	3141.075	16.579	16.189	268.395
95	253.546	.0039	.0002	.0602	4209.104	16.601	16.290	270.437
100	339.302	.0029	.0002	.0602	5638.368	16.618	16.371	272.047

	SINGLE PAYMENT		UNIFORM PAYMENT SERIES				GRADIENT SERIES	
	Compound Amount Factor	Present Worth Factor	Sinking Fund Factor	Capital Recovery Factor	Compound Amount Factor	Present Worth Factor	Gradient Uniform Series	Gradient Present Worth
	Find F Given P F/P	Find P Given F P/F	Find A Given F A/F	Find A Given P A/P	Find F Given A F/A	Find P Given A P/A	Find A Given G A/G	Find P Given G P/G
n								
1	1.070	.9346	1.0000	1.0700	1.000	.935	0	0
2	1.145	.8734	.4831	.5531	2.070	1.808	.483	.873
3	1.225	.8163	.3111	.3811	3.215	2.624	.955	2.506
4	1.311	.7629	.2252	.2952	4.440	3.387	1.416	4.795
5	1.403	.7130	.1739	.2439	5.751	4.100	1.865	7.647
6	1.501	.6663	.1398	.2098	7.153	4.767	2.303	10.978
7	1.606	.6227	.1156	.1856	8.654	5.389	2.730	14.715
8	1.718	.5820	.0975	.1675	10.260	5.971	3.147	18.789
9	1.838	.5439	.0835	.1535	11.978	6.515	3.552	23.140
10	1.967	.5083	.0724	.1424	13.816	7.024	3.946	27.716
11	2.105	.4751	.0634	.1334	15.784	7.499	4.330	32.466
12	2.252	.4440	.0559	.1259	17.888	7.943	4.703	37.351
13	2.410	.4150	.0497	.1197	20.141	8.358	5.065	42.330
14	2.579	.3878	.0443	.1143	22.550	8.745	5.417	47.372
15	2.759	.3624	.0398	.1098	25.129	9.108	5.758	52.446
16	2.952	.3387	.0359	.1059	27.888	9.447	6.090	57.527
17	3.159	.3166	.0324	.1024	30.840	9.763	6.411	62.592
18	3.380	.2959	.0294	.0994	33.999	10.059	6.722	67.622
19	3.617	.2765	.0268	.0968	37.379	10.336	7.024	72.599
20	3.870	.2584	.0244	.0944	40.995	10.594	7.316	77.509
21	4.141	.2415	.0223	.0923	44.865	10.836	7.599	82.339
22	4.430	.2257	.0204	.0904	49.006	11.061	7.872	87.079
23	4.741	.2109	.0187	.0887	53.436	11.272	8.137	91.720
24	5.072	.1971	.0172	.0872	58.177	11.469	8.392	96.255
25	5.427	.1842	.0158	.0858	63.249	11.654	8.639	100.676
26	5.807	.1722	.0146	.0846	68.676	11.826	8.877	104.981
27	6.214	.1609	.0134	.0834	74.484	11.987	9.107	109.166
28	6.649	.1504	.0124	.0824	80.698	12.137	9.329	113.226
29	7.114	.1406	.0114	.0814	87.347	12.278	9.543	117.162
30	7.612	.1314	.0106	.0806	94.461	12.409	9.749	120.972
31	8.145	.1228	.0098	.0798	102.073	12.532	9.947	124.655
32	8.715	.1147	.0091	.0791	110.218	12.647	10.138	128.212
33	9.325	.1072	.0084	.0784	118.933	12.754	10.322	131.643
34	9.978	.1002	.0078	.0778	128.259	12.854	10.499	134.951
35	10.677	.0937	.0072	.0772	138.237	12.948	10.669	138.135
40	14.974	.0668	.0050	.0750	199.635	13.332	11.423	152.293
45	21.002	.0476	.0035	.0735	285.749	13.606	12.036	163.756
50	29.457	.0339	.0025	.0725	406.529	13.801	12.529	172.905
55	41.315	.0242	.0017	.0717	575.929	13.940	12.921	180.124
60	57.946	.0173	.0012	.0712	813.520	14.039	13.232	185.768
65	81.273	.0123	.0009	.0709	1146.755	14.110	13.476	190.145
70	113.989	.0088	.0006	.0706	1614.134	14.160	13.666	193.519
75	159.876	.0063	.0004	.0704	2269.657	14.196	13.814	196.104
80	224.234	.0045	.0003	.0703	3189.063	14.222	13.927	198.075
85	314.500	.0032	.0002	.0702	4478.576	14.240	14.015	199.572
90	441.103	.0023	.0002	.0702	6287.185	14.253	14.081	200.704
95	618.670	.0016	.0001	.0701	8823.854	14.263	14.132	201.558
100	867.716	.0012	.0001	.0701	12381.662	14.269	14.170	202.200

	SINGLE PAYMENT		UNIFORM PAYMENT SERIES				GRADIENT SERIES	
	Compound Amount Factor	Present Worth Factor	Sinking Fund Factor	Capital Recovery Factor	Compound Amount Factor	Present Worth Factor	Gradient Uniform Series	Gradient Present Worth
	Find F Given P F/P	Find P Given F P/F	Find A Given F A/F	Find A Given P A/P	Find F Given A F/A	Find P Given A P/A	Find A Given G A/G	Find P Given G P/G
n								
1	1.080	.9259	1.0000	1.0800	1.000	.926	0	0
2	1.166	.8573	.4808	.5608	2.080	1.783	.481	.857
3	1.260	.7938	.3080	.3880	3.246	2.577	.949	2.445
4	1.360	.7350	.2219	.3019	4.506	3.312	1.404	4.650
5	1.469	.6806	.1705	.2505	5.867	3.993	1.846	7.372
6	1.587	.6302	.1363	.2163	7.336	4.623	2.276	10.523
7	1.714	.5835	.1121	.1921	8.923	5.206	2.694	14.024
8	1.851	.5403	.0940	.1740	10.637	5.747	3.099	17.806
9	1.999	.5002	.0801	.1601	12.488	6.247	3.491	21.808
10	2.159	.4632	.0690	.1490	14.487	6.710	3.871	25.977
11	2.332	.4289	.0601	.1401	16.645	7.139	4.240	30.266
12	2.518	.3971	.0527	.1327	18.977	7.536	4.596	34.634
13	2.720	.3677	.0465	.1265	21.495	7.904	4.940	39.046
14	2.937	.3405	.0413	.1213	24.215	8.244	5.273	43.472
15	3.172	.3152	.0368	.1168	27.152	8.559	5.594	47.886
16	3.426	.2919	.0330	.1130	30.324	8.851	5.905	52.264
17	3.700	.2703	.0296	.1096	33.750	9.122	6.204	56.588
18	3.996	.2502	.0267	.1067	37.450	9.372	6.492	60.843
19	4.316	.2317	.0241	.1041	41.446	9.604	6.770	65.013
20	4.661	.2145	.0219	.1019	45.762	9.818	7.037	69.090
21	5.034	.1987	.0198	.0998	50.423	10.017	7.294	73.063
22	5.437	.1839	.0180	.0980	55.457	10.201	7.541	76.926
23	5.871	.1703	.0164	.0964	60.893	10.371	7.779	80.673
24	6.341	.1577	.0150	.0950	66.765	10.529	8.007	84.300
25	6.848	.1460	.0137	.0937	73.106	10.675	8.225	87.804
26	7.396	.1352	.0125	.0925	79.954	10.810	8.435	91.184
27	7.988	.1252	.0114	.0914	87.351	10.935	8.636	94.439
28	8.627	.1159	.0105	.0905	95.339	11.051	8.829	97.569
29	9.317	.1073	.0096	.0896	103.966	11.158	9.013	100.574
30	10.063	.0994	.0088	.0888	113.283	11.258	9.190	103.456
31	10.868	.0920	.0081	.0881	123.346	11.350	9.358	106.216
32	11.737	.0852	.0075	.0875	134.214	11.435	9.520	108.857
33	12.676	.0789	.0069	.0869	145.951	11.514	9.674	111.382
34	13.690	.0730	.0063	.0863	158.627	11.587	9.821	113.792
35	14.785	.0676	.0058	.0858	172.317	11.655	9.961	116.092
40	21.725	.0460	.0039	.0839	259.057	11.925	10.570	126.042
45	31.920	.0313	.0026	.0826	386.506	12.108	11.045	133.733
50	46.902	.0213	.0017	.0817	573.770	12.233	11.411	139.593
55	68.914	.0145	.0012	.0812	848.923	12.319	11.690	144.006
60	101.257	.0099	.0008	.0808	1253.213	12.377	11.902	147.300
65	148.780	.0067	.0005	.0805	1847.248	12.416	12.060	149.739
70	218.606	.0046	.0004	.0804	2720.080	12.443	12.178	151.533
75	321.205	.0031	.0002	.0802	4002.557	12.461	12.266	152.845
80	471.955	.0021	.0002	.0802	5886.935	12.474	12.330	153.800
85	693.456	.0014	.0001	.0801	8655.706	12.482	12.377	154.492
90	1018.915	.0010	.0001	.0801	12723.939	12.488	12.412	154.993
95	1497.121	.0007	.0001	.0801	18701.507	12.492	12.437	155.352
100	2199.761	.0005		.0800	27484.516	12.494	12.455	155.611

	SINGLE PAYMENT		UNIFORM PAYMENT SERIES				GRADIENT SERIES	
	Compound Amount Factor	Present Worth Factor	Sinking Fund Factor	Capital Recovery Factor	Compound Amount Factor	Present Worth Factor	Gradient Uniform Series	Gradient Present Worth
	Find F Given P F/P	Find P Given F P/F	Find A Given F A/F	Find A Given P A/P	Find F Given A F/A	Find P Given A P/A	Find A Given G A/G	Find P Given G P/G
n								
1	1.100	.9091	1.0000	1.1000	1.000	.909	0	0
2	1.210	.8264	.4762	.5762	2.100	1.736	.476	.826
3	1.331	.7513	.3021	.4021	3.310	2.487	.937	2.329
4	1.464	.6830	.2155	.3155	4.641	3.170	1.381	4.378
5	1.611	.6209	.1638	.2638	6.105	3.791	1.810	6.862
6	1.772	.5645	.1296	.2296	7.716	4.355	2.224	9.684
7	1.949	.5132	.1054	.2054	9.487	4.868	2.622	12.763
8	2.144	.4665	.0874	.1874	11.436	5.335	3.004	16.029
9	2.358	.4241	.0736	.1736	13.579	5.759	3.372	19.421
10	2.594	.3855	.0627	.1627	15.937	6.145	3.725	22.891
11	2.853	.3505	.0540	.1540	18.531	6.495	4.064	26.396
12	3.138	.3186	.0468	.1468	21.384	6.814	4.388	29.901
13	3.452	.2897	.0408	.1408	24.523	7.103	4.699	33.377
14	3.797	.2633	.0357	.1357	27.975	7.367	4.996	36.800
15	4.177	.2394	.0315	.1315	31.772	7.606	5.279	40.152
16	4.595	.2176	.0278	.1278	35.950	7.824	5.549	43.416
17	5.054	.1978	.0247	.1247	40.545	8.022	5.807	46.582
18	5.560	.1799	.0219	.1219	45.599	8.201	6.053	49.640
19	6.116	.1635	.0195	.1195	51.159	8.365	6.286	52.583
20	6.727	.1486	.0175	.1175	57.275	8.514	6.508	55.407
21	7.400	.1351	.0156	.1156	64.002	8.649	6.719	58.110
22	8.140	.1228	.0140	.1140	71.403	8.772	6.919	60.689
23	8.954	.1117	.0126	.1126	79.543	8.883	7.108	63.146
24	9.850	.1015	.0113	.1113	88.497	8.985	7.288	65.481
25	10.835	.0923	.0102	.1102	98.347	9.077	7.458	67.696
26	11.918	.0839	.0092	.1092	109.182	9.161	7.619	69.794
27	13.110	.0763	.0083	.1083	121.100	9.237	7.770	71.777
28	14.421	.0693	.0075	.1075	134.210	9.307	7.914	73.650
29	15.863	.0630	.0067	.1067	148.631	9.370	8.049	75.415
30	17.449	.0573	.0061	.1061	164.494	9.427	8.176	77.077
31	19.194	.0521	.0055	.1055	181.943	9.479	8.296	78.640
32	21.114	.0474	.0050	.1050	201.138	9.526	8.409	80.108
33	23.225	.0431	.0045	.1045	222.252	9.569	8.515	81.486
34	25.548	.0391	.0041	.1041	245.477	9.609	8.615	82.777
35	28.102	.0356	.0037	.1037	271.024	9.644	8.709	83.987
40	45.259	.0221	.0023	.1023	442.593	9.779	9.096	88.953
45	72.890	.0137	.0014	.1014	718.905	9.863	9.374	92.454
50	117.391	.0085	.0009	.1009	1163.909	9.915	9.570	94.889
55	189.059	.0053	.0005	.1005	1880.591	9.947	9.708	96.562
60	304.482	.0033	.0003	.1003	3034.816	9.967	9.802	97.701
65	490.371	.0020	.0002	.1002	4893.707	9.980	9.867	98.471
70	789.747	.0013	.0001	.1001	7887.470	9.987	9.911	98.987
75	1271.895	.0008	.0001	.1001	12708.954	9.992	9.941	99.332
80	2048.400	.0005		.1000	20474.002	9.995	9.961	99.561
85	3298.969	.0003		.1000	32979.690	9.997	9.974	99.712
90	5313.023	.0002		.1000	53120.226	9.998	9.983	99.812
95	8556.676	.0001		.1000	85556.761	9.999	9.989	99.877
100	13780.612	.0001		.1000	137796.123	9.999	9.993	99.920

	SINGLE PAYMENT		UNIFORM PAYMENT SERIES				GRADIENT SERIES	
	Compound Amount Factor	Present Worth Factor	Sinking Fund Factor	Capital Recovery Factor	Compound Amount Factor	Present Worth Factor	Gradient Uniform Series	Gradient Present Worth
	Find F Given P F/P	Find P Given F P/F	Find A Given F A/F	Find A Given P A/P	Find F Given A F/A	Find P Given A P/A	Find A Given G A/G	Find P Given G P/G
n								
1	1.120	.8929	1.0000	1.1200	1.000	.893	0	0
2	1.254	.7972	.4717	.5917	2.120	1.690	.472	.797
3	1.405	.7118	.2963	.4163	3.374	2.402	.925	2.221
4	1.574	.6355	.2092	.3292	4.779	3.037	1.359	4.127
5	1.762	.5674	.1574	.2774	6.353	3.605	1.775	6.397
6	1.974	.5066	.1232	.2432	8.115	4.111	2.172	8.930
7	2.211	.4523	.0991	.2191	10.089	4.564	2.551	11.644
8	2.476	.4039	.0813	.2013	12.300	4.968	2.913	14.471
9	2.773	.3606	.0677	.1877	14.776	5.328	3.257	17.356
10	3.106	.3220	.0570	.1770	17.549	5.650	3.585	20.254
11	3.479	.2875	.0484	.1684	20.655	5.938	3.895	23.129
12	3.896	.2567	.0414	.1614	24.133	6.194	4.190	25.952
13	4.363	.2292	.0357	.1557	28.029	6.424	4.468	28.702
14	4.887	.2046	.0309	.1509	32.393	6.628	4.732	31.362
15	5.474	.1827	.0268	.1468	37.280	6.811	4.980	33.920
16	6.130	.1631	.0234	.1434	42.753	6.974	5.215	36.367
17	6.866	.1456	.0205	.1405	48.884	7.120	5.435	38.697
18	7.690	.1300	.0179	.1379	55.750	7.250	5.643	40.908
19	8.613	.1161	.0158	.1358	63.440	7.366	5.838	42.998
20	9.646	.1037	.0139	.1339	72.052	7.469	6.020	44.968
21	10.804	.0926	.0122	.1322	81.699	7.562	6.191	46.819
22	12.100	.0826	.0108	.1308	92.503	7.645	6.351	48.554
23	13.552	.0738	.0096	.1296	104.603	7.718	6.501	50.178
24	15.179	.0659	.0085	.1285	118.155	7.784	6.641	51.693
25	17.000	.0588	.0075	.1275	133.334	7.843	6.771	53.105
26	19.040	.0525	.0067	.1267	150.334	7.896	6.892	54.418
27	21.325	.0469	.0059	.1259	169.374	7.943	7.005	55.637
28	23.884	.0419	.0052	.1252	190.699	7.984	7.110	56.767
29	26.750	.0374	.0047	.1247	214.583	8.022	7.207	57.814
30	29.960	.0334	.0041	.1241	241.333	8.055	7.297	58.782
31	33.555	.0298	.0037	.1237	271.293	8.085	7.381	59.676
32	37.582	.0266	.0033	.1233	304.848	8.112	7.459	60.501
33	42.092	.0238	.0029	.1229	342.429	8.135	7.530	61.261
34	47.143	.0212	.0026	.1226	384.521	8.157	7.596	61.961
35	52.800	.0189	.0023	.1223	431.663	8.176	7.658	62.605
40	93.051	.0107	.0013	.1213	767.091	8.244	7.899	65.116
45	163.988	.0061	.0007	.1207	1358.230	8.283	8.057	66.734
50	289.002	.0035	.0004	.1204	2400.018	8.304	8.160	67.762
55	509.321	.0020	.0002	.1202	4236.005	8.317	8.225	68.408
60	897.597	.0011	.0001	.1201	7471.641	8.324	8.266	68.810
65	1581.872	.0006	.0001	.1201	13173.937	8.328	8.292	69.058
70	2787.800	.0004		.1200	23223.332	8.330	8.308	69.210
75	4913.056	.0002		.1200	40933.799	8.332	8.318	69.303
80	8658.483	.0001		.1200	72145.692	8.332	8.324	69.359
85	15259.206	.0001		.1200	127151.714	8.333	8.328	69.393
90	26891.934			.1200	224091.118	8.333	8.330	69.414
95	47392.777			.1200	394931.471	8.333	8.331	69.426
100	83522.266			.1200	696010.547	8.333	8.332	69.434

Index